Community-based Natural Resource Management

Community-based Natural Resource Management

ISSUES AND CASES FROM SOUTH ASIA

AJIT MENON
PRAVEEN SINGH
ESHA SHAH
SHARACHCHANDRA LÉLÉ
SUHAS PARANJAPE
K.J. JOY

Los Angeles | London | New Delhi
Singapore | Washington DC | Melbourne

First published in 2007 by

 SAGE Publications India Pvt Ltd
B1/I-1 Mohan Cooperative Industrial Area
Mathura Road, New Delhi 110 044, India
www.sagepub.in

SAGE Publications Inc
2455 Teller Road
Thousand Oaks, California 91320, USA

SAGE Publications Ltd
1 Oliver's Yard
55 City Road
London EC1Y 1SP, United Kingdom

SAGE Publications Asia-Pacific Pte Ltd
3 Church Street
#10-04 Samsung Hub
Singapore 049483

Published by Vivek Mehra for SAGE Publications India Pvt Ltd, typeset in 10.5/13pt Minion by Star Compugraphics Private Limited, Delhi and printed at Sai Print-o-Pack, New Delhi.

Library of Congress Cataloging-in-Publication Data

Community-based natural resource management: issues and cases from South Asia/ Ajit Menon ... [et al.].
    p. cm.
  Includes bibliographical references and index.
    1. Natural resources—Co-management—South Asia. I. Menon, Ajit, 1966–

HC430.6.Z65C64   333.70954—dc22   2007   2007028910

ISBN: 978-07-619-3574-2 (PB)

The SAGE Team: Sugata Ghosh, Neha Kohli and Rajib Chatterjee

Note regarding maps:
1. The maps on pages 26, 33, 85, 119, 128, 162, 163, 198, 209 and 255 are recreated from maps available in the public domain and are not to scale.

Contents

List of Tables

List of Figures

List of Boxes

List of Abbreviations

AGY	Adarsh Gaon Yojana
ASAG	Ahmedabad Study Action Group
BG-SRDP	Bhutan–German Sustainable Renewable Natural Resource Development Project
BLP	block-level planning
CBNRM	community-based natural resource management
CCTs	continuous contour trenches
CFM	community forest management
CFMGs	community forestry management groups
CPR	common property resource
CWSS	Community Water Supply and Sanitation Programme
DFID	Department for International Development
DPAP	Drought Prone Areas Programme
DTLVS	Doodha Toli Lok Vikas Sansthan
DYT	Dzongkhag Yargye Tshogchung
EGS	Employment Guarantee Scheme
FMIS	farmer-managed irrigation systems
FYM	farmyard manure
GC	Gono Chetona
GRWSSP	Ghogha Regional Water Supply and Sanitation Programme
GSS	Gono Shahajjo Shangstha
GWSSB	Gujarat Water Supply and Sewerage Board

GYT	Geog Yargye Tshogchung
HYV	high-yielding variety
IMT	irrigation management transfer
IWDP	Integrated Wasteland Development Programme
JFM	joint forest management
MMD	mahila mangal dal
NEMAP	National Environment Management Action Plan
NRSP	National Rural Support Programme
NWDP	National Watershed Development Programme
NWDPRA	National Watershed Development Programme for Rainfed Areas
PAWDI	People's Action for Watershed Development
PIA	project implementation agency
PIM	participatory irrigation management
PHWA	per household watershed area
PRA	participatory rural appraisal
RNR	renewable natural resource
RRA	rapid rural appraisal
RWSSP	Rural Water Supply and Sanitation Project
SB	Swanirwar Bangladesh
SC	Scheduled Castes
ST	Scheduled Tribes
SEMP	Sustainable Environment Management Plan
SHGs	self-help groups
SLRC	Sustainable Livelihoods in Riverine Charlands
TBS	Tarun Bharat Sangh
TDH	Terre Des Hommes
VP	van panchayat
YKGPVS	Yashwant Krishi Gram and Panlot Vikas Sanstha
WAA	water augmentation activities
WASMO	Water and Sanitation Management Organisation

Preface and Acknowledgements

Undertaking this study on community-based natural resource management (CBNRM) in South Asia was for all of us an interesting and
necessary challenge. Interesting because as researchers, it provided us
an opportunity to visit and explore different regions of South Asia
and observe CBNRM initiatives that had received a considerable
amount of 'good' press. Necessary because so much has been made of
the potential role of civil society and NGOs as a means towards a
more decentralised, democratic and participatory form of development, yet so little attention has been paid to specific initiatives. Hence,
we felt the need to take it upon ourselves to study some of these initiatives, hopefully with a more critical and analytical perspective.

This book is the product of our endeavour, an endeavour that started
with lengthy discussions as to how best to approach the study. Being a
mix of researchers and activists, both individually and collectively, as
well as having widely different academic backgrounds, the task at times
was not an easy one, often being confronted by both ontological and
epistemological differences. Yet, our heated discussions and often irre-
concilable differences, helped us in the long-run produce what we
hope is a better finished product. Our other hope is that the book will
provide thick descriptive and analytical insights into the what, how
and why of CBNRM in South Asia, based of course on our selected
case studies.

While the book is a collective product, we ventured into the field in
groups of two to four people based primarily on our own interests.

Chapter 2 on Hivre Bazar is authored by Ajit Menon, Praveen Singh, K.J. Joy and Suhas Paranjape; Chapter 3 on Utthan by Suhas Paranjape and Esha Shah; Chapter 4 on Tarun Bharat Sangh by Esha Shah and Praveen Singh; Chapter 5 on the Lingmuteychhu Watershed in Bhutan by Ajit Menon and K.J. Joy; Chapter 6 on the Doodha Toli region by Sharachchandra Lélé and Praveen Singh; and Chapter 7 on the Bangladesh Charlands by Praveen Singh and Suhas Paranjape. The case study chapters are stand-alone chapters both in their descriptive and analytical content bound together by a common argument and frame of reference detailed in the introduction.

The making of this book would not have been possible without the help of organisations and individuals who helped us in the course of our research. Gathering the literature on CBNRM was a lengthy and difficult task given the fact that much of it is unpublished, and hence we pestered a number of organisations in the course of that undertaking. In particular, we would like to mention Society for Promotion of Wasteland Development (SPWD); the Ford Foundation; International Development Reseach Centre, New Delhi; Professional Assistance for Development Action (PRADAN); the Aga Khan Rural Support Programme (India) or AKRSP (I); the International Conversation Union (IUCN), Bangladesh; International Water Management Institute, Colombo; Centre for Science and Environment (CSE); and CARE Nepal. We also requested and got help from many colleagues in India and other South Asian countries, including Atiq Rahman, Usman Iftikhar, Shekhar Pathak, Harnath Jagawat, Apoorva Oza, Amita Shah, Shilp Verma, Atiur Rahman, Mokhlesur Rahman, Shireen Kamal Sayeed, Philip Gain, Neelima Khaitan, Irfan Maqbool, S.P. Wani, Nandini Sundar, Madhu Sarin and C.M. Wijeratna. We requested Ajaya Dixit and Ram Kumar Sharma from Nepal, Mushtaq Gaadi from Pakistan and L.P. Dayananda from Sri Lanka to put together a review document for their respective countries.

Local organisations, namely, Yashwant Krishi Gram and Watershed Development Trust in Hivre Bazar; Tarun Bharat Sangh (TBS) in Gopalpura; Renewable Natural Resource Research Centre (RNRRC), Bajo, Bhutan; Gono Chetona (GC) in Bangladesh; Utthan in Nathugarh; and Doodha Toli Lok Vikas Sansthan (DTLVS) in Uttarakhand, helped us immensely in the course of our fieldwork.

In particular, we would like to thank Popatrao Pawar, Rajendra Singh, Sangay Duba, Aita Kumar Bhujel, Ira Rahman, Nafisa Barot and Sachchidanand Bharati, and of course the various staff and volunteers in these organisations. We would also like to acknowledge our debt to the different communities in all these locations who tolerated our intrusions and extended their hospitality generously.

We organised a workshop in December 2005 to discuss the first draft of what was then a report. This workshop, attended by a total of about 45 persons from the case study organisations, and the practitioner, activist, donor, policymaker and academic communities, provided us with a rich set of comments and ideas that helped us revise the draft very substantially. We would like to thank all the participants in this workshop, and especially the discussants and session chairs who included, Sumi Krishna, Neelima Khaitan, Sara Ahmed, C.R. Bijoy, Sudarshan Iyengar, Amita Shah, Liz Fajber, R. Rajesh, Emmanuel Theophilus and others. Our colleagues at the Centre for Interdisciplinary Studies in Environment and Development (CISED), especially Priya and Rinki, gave comments on the whole report or various chapters. Shrini generously lent material. Santosh patiently helped us make the maps, and with GIS related work. The support staff in CISED, most notably Anand and Ganesha, provided back-up support throughout the research project. We would also like to mention the help that the Society for Promoting Participative Ecosystem Management (SOPPECOM), particularly Ravi Pomane, gave us during our study in Hivre Bazar.

The conceptual and analytical framework of this book owes a lot to the intellectual contributions of other scholars, many of whom have undertaken similar scholarly forays. There are of course too many of them to mention but the work of Amita Baviskar, David Mosse, Sangita Kamat, Maxine Weisgrau, Tania Murray Li, Nandini Sundar, Roger Jeffery, Arun Agarwal, Norman Uphoff, Rajni Kothari, Anil Agarwal, Nirmal Sengupta and Anupam Mishra all helped shape the direction of our research.

Finally, we would like to acknowledge the financial support provided by IDRC, which made this study possible, and would also like to thank Liz Fajber at IDRC, New Delhi for her constant support and encouragement.

1

Introduction

Background

Since the 1990s, the concept of community-based natural resource management (CBNRM) has come to the forefront of rural development policy in developing countries. Governments across South and South-East Asia, Africa and Latin America have adopted and implemented CBNRM in various ways. These include programmes in individual sectors such as forestry, irrigation or wildlife management, or multi-sectoral programmes such as watershed development or rural livelihoods development, with or without donor support, with varying emphasis on conservation or local livelihoods, with statutory backing or in an ad-hoc manner, through non-government organisations (NGOs) or directly through state agencies and so on.

Broadly speaking, the umbrella term CBNRM emphasises that involving, if not privileging, local communities is essential for 'successful' natural resource management (NRM), and that doing so can simultaneously ensure environmental sustainability, social justice and development efficacy. The faith being placed in CBNRM is part of a wider shift away from state-driven development towards a more 'communitarian' and civil society-driven development. The concept and practice of CBNRM is backed by various diverse academic discourses, ranging from common property theory to traditional knowledge to decentralisation.

Perhaps most significant about these developments, and certainly most important to us, is that in this process of 'mainstreaming' community-based development and CBNRM, the role of NGOs is almost ubiquitous. Not only are NGOs themselves implementing CBNRM, but many state-driven initiatives are operating through NGOs as well. The expansion of NGOs in the 1970s and 1980s throughout much of the world was seen as an opportunity for civil society to offer 'alternative' forms of development and as a means to help democratise the state. However, the mushrooming of NGOs thereafter has not only resulted in a significant diversity of NGOs in terms of their type and priorities, but also, some argue, in the content of their alternative discourses of development and their interest in concerns of social justice and structural change being watered down (Fisher, 1997). It is this mainstreaming of NGO-driven development that forms the context of our study of CBNRM.

Such an enquiry is important because both CBNRM and more so NGO-driven CBNRM are at a crossroads. A large number of critiques have emerged that call for rethinking the initial enthusiasm about CBNRM and the form it is increasingly taking. While the concept of community-based development itself has rarely been discarded, several critiques have seriously questioned the design, content and implementation of community-driven development. Three main strands of criticism have been mounted against community-driven development in general and CBNRM in particular. The first is that community-based development is constrained by wider hegemonic discourses and practices of development and hence allows for limited manoeuvrability and possibilities for communities to articulate their own agendas (Chatterjee, 1998; Manor, 1999). The second critique pertains to the 'community' in CBNRM and the unproblematic way in which community is idealised as a harmonious and symbiotic entity glossing over internal differentiations of class, caste, gender and race and the micro-politics that arises as a result of this differentiation (Agrawal,

2001a; Baviskar, 2002; Ferguson, 1990; Manor, 1999; Mosse, 2003a; Rossi, 2004; Sundar et al., 2001). A third critique that follows from these two and partly explains them as well pertains to the fact that most CBNRM seems to be implemented in project mode by the implementing agencies (often NGOs). As Mosse (2003a, 2005) and Baviskar (2002) have pointed out, project implementing agencies are driven by the exigencies of meeting targets and delivering outputs. As a result, the *process* of CBNRM is often compromised resulting in the micropolitics of participation being ignored.

This study hopes to contribute to a better understanding of NGO-driven CBNRM, using these critiques as a platform from which to examine how NGOs 'practice development'. While there is a significant critical literature on NGOs, that at least partly applies the critiques discussed earlier to NGOs, there are few detailed studies (especially in the South Asian context) that examine NGO-driven CBNRM. Existing critiques, moreover, are often case specific and also partial, focusing on particular aspects and driven by particular concerns such as (say) the limits to participation, the hierarchical nature of communities, etc., and often do not address other concerns such as sustainability that too might constitute 'alternative' notions of development. Moreover, if indeed NGO-driven CBNRM has its limits, as we too argue, there is a need for a more detailed and nuanced explanation for this that captures diversity within NGO initiatives as well. Detailed case studies are necessary to understand and analyse the nature of NGO visions, the manner in which they function, the constraints (and opportunities) they face and the possibilities that such experiments offer (or not). Hence, our main purpose is to understand in more detail not only the nature of different NGO-driven CBNRM experiments, but also the manifold reasons behind the visions they set themselves and the 'outcomes' they achieve.

Some would argue that a detailed 'ethnography of development practices' needs to be carried out to understand these issues fully (Mosse, 2003a). While we believe such an approach would be useful, our study is based on detailed case studies of six seemingly innovative CBNRM initiatives that might help unpack their visions, workings and outcomes. In that sense, we attempt to cover a wider terrain than a single ethnographic study might allow us to cover. Due to the ubiquitous time and resource constraints, moreover, our fieldwork cannot be as detailed as an ethnography would be.

The remaining part of this chapter traces the genealogy or trajector-
ies of the concept and practice of CBNRM, including the definitional
maze the term is entangled in, followed by a further detailed critique
of it. The rationale for the selection of case studies, scope of the study,
research questions and methodology are discussed subsequently.

Emergence of CBNRM

In the late 1970s and early 1980s, a variety of CBNRM experiments
and initiatives emerged across South Asia that provided the inspiration
for subsequent efforts. This early set emerged for various reasons and
took various forms. In many cases, they came out of disillusionment
with the developmental state. Social movements emerged which
challenged the authority of the state and highlighted the need for more
decentralised decision-making that would give a voice to local communities (Kothari, 1989). Some of these social movements, such as Chipko
in the Himalayas, placed the 'environment' at the centre stage. They
blamed the state for ecological destruction and sought to redress this
by asking for greater local control over forests. Similarly, in Orissa,
many village communities took control of patches of forests that they
were using and that were not being managed well by the forest
department. But other movements and smaller initiatives, such as the
Pani Panchayat in Maharashtra, emerged more out of concern for
inadequate or failed development efforts of the state. Rather than
excluding the state entirely, they asked for changes in the developmental process that would enable local communities to play a more
active role in development programmes.

Several other efforts were initiated by 'outside agents' often out of a
sense of voluntarism and sometimes due to 'practical' managerial
necessities. Individuals and NGOs, inspired by ideas of welfare work
or more radical thinking, took on the task of what was perceived of as
a different mode of rural development. They mobilised resources and
organised communities to help define and implement developmental
activities of various kinds, from literacy to irrigation tank rehabilitation
to soil conservation to improved agricultural practices. Some focused

on specific concerns such as women's issues (Gupta, 1999: 28–31; van Koppen et al., 2001) or tribal development (Gupta, 1999; Paranjape et al., 1997; Pathak and Gour-Broome, 2000; Sinha and Sinha, 1996; Vohra, 1990) and sought to create space for these marginalised groups to have a say in NRM. On the other hand, some of the most famous initiatives were initiated by innovative government officials. The famous Sukhomajri experiment in Haryana was led by a technocrat who was concerned over the silting of a lake, for which he got soil conservation structures constructed on the slopes of the catchment (Chopra et al., 1988; Seckler and Joshi, 1981; SPWD, 1984). Local villagers were then involved in order to regulate their grazing activities on the slopes and to set up sharing mechanisms to manage the water-harvesting structures. The success of the experiment led to the setting up of many such 'hill resource management societies' with official support. Similarly, a forest officer in Arabari, West Bengal, striking a deal with local communities so as to save the plantations of the forest department eventually led to the now-famous concept of joint forest management (JFM) (Chatterji, 1996; Corbridge and Jewitt, 1997; Correa, 1996; Deb and Malhotra, 1993; Jeffery and Sundar, 1999; Kumar et al., 1999; Lélé, 1999; Lélé et al., 2005; Ravindranath et al., 2000; Sarin, 1995; Sarin et al., 2003b; Saxena et al., 1997; Sundar et al., 2001).

In the period following these earlier movements and initiatives, much older 'traditional' systems of community management were 'rediscovered'. Van panchayats (VPs) in the Kumaon region of the Indian Himalayas were found to be a longstanding example of a state-recognised but highly decentralised forest management programme (Agrawal, 1999; Agrawal, 2001b; Ballabh and Singh, 1988; Sarin, 2001b). Many communities in Orissa copied the examples set by decades-old community protection initiatives in their neighbourhood to launch their own efforts (Conroy, 2001; Conroy et al., 1998; Kohlin, 1998; Pattanaik, 2002). Traditions of farmer-managed canal irrigation systems in the north (*kuhls* of Himachal Pradesh or *kuhlos* of Nepal) (Baker, 1997; Pradhan, 2003; Yoder, 1994; Zwarteween and Neupane, 1996) as well as farmer-managed tank irrigation systems in the south were studied with renewed vigour.

Although academic arguments in favour of community-based de-velopment in general existed as early as the 1950s, the major discourses that directly or indirectly lent support to CBNRM emerged to a large extent as a result of 'lived experiences'. One major academic discourse that emerged in the 1980s highlighted the limits of the post-colonial state in environmental management. Based to a large extent on evidence from forestry, much of this discourse has highlighted continuities in underlying political and ideological imperatives of colonial and post-colonial development and its negative impact on environmental policies. The thrust of the argument has been that the bureaucratic centralisation of power in order to prioritise industrial and commercial needs has resulted in the alienation of local communities from control over common resources (Alvares, 1979; Gadgil and Guha, 1992; Guha, 1989; Mishra, 1993; Mukundan, 1988; Nadkarni et al., 1989; Pathak, 1994; Sengupta, 1991; Shankari, 1991; Shiva, 1991). Although critics of the post-colonial state differ in terms of the prescriptions offered, there is a common underlying contention that local communities must play an important role in environmental management. Agarwal and Narain (1989) in their book *Towards Green Villages*, perhaps go furthest in their call to hand over control (and property rights) to village communities. Activists too have often argued on similar lines; for example, many *adivasi* (tribal) groups have spoken about self-rule and how tribal communities can manage their own resources best (Rahul, 1997).

A second stream of thought, also critical of the state's role in NRM, has been more concerned with the manner in which development planning works and how it underplays the role of local communities in this process. Scholarly works like 'farmer first' highlighted the manner in which development planning has privileged the voice and knowledge of the development planner and silenced, to a large extent, the voice and knowledge of local communities (Chambers, 1983; Chambers et al., 1989; Thompson and Scoones, 1994). These scholars have contended that the failure of many development projects, programmes and policies in various sectors, particularly agriculture and rural development, is the result of an overly centralised, bureaucratic and technocratic approach to development. Chambers also highlights the importance of participatory techniques such as participatory rural appraisal (PRA) and rapid rural appraisal (RRA) as a means by which the local's voice can be articulated. What sets this discourse apart from the previous one is that it is less ideologically rooted in a critique of the post-colonial state and more concerned with changing the way the state functions. Hence, it does not call for a retreat of the state: external support in the form of policies, funds and expertise is acknowledged to be crucial for development. Rather, the emphasis is on changing the development process to make it more participatory and give greater control to local people.

A third stream in the literature is the one on 'traditional knowledge', a discourse which goes beyond just a critique of development planning. In the decade of the 1980s, as the pathology of modernity was challenged and unmasked, and the movement against big dams sparked by the Narmada agitation emerged, a significant number of publications sought to highlight the environmental soundness and cultural embeddedness of traditional, indigenous or local knowledge systems (Agarwal and Narain, 1997; Mukundan, 1988; Reddy, 1991; Shankari, 1991). Traditional or indigenous resource management systems constructed, managed and maintained by local communities were thus seen as a viable alternative to allegedly disastrous modern technology such as large dams. Two congresses on traditional sciences and technologies were held in the decade of the 1990s to acknowledge and highlight the importance of the non-Western scientific heritage. Related to this was the discourse on 'appropriate technology' and 'small is beautiful' that located the failure of development processes (in terms of inequity and unsustainability) in the failure to match technology with local needs. It focused on the need to develop innovative technologies that blend local knowledge with modern scientific methods to make them socially, economically and ecologically more viable (Chambers et al., 1989; Reddy, 1999, 2004).

Much of this discourse fits well into (but is not necessarily part of) a wider critique aimed at highlighting the limits of 'development rationalism' and the stifling of cultural plurality. The main thrust of this critique is that the modern state by over-centralising has ignored local cultures and stymied cultural plurality (Chatterjee, 1998; Coward, 1980; Kothari, 1988; Ostrom, 1990, 1992). The emergence of 'community' and 'community development' especially in the work of social anthropologists has been a way not only to highlight the importance of community in development, but also a way to imagine a wider process of democratic empowerment. Community development as imagined by Gandhians emphasised self-sufficiency of village communities based on absolute human needs. Ecological philosophies of the 1970s, such as Schumacher's (1975) influential work on 'small is beautiful', also led to more communitarian thinking.

At the same time, there are other discourses supporting CBNRM that are more grounded in new institutional economics and less in alternative imaginations of the community. Most notable of these is literature that emerged as a response to Hardin's proposition that the atomistic nature of human behaviour and the indivisibility of the commons would inevitably lead to a 'tragedy of the commons'. This 'collective action' literature demonstrated that in fact such a tragedy was neither inevitable in theory nor necessarily obtained in practice (Berkes, 1989; Bromley, 1992; McKay and Acheson, 1987). Numerous case studies brought to light the existence of old and new institutional arrangements for community management of forests, fisheries, pastures, tanks, etc., and highlighted the cultural specificity and ecological sophistication of these arrangements. As evidence for the existence of common property management grew, attention shifted to mapping out the conditions under which collective action would take place (Agrawal, 2001a; Agrawal and Ostrom, 2001; Baland and Platteau, 1999; Ostrom, 1990, 1992; Wade, 1988). Much of this literature implicitly assumes that the role of the state should be limited to facilitating community control.

What emerges from these streams of lived experiences and thought is a fairly compelling set of arguments as to why CBNRM is desirable and workable. First, there is the normative argument that greater (if not complete) control of the local environment and natural resources is the right of local communities (a right that was taken away by colonial and post-colonial governments) and needs to be restored.[1] Second, there are several instrumental arguments why a community-based approach will be more effective (because it mobilises local knowledge and skills, ensures greater accountability),[2] result in greater conservation (because assigning clear rights to specific local groups will close open-access situations and these groups can protect and monitor the resource more effectively) and even more equitable outcomes (because they are more democratic).

By the 1990s, the stage was thus well set to 'mainstream' CBNRM. Different discourses suggested the perceived advantages of CBNRM and concrete examples on the ground offered testimony to its possibilities. We now describe how CBNRM assumed greater importance in official NRM policy in South Asia.

Mainstreaming CBNRM

Starting around 1990, governments across South Asia began to pay serious attention to the concept of CBNRM. This was not the first time that communities were discovered by the state. In the 1950s and 1960s, community development programmes were very much part of development planning (Rudolph and Rudolph, 1987). What was new about the trend in the 1990s was the fact that sectors that were traditionally state-controlled—forests, water—were now being opened up to 'community participation', supported both by policy pronouncements and financial backing from international donor agencies.

The joint forest management (JFM) policy emerged in India in the 1990s. Similarly, the community forest management (CFM) policy emerged in Nepal in 1992. Social forestry emerged in Bangladesh in 1994 and was given legal backing with the passing of the Forest (Amendment) Act 2000; this was followed by the draft 'Social Forestry Rules'.[3] Simultaneously, in the decades of the 1980s and 1990s, heavily-funded programmes emerged for restructuring irrigation policy in South Asia. A community-based tank modernisation programme in Tamil Nadu and Andhra Pradesh in India was initiated with massive support from the European Commission in the 1980s (see CWR, 2000). Policies for the management of small irrigation tanks are now being changed in several other states in India. Participatory irrigation management (PIM) for canal irrigation was initiated in India in the decade of the 1990s with funding from the World Bank (Hooja et al., 2002; Joshi and Hooja, 2000). With regard to watershed development, India has substantially revised its guidelines in 1995 so as to give much greater prominence to community participation (GOI, 2001).

These policy changes under the rubric of democratic decentralisation were translated into practice through a large number of programmes and projects in all these sectors and countries. By 2000, JFM programmes had been implemented in at least 26 states in India. CFM had been implemented in most districts of Nepal. Moreover, due significantly to donor conditionalities, several states in India began implementing PIM programmes. Subsequently, legal frameworks for the formation of water user groups and the transfer of irrigation infrastructure to these user groups have been framed and implemented in some states (see various articles in Hooja et al., 2002). This process has also been in full swing in Pakistan. Pilot PIM efforts are on in Sri Lanka. In Nepal, farmer-managed irrigation systems (FMIS) account for more than 60 per cent of total irrigated area though there is no legal backing for water user groups per sé. Participatory watershed development programmes are being implemented across the length and breadth of the drier parts of India (Kerr et al., 2002). While the protected area policy seems to have gone the way of eco-development, that is, it seeks to move communities out of protected areas, there are still those who continue to advocate joint protected area management (Sarkar et al., 1995).

Nepal has a number of CBNRM strategies, mostly funded by donors which go back to even the 1970s. The first of its kind was the UNICEF-funded Community Water Supply and Sanitation Programme (CWSS). Subsequently, the government internalised the modalities of this programme and started the Rural Water Supply and Sanitation Project (RWSSP) that was funded by the Finnish government. Recently, an Asian Development Bank (ADB)-funded 'Community-based Rural Drinking Water Project' has been initiated and is expected to benefit more than five million people (for details on drinking water and sanitation projects in Nepal, see Dixit, 2000). In the forestry sector, besides CFM, a community-managed leasehold forestry programme has been operationalised (IFAD, 2003; Yadav and Dhakal, 2000).

Bangladesh has a number of programmes under the forestry sector, mostly ADB funded, including, Forestry Sector Project (1998–2004), Community Forestry Project (1981–88), Upazila Afforestation and Nursery Development Project (1989–96) and Sundarban Biodiversity Conservation Project (1999–2006) (Khan et al., 2004). In Pakistan, the National Rural Support Programme (NRSP), modelled on the well-known Aga Khan Rural Support Programme, was established by the government to undertake community development efforts across the country. The NRSP has been granted Pakistani Rs 10 billion over 10 years (Runnalls, 1995).

There are, however, important differences with regard to the government approach to CBNRM in South Asian countries. Bhutan, for example, has a CBNRM policy framework in place that encourages community-based NRM. This policy document advocates a fresh assessment of laws so that they are made conducive to CBNRM (DRDS, 2002). In Nepal, this has already happened partly with the Forest Act, 1993, lending support to CFM (Belbase and Regmi, 2002; Bhatta, 2002a, 2002b; Chakraborty, 1998; Dougill et al., 2001; Gilmour, 2003; Tiwari, 2002; Upadhyaya, 2002). However, there is no legal support in Nepal for community-based irrigation management (Pradhan, 2003; Pradhan and Gautam, 2002). In India, JFM is not supported in the Forest Act, but PIM as highlighted earlier, because water is a state subject, is legally supported in some states.

As suggested earlier, many international donors and lending agencies have played a key role in developing and implementing CBNRM-related policies. They have not only supported CBNRM initiatives, but driven them as well. The British Department for International Development (DFID) was perhaps one of the earliest to support 'participatory management' in its forest sector programmes from 1991 onwards, but the concept was quickly picked up by several other agencies, including lending agencies such as the World Bank and more recently the Japanese Bank for International Cooperation. While exact figures on the level of total funding are hard to come by, the amounts involved are very substantial: the World Bank alone has lent US$ 460 million over the period 1992–2000 for JFM-based forestry projects in India (Kumar et al., 1999) and has also supported PIM and irrigation tank 'rejuvenation' projects in several states.

The World Bank has, in fact, picked up and adopted the concept through the establishment of a common property resource (CPR) related network. To the World Bank, CBNRM is very much linked to building up social capital. In Bangladesh, the ADB has invested Taka 5,844.75 million (out of the total investment of Taka 12,472.25 million) in just six major social forestry projects.[4] The role of donor agencies seems as important in Nepal and Bhutan as well. In the case of the former, many of the government programmes discussed earlier are supported by donors. In the latter, donor funding has and continues to be pivotal to the promotion of CBNRM. On the whole, therefore, international aid agencies play a very significant role in propagating and designing these programmes. Of course, not all programmes are externally funded. In India, the central government spends at least Rs 2,000 crore ($400 million) of its own money every year on participatory watershed development programmes (Farrington et al., 1999).

In parallel to these state-initiated programmes and projects, the 1990s also saw other state initiatives in political devolution. In India, the central government passed the 73rd and 74th Amendments to the Constitution in 1992, thereby requiring the state governments to create a statutory three-tier local self-government structure down to the village (*gram*) level. By the mid-1990s, most states had complied. Several natural resources, including minor forest produce, small water bodies, etc., were to be brought under the jurisdiction of these bodies. The Indian Government also passed a Panchayats (Extension to the Scheduled Areas) Act (PESA) in 1996, which empowered *gram sabha*s in 5th Schedule areas to have the right to decide upon or veto development projects within its jurisdiction (Lélé, 2005). Similar initiatives for political devolution have also been taken up in other countries in the region. For instance, the Bangladesh Government has set up a Public Administration Reforms Commission. The draft report of the Local Government Commission was reviewed by the cabinet of ministers and a new four-tier local government system may be put in place once the relevant legislation is passed by Parliament (UNDP, 1997). These forms of decentralisation are part of wider programmes of devolution aimed at improving the overall voice of local communities within the political process and with regard to NRM. Donor involvement in political devolution (as against development implementation) is somewhat less than in CBNRM-type projects, perhaps because there are serious limits to how much money can be pumped into what is essentially a political process, not involving building of check dams or planting of trees.

Non-governmental organisations have played a prominent role in the mainstreaming of CBNRM. At one level, while governments were adopting the concept of CBNRM in various ways, the stream of civil society-driven CBNRM experiments that had emerged in the 1980s continued to expand and at times mingled somewhat with the state-initiated efforts. Hundreds of NGOs and grassroots voluntary groups attempted to replicate the early successes of Sukhomajri, Pani Panchayat and Ralegaon Siddi: somewhat larger-scale efforts by the Aga Khan Rural Support Programme in Pakistan and India, the Chakriya Vikas Pranali initiative in Jharkhand, the irrigation tank rejuvenation work of DHAN Foundation in Tamil Nadu and several integrated mountain development programmes and projects in the Himalayan region of Nepal, Bhutan and India.[5] More recently, some experiments have focused on conservation of biodiversity through enhancement of local livelihoods. These 'enterprise-based conservation efforts' include some that are based on non-timber forest product processing, and marketing, such as the ATREE-VGKK experiment (Bawa et al., 1999; Lélé et al., 1998; Lélé et al., 2004), and some based on eco-tourism (Bookbinder et al., 1998). While reliable estimates of the numbers of such efforts are not available, they are probably in the thousands and include many individual village-level initiatives, several efforts covering a few neighbouring or tens of villages to a few programmes that cover several hundreds of villages across different regions. Second, in many of the donor-funded and state-implemented CBNRM initiatives, NGOs have continued to play key roles, for example, as community organisers in JFM or as members of watershed development teams at the village-level in watershed programmes. For example, the mission-mode of watershed development in Jhabua district of Madhya Pradesh and the 'livelihoods approach-based projects' in Orissa and Andhra Pradesh involve NGOs as project implementers. In that sense, NGOs are now very much central to the discourse of CBNRM.

Defining CBNRM and the Scope of the Study

As evident from this discussion, the term CBNRM can be applied to a rather broad set of initiatives and is potentially confusing as it coexists with a number of other terms and acronyms coined by various analysts.

These include earlier terms such as CPR management, co-management, collaborative management or participatory resource management as well as more recent ones such as decentralised NRM (Ramakrishnan et al., 2002), democratic decentralisation of natural resources (Ribot, 2002) and decentralised governance of natural resources (Lélé, 2005). Before we get into a discussion of the critiques of CBNRM, it would be useful to come up with at least an approximate taxonomy so as to be clear as to which kinds of experiments or programmes one is talking about and which ones we would later be focusing on in this study.

Broadly speaking, any situation where the local community is involved in some manner in the management of natural resources in its immediate environment could be called a case of *community-based natural resource management*. Within South Asia alone, such a term would encompass a wide variety of initiatives and practices that have emerged over the past two to three decades. One can think of these initiatives falling into four broad categories:

(a) *Traditional systems* of resource management that continue to exist if not flourish, for example, irrigation *kuhls* of Himachal Pradesh or *kuhlos* of Nepal, minor irrigation tanks of peninsular India, the *phad* system and *malgujari* tanks in Maharashtra, tribal forest management systems of north-eastern India, *van panchayats* of Kumaon and customary fisheries management systems on the Sri Lankan and south Indian coast.

(b) Individual experiments typically initiated by *voluntary efforts*, activist groups or even local communities with or without indirect support from the state and other sources. These range from entirely self-initiated forest protection groups such as those in Orissa to NGO-initiated tank renovation projects in Tamil Nadu to more multi-sectoral watershed development projects taken up by various NGOs.

(c) A number of government-implemented programmes for *sectoral decentralisation* of NRM in forestry and irrigation, including JFM and PIM in India, CFM in Nepal, irrigation management transfer (IMT) in Sri Lanka and Pakistan and

participatory management of protected areas in Pakistan and
Nepal.

(*d*) A few (and mostly tentative) state-initiated efforts at *decentralisation of government as a whole*, including devolution of
control over natural resources, as has been attempted in parts
of India under the recent panchayati raj and PESA legislations
and earlier under the 6th Schedule legislation in the north-
eastern states of India.

Admittedly, the boundaries are fuzzy and so there is considerable
disagreement over which categories are significant and which category
or categories the term CBNRM refers to. For instance, Uphoff (1998:
1–3) makes a clear distinction between the first category and the rest,
calling the first one 'community management' (CNRM) which 'refers
to [situations of] communities having full and generally autonomous
responsibility for the protection and use of natural resources' and
pointing out that this situation obtains only in special cases. He seems
to include the remaining three strands in CBNRM, but others differ.
Ribot combines the second and third categories under CBNRM and
distinguishes them from democratic decentralisation (which corres-
ponds to our fourth category), arguing that the difference between
an explicit process of devolution of political power, on the one hand,
and NGO-driven experiments or even state-led sectoral programmes
that amount to only administrative decentralisation, on the other,
makes a significant difference in terms of outcomes (Ribot, 2002). Yet
others (Ramakrishnan et al., 2002) club democratic devolution programmes along with sectoral programmes when studying what they
call decentralised natural resource management (DNRM).

While this categorisation is useful in a heuristic manner and could
serve as a way to undertake a comparative study of 'different' types of
CBNRM, this is not our intent. Instead, our interest is to study NGO-
driven CBNRM because it is that form of CBNRM that is increasingly
being mainstreamed. The increasing prominence of NGO-driven
CBNRM is at least partly due to the fact that state-driven CBNRM
has fallen short on a number of counts. But equally important, as we
suggested at the outset, is that the discursive terrain in which development is envisaged has increasingly focused on the role of 'civil society'.

Our interest is to see what form such civil society initiatives have taken under NGO-driven CBNRM. We do this moreover keeping in mind four thematic (normative) concerns we believe are central to CBNRM, namely, livelihood enhancement, sustainability, equity and democratic decentralisation. We elaborate on these concerns later on.

Situating the Study: NGO-Driven CBNRM

Non-governmental organisations throughout the world, at some point of time, were envisaged to play an important role in seeking alternatives to development and a means to encourage processes of democratisation within the state (Fisher, 1997: 445). This was perhaps most noticeably the case in Latin America. As Bebbington (1997: 118) argues, a number of different types of NGOs emerged in Latin America in the 1960s, 1970s and 1980s ranging from church-based groups to groups linked more closely with political parties who drew their ideological inspiration from a number of sources ranging from Marxism to liberation theology to Freire's 'education for critical consciousness'. In South Asia, and more specifically India, a diverse number of leftist movements, social movements and people's science movements also gained prominence at this time (Mencher, 1999: 2081) These movements were innovative, demanded space for their innovations, sought and emphasised alternative ways of thinking as well as of grass-roots struggle.

The scenario, however, has changed significantly. As Bebbington (1997: 124) argues in the context of Latin America, many NGOs are now very much part of state-funded programmes or have agendas which are far less 'confrontational' in nature. Moreover, given the changed character of the state and the increasing role of the market in a more 'open' and 'competitive' economy implies that the role of NGOs in making demands on the state has become somewhat redundant given the different role envisaged for NGOs in a new political economy.

A critical study of NGO-driven CBNRM should take cognisance of these developments and take off from the existing critiques of community-based development that at least partially apply to it. The first major critique of community-based development is that it is framed in the context of wider hegemonic discourses and practices of development. Grounded in Foucauldian ideas of development as a discursive formation, it highlights the fact that practices and perceptions of development are rooted in a certain historically rooted regime of rationality, and this regime works as a structure of knowledge, allowing at certain historical moments, certain events and patterns of agency (Rossi, 2004). This means that envisaging alternative forms of development in the context first of state-centred development and more recently a neo-liberal paradigm of development has its own limits, be it in terms of limited devolution to communities, the privileging of hegemonic market-based ideologies or the manifestation of what Ferguson calls 'depoliticised' development—all of which prevent the emergence of new forms of democratisation (Ferguson, 1990).

Thus far, such a critique has been applied much more to state-implemented CBNRM. A significant literature exists, for example, on JFM and PIM in India that highlights the limited powers given to communities and the fact that besides the 'new' concern for environmental management, the state's thinking continues to be very much situated in old state-centred thinking (Kolavalli, 1995; Sundar et al., 2001). Even initiatives celebrated in the media, such as the Rajiv Gandhi Watershed Mission in Madhya Pradesh, have come in for criticism from scholars (Baviskar, 2002). There is also evidence to suggest that even older community-initiated collective action is being stifled by excessive government intervention under the guise of the new 'joint' management programmes. Experiments with CFM in Orissa, for example, have been incorporated within official JFM strategies, hence reducing the autonomy of community involvement in the management of forests and increasing the role of forest department staff (Edmunds et al., 2003). Similar experiences exist with regard to the 'institutionalisation' of participation in tank management. State-initiated devolution efforts are generally in worse shape—while local bodies may be in place and even implementing developmental programmes, control of natural resources has almost never been handed over to them (Lélé, 2005).[6]

This critique, however, applies to NGOs as well. Kamat, using the example of activists who started taking up developmental work in Maharashtra, highlights the manner in which the NGO developmental agenda is shaped and directed by wider hegemonic (apolitical) discourses of development resulting in substantive redistributive forms of development being sidelined, something that a priori at least appears to hold true for the vast majority of NGO-initiated CBNRM experiments (Kamat, 2002). Other critiques have also highlighted the dangers of strategies such as enterprise-based conservation where nature becomes a commodity and communities increasingly see their environment in individualistic market-oriented terms, very much part of the neo-liberal agenda (Anderson, 2000).

The second critique pertains to the 'community' in CBNRM. The discursive limits to CBNRM highlighted earlier are central to the manner in which 'community' has been problematised. Many scholars have argued that in current development practice, community is visualised as primarily a rational, economic space that gives cultural and historical aspects peripheral importance. Not only culture but even mediating categories such as class, caste, gender and race are poorly understood in forming community. Oblivious of internal differentiation and history and the cultural context of community, and even the state's influence in structuring social relations and community space, contemporary development practice is seen to be based upon a conceptualisation of community as a small, locally situated, harmonious and autonomous social (in fact, economic) formation (Agrawal and Ostrom, 2001; Baviskar, 2002; Ferguson, 1990; Manor, 1999; Mosse, 2003a, 2003b; Rossi, 2004; Sundar et al., 2001).

While Li (1996) has highlighted how 'idealised' notions of community are a means by which to highlight the need for decentralisation and the privileging of the local community, our concern here is with what this implies in terms of the working of CBNRM. A number of points need to be mentioned: (i) communities are often envisaged as communities of shared understanding (Agrawal and Gibson, 1999; Agrawal and Ostrom, 2001) and hence internal differentiations are ignored; (ii) imagining such communities of shared understanding results in certain voices within the community being privileged over others; (iii) the priorities of the community (in a differentiated sense) are not adequately examined and (iv) communities are viewed as 'autonomous' and in opposition to the state often resulting in simplified claims that community-based management is the solution. One dimension of this critique is elite capture and the manner in which women and socially-disenfranchised castes and classes are routinely excluded from the space of community participation (Agrawal, 2001a; Agrawal and Gibson, 1999; Baland and Platteau, 1996; Harriss and Alavi, 1989; Manor, 2004; Mollinga, 2002). To what extent questions of *whose resource management* and *what kind of resource management* are problematised in CBNRM is open to question.

The results of such simplified assumptions are visible in the actual practice of CBNRM at a number of levels. Community-based management is often envisaged at the scale of the community alone and attempts are primarily made to forge community-level NRM groups. In the case of government schemes such as PIM and JFM, the community is envisaged largely in terms of user groups that invariably have government officials as ex-officio members.[7] The initiatives of NGOs are rarely very different. Collective action is privileged and conflicting claims and needs are sought to be managed. The user groups themselves have limited internal democracy and by their very nature are less than representative of the community. Furthermore, although 'participatory' methodologies are often used, it is very often the case that exercises such as resource mapping pay inadequate attention to differential and competing needs. As a result, little attempt is made to engage with heterogeneous needs of the community or to address inequities (Jairath, 1999; Meinzen-Dick and Palanisami, 2001; Mosse, 1999).

The third critique that links the prior two is that of the pressures of projectised implementation through large external agencies and the manner in which the process works on the ground. Mosse (2003a, 2005) and Baviskar (2002), for example, highlight the manner in which targets and achievements often drive the nature of intervention. While Mosse highlights the manner in which specific project decisions are thrust upon communities due to the pressures on staff in NGOs to deliver desirable outcomes, Baviskar also focuses on how particular villages are selected to make the 'process' more workable. Whether or not these constraints are endemic to community-based management requires, as Mosse correctly argues, more ethnographies of development.

There are a number of reasons, despite these critiques and their applicability to NGO-driven development and CBNRM more particularly, that more detailed, comparative studies are required. First, there is still a relatively scant literature on NGO-driven CBNRM in South Asia. Much of the literature on CBNRM in South Asia, in fact, is 'grey' unpublished literature that invariably is more descriptive than analytical. Second, NGO-driven CBNRM is at best an umbrella term that describes a vast array of experiments that could be very different (Fisher, 1997; Mencher, 1999; Vohra, 1990) hence the danger of generalising based on a small sample of studies.[8] Third, there are a number of NGOs which appear to be cognisant of the critiques of NGO-driven development and which have ostensibly tried to grapple with them. It is not obvious that all practitioners of CBNRM have confined themselves to 'mainstream' ideas of development or are confining themselves to the neo-liberal paradigm of development more recently, blindly rejecting any role for the state or blindly embracing the market. Fourth, there is still a need for 'a more tempered and more nuanced analysis' (Weisgrau, 1997: 3) with regard to the possibilities/limits of NGO-driven CBNRM that explores the reasons for the types of interventions that NGOs make and the nature of outcomes.

As Mosse has most convincingly argued in the context of tank irrigation in Tamil Nadu (Mosse, 1997, 1999, 2003a, 2003b), but equally applicable to other regions and resources, communities are culturally and historically rooted formations. Hence, the relationship between communities and the environment is socially constructed in different ways in different landscapes. It is critical to remember this in trying to understand the impact of NGO intervention in particular landscapes. In other words, NGO-driven CBNRM needs to be examined keeping in mind the historical and cultural dynamics of community relationships to the natural resource base and assessed keeping in mind the critiques discussed here.

Framework, Research Questions and Methodology

We hope, therefore, that our study can unpackage in more detail the working of NGO-driven CBNRM and the factors that shape these workings. We have chosen six case studies that have some definite elements that can be identified as 'innovative' and have also been looked upon as possible 'models' on which to 'design' further CBNRM interventions. In other words, these experiments are prima facie interesting in certain ways. By unpacking the details of these different experiments, we hope to capture the working of CBNRM, how NGOs grapple with possible micro constraints and macro discourses of development in practice (if at all) and assess in a wider sense the potential and limits of NGO-driven CBNRM.

In undertaking our studies of different CBNRM experiments it is necessary to also look at how we assess particular experiments. It is our contention that most CBNRM initiatives envisage different forms of development with emphases on four main issues: livelihood enhancement, sustainability, equity and democratic decentralisation. All of these concerns can no doubt have multiple meanings. Different experiments place varied emphasis on these four 'normative' concerns, yet all of them aim at improving the overall quality of life of all actors given specific historical and cultural boundaries. We try to explore how the various project implementers grapple with these concerns and what they mean by them.

It is important to point out here, however, that we have our own understanding of these concerns. The four concerns of livelihood enhancement, sustainability, equity and democratic decentralisation, we believe, should be intrinsically linked. Livelihood enhancement needs to be ecologically sustainable which means that not only should CBNRM initiatives aim to regenerate resources, but also ensure that mechanisms are in place to ensure that they are not over-exploited. Equally important, livelihood enhancement should be cognisant of distributive/equity concerns. Many experiments assume that interventions will help the poor and leave it at that. But we are concerned that the marginalised are 'targeted' more directly and that structural forms of inequality should be addressed. This is, moreover, central to our understanding of democratic decentralisation. Livelihood enchancement should be achieved not only by getting the 'community' to participate, or by achieving greater decentralisation for the community, but also through a process of democratisation in which the marginalised have an increasing role to play in the domain of decision-making.

We have a number of research questions in mind:

(*a*) What have been the outcomes of different NGO-driven CBNRM initiatives in terms of addressing issues of livelihood enhancement, sustainability, equity and democratic decentralisation?

(*b*) What are the factors that influence the nature of NGO interventions and the manner in which these issues are addressed and materialise?

(*c*) What does our analysis of the different cases tell us comparatively and more generally about NGO-driven CBNRM in South Asia?

The overarching aim of our enquiry is to undertake a substantive analysis of a number of NGO-driven CBNRM experiments, understand their outcomes and the multiple factors that shape these outcomes (the first and second objectives). In particular, we explore how the visions of the different project implementers (NGOs), their strategies (shaped by the socio-ecological context and concern for 'community formation') and state policies affect both choices made and the outcomes (the second objective). These case studies will allow us to say something about the direction NGO-driven CBNRM seems to have taken over the last 10–15 years and its possibilities and limits (the third objective).

Our study does not, therefore, engage with other forms of CBNRM (detailed earlier) or assess the role of social/political movements in development. This is no doubt a lacuna in our study, but one that is partly deliberate. Given the fact that NGO-driven CBNRM accounts not only for a large percentage of CBNRM cases, but also a significant amount of money, we felt there was an urgent need to study it in detail.

Our methodology combines secondary analysis of a vast literature on CBNRM with fieldwork. We make use of the more academic literature on community-based development to contextualise our own study. Chapters 2–7 are individual case studies that have been selected because they are in various degrees considered to be 'innovative' or 'successful' initiatives.[9] The emphasis in the case studies is on understanding the social and ecological context that affect community formation and the nature of CBNRM. The outcome of the CBNRM initiative is understood with reference to the strategies and approaches adopted by the initiating agencies, which are compared, evaluated and interpreted in the light of the actions, expectations and choices made by various sections of rural people. We have made use of a number of sources of information. Information about the CBNRM interventions was collected through published and unpublished reports, both from the agency itself and at times from academic studies. Overall information about the village (population, land use, cropping pattern, land ownership, etc.) was obtained from a variety of data sources such as the census and village-level records. Many of the implementing NGOs also shared with us any data they had. We undertook transect walks to traverse the landscape and view for ourselves the types of interventions made.

The main material is, however, derived from narratives obtained from various sections of the village(s) in light of the CBNRM initiative. We conducted focus group discussions at different scales: with villagers as a whole, with selected 'communities' within the village (landed, landless, Dalits, etc.) and with organisations such as self-help groups (SHGs) and user groups. More detailed discussions were conducted with individuals, selected based on a number of criteria: their location in the watershed (or other ecosystem), their social and economic status, whether they played a prominent role in the initiative, whether they had critiques of the initiative, etc. However, as highlighted earlier, the case studies are not detailed ethnographies. Rather, they are attempts to situate the CBNRM experiment in the wider socio-historical and cultural as well as political economy context of rural life. The limited amount of time available for the fieldwork is likely to prove a significant limitation of the study. We still hope that the insights generated by intensive interaction with villagers[10] over a period of 12–15 days for each case study provides enough substantive material for raising questions and identifying issues with respect to larger processes concerning CBNRM initiatives.

The six case studies that we have chosen are: (*i*) watershed development in Hivre Bazar, Maharashtra, (*ii*) Utthan's watershed development work in Nathugadh village in Gujarat, (*iii*) Tarun Bharat Sangh's

(TBS) revival of water-harvesting structures in Gopalpura, Rajasthan, (*iv*) RNR Research Centre Bajo's watershed development programme in the Lingmuteychhu Watershed in Bhutan, (*v*) the multi-sectoral work of Doodha Toli Lok Vikas Sansthan (DTLVS) in Paudi Garhwal, Uttarakhand and (*vi*) the sustainable environment management programme implemented by Gono Chetona (GC) in the chars in northern Bangladesh. We tried to choose initiatives that were different in terms of the project implementers. Although our focus was almost totally on NGO-based initiatives, barring the Lingmuteychhu case in Bhutan, the nature of the NGOs were significantly different. For example, Yashwant Krishi Gram and Watershed Development Trust in Hivre Bazar was a local village-based NGO overlapping with the gram panchayat and the DTLVS a relatively small-scale NGO working with limited budgetary support. On the other hand, TBS, Utthan and GC are bigger NGOs with substantial funding from donor agencies. Finally, the Bhutan case study provides a contrast as the implementing agency is a government-funded research centre. In the context of Bhutan, renewable natural resource (RNR) research centres have 'autonomy' in similar ways that NGOs have elsewhere in South Asia.[11] Moreover, in the Bhutan context, NGOs have a relatively minor role to play. Basic information on the location and nature of intervention are given in Table 1.1. Their geographic location is indicated in Figure 1.1.

The selection of these case studies was also based on a couple of other criteria: (*i*) regional diversity across South Asia and (*ii*) a cross-section of initiatives in terms of agro-climatic conditions. During the course of our fieldwork, what also emerged was that there was diversity in terms of the scale of our study: some were village-based enquiries, others involved two villages and a couple a number of villages (Bhutan and Bangladesh).

In Chapter 8, we try to draw comparisons across case studies and locate them in a wider analysis of NGO-driven CBNRM and its limits. We do this to comment upon the practice of CBNRM and to highlight issues that need further research.

TABLE 1.1

Summary of Case Study Locations, Agro-ecological Conditions and Intervention Areas

Name of the Case Study	Location	Agro-ecological Conditions	Type of Initiative
Yashwant Vikas Sangathan, Hivre Bazar	Hivre Bazar, Ahmednagar district, Maharashtra	Semi-arid tropical, undulating	Integrated watershed development
Utthan, Nathugadh	Ghogha taluka, Bhavnagar district, Gujarat	Semi-arid tropical	Integrated watershed development, piped drinking water supply
Tarun Bharat Sangh, Gopalpura	Gopalpura, Alwar district, Rajasthan	Semi-arid tropical, undulating	Water harvesting
RNR Research Centre, Bajothang, Lingmuteychhu Watershed	West-central Bhutan	Temperate, mountainous	Forestry, improvement in agricultural and animal husbandry, marketing
Doodha Toli Lok Vikas Sansthan	Thalisain tehsil, Paudi Garhwal district, Uttarakhand	Sub-temperate, mountainous	Forestry, horticulture (walnut), drinking water, fodder
Gono Chetona, Jamuna–Brahmaputra, Chars	Char villages in Gaibanda and Jamalpur districts, Bangladesh	Tropical floodplains	Flood proofing, improved agriculture, horticulture, poultry farming, animal husbandry, sanitation

FIGURE 1.1
Location of the Six Case Studies

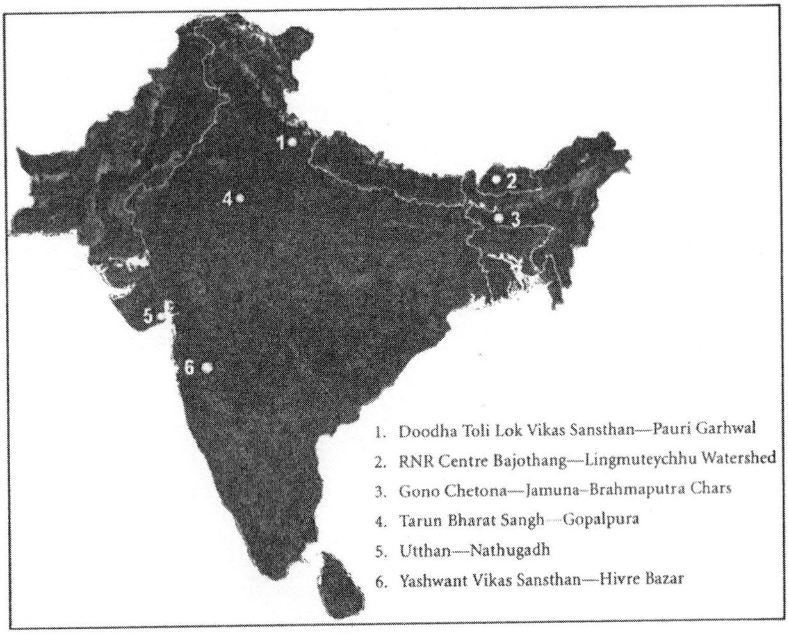

1. Doodha Toli Lok Vikas Sansthan—Pauri Garhwal
2. RNR Centre Bajothang—Lingmuteychhu Watershed
3. Gono Chetona—Jamuna–Brahmaputra Chars
4. Tarun Bharat Sangh—Gopalpura
5. Utthan—Nathugadh
6. Yashwant Vikas Sansthan—Hivre Bazar

Notes

1. Related to this is the normative goal of ensuring more cultural diversity or protection of the identities of certain ethnic groups, or more generally maintaining
   a certain lifestyle (Uphoff, 1998).
2. For example, Korten (1986) talks of three reasons for getting into CB(N)RM: (i)
   local people can adapt the centralised design to suit local conditions, (ii) local
   people can mobilise local resources and (iii) increases local accountability.
3. For a critique of Bangladesh forestry, see Gain (2002).
4. The six projects are the CHT Development Project, Community Forestry
   Development Project, Forestry Sector Project, Biodiversity Conservation in the
   Sundarban Reserve Project, Coastal Greenbelt Project, and Upazila Afforestation
   and Nursery Development Project (Khan et al., 2004)
5. See for Sukhomajri (Chopra et al., 1988; Seckler and Joshi, 1981; SPWD, 1984);
   for Pani Panchayat (Apte, 2001; Pangare and Lokur, 1996; Salunkhe et al.,
   2000); for Ralegaon Siddhi (Anonymous, n.d.c; Antia and Kadekodi, 2002;
   Gunjal and Deshmukh, 1998); for Aga Khan Rural Support Programme, Chitral
   (Gloekler and Seeley, 2003; Husain, 1992; Hussein and Plateau, 2003; Najam,
   2003; Wood and Malik, 2003; Wood and Shakil, 2003); for Aga Khan Rural Support

Programme, India (Agrawal, 1999a, 1999b; Joshi et al., 2004; Parthasarathy et al., 1994; Shah, 1996; Shah et al., 1994a, 1994b; Shah and Shah, 1999; Sinha and Sinha, 1996; Underwood, 1997); for Chakriya Vikas Pranali (Kadekodi et al., 1991; Kumar, 1993; Roy et al., 2001); for DHAN Foundation (DHAN Foundation, 2004).

6. There is a wider critical literature on state-implemented programmes that highlights other limitations, namely:

  (i) biophysical impacts over longer temporal and larger spatial scales are poorly understood and could be negative,

  (ii) institutional structure is unclear as multiple ad-hoc user groups proliferate while local self-government institutions are left out and

  (iii) fiscal efficiency and sustainability are highly questionable since funds are borrowed and spent by line departments (Jairath, 1999; Jeffery and Sundar, 1999; Kerr et al., 2002; Lélé et al., 2005; Saxena et al., 1997; Sundar et al., 2001).

7. Manor (2004) focuses on how the second wave of decentralisation, namely that of user groups, is intended to happen outside the political process and hence breeds 'stakeholderism' and not actual empowerment of marginal people.

8. There is already a fairly vast literature that attempts to characterise NGOs in terms of their compostion, size, sources of funding, area of focus and so on (Fisher, 1997; Mencher, 1999; Uphoff, 1998). While such typologies are useful in classifying NGOs according to different characteristics, we have not explicitly made use of such categories in selecting our case studies.

9. All the case study chapters were sent to the concerned organisations for their comments and feedback. Accordingly, we have made changes to the chapters where deemed necessary to address mostly 'factual' inaccuracies. The usual disclaimers, however, apply.

10. While the names of villagers whom we interviewed and quoted in this report have been changed to protect their identity, village names, names of implementing agencies, their staff and representatives of village-level formal institutions have not been changed.

11. We are not claiming that NGOs are, in fact, autonomous, that is, that they are not affected by wider discourses and practices of development. However, it is this claim that provides the main justification for mainstreaming NGO-driven CBNRM and hence we are examining that claim.

2

Hivre Bazar

A 'MODEL' WATERSHED EXPERIMENT?

Introduction

Hivre Bazar, a village in Nagar taluka of Ahmednagar district, was chosen as an Adarsh Gaon Yojana (AGY, literally meaning Model Village Scheme) village in the early 1990s. The AGY was conceived by the government of Maharashtra to replicate the experiment in Ralegaon Siddhi, a very prominent and much cited example of participatory and sustainable watershed development. Hivre Bazar today is also considered a 'success' story like Ralegaon Siddhi and has become a 'model' village visited regularly by bureaucrats, development practitioners and even villagers from other parts of the country.

What sets Hivre Bazar apart from many of our other case studies is the fact that it is an initiative that has its roots partly at least in state support for a CBNRM type of approach. As importantly, at another level, it is an example of a village community being in the forefront of decentralised NRM without the intervention of an 'outsider' NGO. While the former is important because it suggests that the state, with a far bigger reach, can play a proactive role of sorts in promoting CBNRM, the latter is important in the search for community-based initiatives that are driven by the 'community'.

This chapter, like the others, is aimed at critically examining the emergence and functioning of a particular CBNRM experiment. Our analysis, in particular, focuses on how efforts were made to create a collective consciousness through both social controls and economic promises of a better future. We examine not only how this collective consciousness was brought about, but how this has shaped the manner in which livelihood enhancement has been envisaged and carried out.

Emergence of the AGY

Maharashtra is divided into nine agro-ecological zones that vary widely in terms of rainfall (Joy et al., 2004). The Sahyadri mountain range separates coastal Maharashtra from the Deccan Plateau. While the western face and mountain tops of the Sahyadris receive rainfall in excess of 2,500 mm, there is a sharp drop in rainfall immediately east of the Sahyadris. The entire stretch encompassing Dhule, Ahmednagar, western parts of Pune, Satara, Sangli and Kolhapur districts constitutes this rain shadow area, with annual rainfall averaging around 500 mm. Ahmednagar district, which includes both Hivre Bazar and Ralegaon Siddhi (the inspiration behind Hivre Bazar), has an average annual rainfall of about 590 mm, but with high variability, ranging from as low as 225 mm to above 900 mm (Brahme, 1983). The district is sharply divided between the plains that have a significant amount of irrigation and a number of sugarcane factories and the hillier regions that fall in the scarcity zone. Hivre Bazar lies in the latter portion of the district.

All discussions about drought and drought proofing in Maharashtra take 1972 as a landmark, the year in which Ahmednagar district barely received 225 mm of rainfall. The extent of disruption to livelihoods in 1972 has been unmatched and hence has left an indelible imprint on Maharashtra's economy as well as society.[1] Immediately after the drought, the Dushkal Nirmulan Samiti (the Drought Eradication Committee) was formed. It included 'experts' as well as social and political activists. The Samiti influenced public consciousness and state policy to the extent that drought-proofing measures became central to the discourse. Rural development strategies and schemes began to increasingly incorporate soil and water conservation components.

The 1980s saw some interesting and pioneering efforts in Maharashtra of CBNRM, especially with regard to water. Ralegaon Siddhi in Ahmednagar district (see Antia and Kadekodi, 2002), Adgaon in Aurangabad district (see Anonymous, n.d.a) and the Pani Panchayat movement in Pune district (see Pangare and Lokur, 1996) are all examples of this. In the early 1990s, the watershed-based approach had begun to be mainstreamed and many experiments were launched by the government, donor agencies and NGOs that focused on community participation and also aimed at improving the sustainability of development interventions, though they were not yet strictly based on watershed concepts. Over the next decade, a spate of such programmes emerged: apart from central government programmes like Drought Prone Areas Programme (DPAP) and National Watershed Development Project for Rainfed Areas (NWDPRA) implemented either through line departments or NGOs, there were some other notable interventions including the Indo-German Watershed Programme started in Pimpalgaon Wagha (a village close to Hivre Bazar), Manavlok's work in Beed district, Action for Agricultural Renewal in Maharashtra or AFARM's work in the Marathwada region, and Gomukh Trust's work in Kolwan Valley of Pune district (Joy et al., 2004). However, it was the Ralegaon Siddhi experiment in particular, as suggested earlier, that led to the formulation of the AGY and the direction that watershed development was to take in Hivre Bazar.

Ralegaon Siddhi, a village in the Ahmednagar district of Maharashtra, suffered from problems of drought, ecological degradation and social problems such as liquor addiction. It witnessed a turnaround in its fortunes because of the inspiring and entrepreneurial leadership provided by Anna Hazare, a retired army driver. The villagers formed their own organisations/institutions to manage the affairs of the village and to regenerate the natural resource base. Most of the investments came through government schemes and departments (like soil con-servation, social forestry, etc.). Anna Hazare saw to it that the schemes were implemented properly. Efforts were also made to regulate resource use: the decision not to allow individual wells behind check dams and encouragement given to collective wells are often cited as examples of this.[2] Social norms and regulations like a ban on alcohol and tobacco were introduced. Ralegaon Siddhi is often projected as a good model of what could be an 'ideal' village, more in line with the Gandhian concept of *gram swarajya* (village self-rule) with the assumption that the solution to rural India's problems lie in creating more 'ideal' villages like Ralegaon Siddhi.

The roots of the AGY are generally traced to a meeting called in 1992, by the then chief minister of Maharashtra, of eminent people to discuss how best to celebrate the golden jubilee of the 1942 Quit India Move-ment. Achyutrao Patwardhan, a renowned and well-respected socialist thinker, urged the chief minister to start the AGY so that the success of Ralegaon Siddhi could be replicated throughout the state, starting a large-scale process of social transformation (Warghade, 2003). The chief minister followed through on this suggestion and the AGY was started with Anna Hazare as the president of the AGY committee. Under the AGY, which started as more than a watershed programme but became only that eventually, one village in each taluka was to be selected. A local NGO selected by the AGY committee was to be the implementing agency. Given the roots of the AGY, it is not surprising that the pre-conditions for being selected as an AGY village were the four *bandis* or bans and *shramdaan* (voluntary labour) implemented in Ralegaon Siddhi as well as the presence of certain physical conditions.[3] That apart, what was unique about the AGY was that it involved a govern-ment attempt to support community-led and NGO-assisted devel-opment with community participation before it had been mainstreamed within official guidelines elsewhere.[4]

Along with the NGO, local institutions were also supposed to play a prominent role. The gram sabha (the village assembly) had a say in the identification and selection of the NGO too. The gram sabha was to be involved by the NGO in all aspects of implementation. In fact, the NGO had the role of assisting the gram sabha in the imple-mentation of the programme. The entire development work in the village was the responsibility of the NGO: planning of activities according to the project requirements, maintaining accounts, creating awareness among the villagers about their social responsibilities, organising shramdaan, organising gram sabha meetings and providing technical support. Though the finances were to be maintained by the NGO, the NGO had to present financial statements in the gram sabha. Also, one person from the village (to be selected by the gram sabha) along with an NGO representative and a technical person, were made the joint signatories of the bank account created specially for each village.

Brief Profile of Hivre Bazar

Hivre Bazar is located in Ahmednagar district about 16 km from the district headquarters (see Figure 2.1). Hivre Bazar is bounded by hillocks on three sides and in that sense approximates an ideal watershed. The hills separate this village from two of its neighbouring villages, Pimpalgaon Wagha and Daithane Gunjal. The main village settlement is situated at the bottom of the hills, towards the upper reaches (south) of the watershed while the better agricultural lands are spread towards the northern side. In the last 15–20 years, a number of people have shifted their habitation to other parts of the village and hence a number of *basti*s (hamlets) have come up.[5]

The village is divided into three sub-watersheds. The largest of these drains the hills on the southern and western side. The second sub-watershed drains the hills of the western side and the third drains the hills of the eastern side. Agricultural lands on the upper reaches (southern side) of the village comprise top soils that have a thin admixture of gravel and pebbles washed down from the hills. The lower reaches of the watershed (on the northern side), on the other hand, have soils with better water retention qualities. Availability of water in these lower reaches is also higher than in any other part of the village, especially after the construction of a check dam in 1982. The best lands in the village are said to lie in the northern and middle portion of the plain area comprising the Phupata, Padir and Gohad bastis. The total

FIGURE 2.1
Location of Hivre Bazar

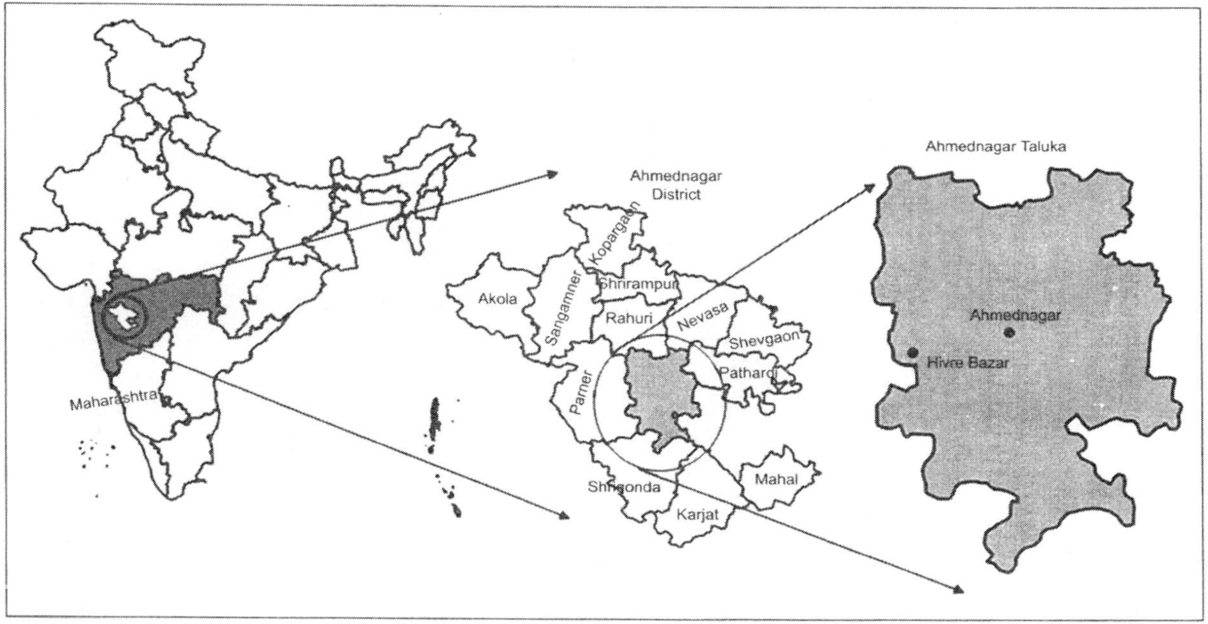

geographical area of the village is about 977 hectares (ha). Out of this, about 795 ha (about 81 per cent) has been classed as cultivable. The remaining area is non-agricultural, including 70 ha forest (about 7 per cent) and wastelands (44 ha, about 4.5 per cent). *Jowar* (sorghum) and *bajra* (pearl millet) are the main crops.

According to the 2001 Census, there are 217 households in Hivre Bazar with a total population of 1,141—589 males and 552 females. There are 69 Scheduled Caste (SC) and four Scheduled Tribe (ST) persons, respectively. Out of a total workforce (main and marginal) of 694, 598 are engaged in agriculture (530 as cultivators and 68 as agricultural labourers). All castes are primarily engaged in agriculture today with bajra and jowar being the major kharif and rabi crops, respectively.

The village is fairly homogeneous in its social composition. The Maratha-Kunbi caste is the dominant community in terms of both number of households (more than 180 of the 200 odd households in the village) and size of land holdings. All Maratha-Kunbi families own land though operational holdings vary greatly in size, from less than a hectare to approximately 20 ha. The Maratha-Kunbi caste is composed of clans known as *bhavki*s in Marathi and often designated by their surnames. In Hivre Bazar, the Thanges are numerically the largest clan with more than 90 households. The other major Maratha clans in the village are the Padirs (about 30 households) and the Pawars (about 25 households). Dalits have no land though some of them may be sharecroppers.

Background to Popatrao's Emergence

Understanding the AGY and CBNRM in Hivre Bazar requires tracing the genealogy of development intervention within the village as seen by the major actors. All of them see the village's history as divided into two parts, before and after the arrival of Popatrao Pawar, the present *sarpanch* (elected head of the institution at the lowest rung of the panchayati raj structure) of the village. The importance of the AGY

and Ralegaon Siddhi to Hivre Bazar need to be examined in the context
of developments within Hivre Bazar that had taken place in actual
fact or at least are very much part of the 'story' behind Popatrao's arrival,
and which paved the way for the emergence of his leadership.[6]

Hivre Bazar, as the story goes, has two kinds of claim to historical
fame. It was an important marketplace within the Maratha kingdom.
It lay very close to the border that later separated the Maratha and
Nizam territories (Warghade, 2003). It was situated on the old trade
route that started from Nalasopara and the Kanheri caves and continued onward through Naneghat and Junnar on to Ahmednagar. It
was only after the advent of the British that other routes came into
prominence. Older villagers speak with pride of the trade that went
on, especially in horses, swords and shields. The other claim to fame
is that it was a village that produced wrestlers. The trade produced
prosperity, however modest it may seem by today's standards, where
'milk and curds were plenty'. The village boasted not only a tradition
of wrestlers, but also *ustads* (coaches) who trained wrestlers. Popatrao's
father, Baguji Pawar was one such wrestler. The villagers, especially
the elders, pride themselves on these two traditions and also say that
there was very little crime, poverty or drunkenness in the village earlier.

The fall from grace again is put down to 1972 and its aftermath.
People link the drought to the economic and social deterioration that
was to follow. A commonly held opinion is that with the arrival of
government money for drought relief and drought mitigation measures that poured in after the drought, local elites scrambled to get a
larger share. Many villagers narrated how internal fighting increased
and how social cleavages emerged as a result of trying to corner money
meant for government relief and development work.

Finally, what is very much part of the collective memory of Hivre
Bazar is the high incidence of drinking and the emergence of illegal
local breweries. It is said that some moneylenders were behind the
entry and prevalence of liquor in the village (Warghade, 2003). They
first drew the wrestlers under their influence and got them habituated
to drinking and later got some of the unemployed persons from lower
castes to set up illegal distilleries to serve the demand. It was a spiral

that grew and finally earned Hivre Bazar the reputation of being a drunkard's village. The situation in the village was so bad that government officials feared starting any programme in the village. Even the teachers in the government primary school considered Hivre Bazar a 'punishment posting'.

This was paralleled by the emergence of the *gram panchayats* (institution at the lowest rung of the panchayati raj institutions or PRIs) and a politics that was turning factional at all levels. Apparently, the leadership in the village (sarpanch and elders) was ineffective in controlling the different factions and the sarpanch's post was a much sought after one because of the money involved. A number of people we met said that they shifted to the hamlets because of the situation prevailing in the village.[7] Clan rivalries are not a new thing, but oral history accounts suggest that earlier there was a greater spirit of accommodation and mutual adjustment between the clans and clan elites/elders kept rivalries in check. The politicisation of village affairs through the gram panchayat elections and the competition for spoils that waited as a reward exacerbated the rivalries and the situation deteriorated rapidly in the 1980s. Things reached a climax in 1982 when a villager succumbed to his injuries in an intra-village fight. The sarpanch during that time, Ganesh Pawar, was also allegedly involved in this fight. Every panchayat election was a virtual war, and approached as such. There was no sign of that much vaunted village unity. The 1989 elections to the gram panchayat were to be held in this tense atmosphere. Complaints by the various factions had already been lodged with the police and people feared that violence was just round the corner.

But there was also another constituency that was growing. Opinion was gaining ground within Hivre Bazar that something needed to be done to change the situation. While their fathers and uncles plotted and fought their factional wars, the youth had become increasingly disillusioned with the turn that events had taken. The very old still remembered the village as it once was and could be, and hence the youth and elders came together to search for a new alternative, for a leadership that could take them away from the mess that the village had gotten into. They were soon to find the man they were looking for.

The Arrival of Popatrao Pawar

The man who was soon to answer their call was Popatrao Pawar, son of the renowned wrestler and ustad Baguji Pawar. He was born in 1960 in Kedgaon in his mother's parental home, had studied till fourth standard in Hivre Bazar and then was sent to Kedgaon by his parents in view of the 'poor atmosphere' in Hivre Bazar. He used to visit the village regularly and was quite popular among the youth because of his cricketing talents. Moreover, he was not identified with any of the factions within the village. The higher education he received in the city also helped to improve his image amongst the people of Hivre Bazar. Popatrao was at that time completing his M.Com. degree and in search of a job.

With the threat of imminent violence hanging over the village, the youth of the village requested Popatrao Pawar to return to the village and stand for gram panchayat elections. Popatrao was a reluctant candidate and even his close friends and family members discouraged him from taking up the post. Finally, he agreed to run for office, but he did so by calling a gram sabha and telling the villagers in a long speech that he was accepting the invitation not for the power it would bring, but for the possibilities of improvements that he saw for the village. In this gram sabha he also declared that if the village did not bury its differences within six months, he would renounce his post and return to Ahmednagar. He stood and was elected unopposed to the gram panchayat and as sarpanch in 1989.

Popatrao Pawar was quite aware of the political and social situation in the village. He was also aware of the fact that in order to change the situation in the village for the better he needed to garner the support (albeit passive perhaps) of his fellow villagers. He knew that he had to prove himself true to his gram sabha speech. It is not surprising, therefore, that his initial approach as sarpanch was one of choosing issues which faced least possible resistance and were felt needs that had the potential to garner large support within the village. The first issue he took up was that of drinking water. There were two handpumps in the village that were not functioning and women had to walk 2–3 km to fetch water. In 1990, he managed to get the panchayat *samiti*[8] at the taluka level to install new hand pumps in place of the two that were beyond repair (Warghade, 2003).

But what really got the village firmly behind him was the interest he showed in the village school and the tenacity with which he pursued and continues to pursue the issue. Before he became the sarpanch, the teachers of the village school were not interested in teaching and were mostly on leave, the students did not attend classes and the school building was in a bad shape. One of the first things he did in 1989 was to get these teachers transferred and get honest teachers appointed. He also got the school building renovated with villagers (mostly youth) undertaking shramdaan. The school had a good playground, but a few dilapidated structures and a *masjid* (Muslim place of worship) stood on it. He appealed to the owners, and the owners along with Banabhai Sayyad who owned the masjid, readily made way for the playground. In 1993, the school got permission to start teaching students in the fifth standard. In 1994, the school got sanction to teach students upto the seventh standard. Popatrao Pawar also tried to start a high school and despite the fact that the district administration did not give financial support for a new building or teachers' salaries, he organised the village youth to start teaching the students in makeshift buildings. After the good performance of the school, the district administration granted it financial support. Today its students have won many awards, the school has won the model school award for the district and one of its teachers, Rohidas Padir, has been adjudged a model teacher for the district (Warghade, 2003). The other major activity undertaken was the renovation of the village temple.[9] Interestingly, one of the early initiatives of tree planting through a grant from the social forestry department (again through shramdaan) failed because the tree saplings were uprooted.

It was also at this time that the sarpanch consulted more with his fellow villagers to address concerns they felt most important given their precarious livelihood situation. Of particular concern to all villagers was water scarcity and agricultural productivity (more so for the landed). Scarcity of water often meant only one crop and often poor productivity at that. The landless, marginal and small farmers were the worst affected. Migration for work was, therefore, the easiest (and maybe the only) option for most people (landless, small and marginal farmers) in the village. The Employment Guarantee Scheme (EGS)[10] and wage labour outside of the village was a common way to earn a livelihood even for medium farmers. Drinking water too was in short supply during summers and people had to wait for government tankers. In the absence of any vegetative cover and any structures to harvest water (the two water-harvesting structures built in 1972 and 1982 had structural problems and could not hold water for long), all the rainwater used to flow away. So even in a year of normal rainfall, wells used to go dry by December or January. Fodder and fuelwood were in short supply and there were no attempts to regulate open grazing and the cutting of trees.

By early 1990, Popatrao was increasingly being influenced by Ralegaon Siddhi and Anna Hazare. He had started looking for different sources of support for watershed development in the village. To start with, he approached the Indo-German Watershed Programme that was active in the neighbouring village (Pimpalgaon Wagha), but was reportedly turned down because Hivre Bazar had a bad reputation (Warghade, 2003). In 1993, the social forestry department in Ahmednagar district came to his help. Seventy hectares of forest land was treated with 40,000 contour trenches. The area was planted with various kinds of indigenous trees. Grazing was banned in the area, though people were allowed to graze their animals on adjacent private wastelands. In the gram sabha it was decided that everyone wanting fodder from the forest should pay Rs 100 to the village fund. They would then be free to collect fodder so long as they did it with the help of a sickle and carried it away by head load. That year the fodder served villagers through the four most difficult months and the village fund was richer by Rs 30,000. The trenches had a delayed but visible effect. Though nothing happened in the monsoon months, by the end of the monsoon, many adjacent wells were still full and the area they served increased from 20 ha to about 70 ha. This was a visible demonstration of the promise that watershed development held.

Thus, the material and social conditions existing at that time in Hivre Bazar helped in the emergence of Popatrao's leadership. A visit to Ralegaon Siddhi helped impress upon people the possibilities that existed through watershed development initiatives. Promises of tangible benefits played a major role in forging a community in Hivre Bazar. Promises of increased agricultural productivity, better availability of drinking water and more employment opportunities, all visible in Ralegaon Siddhi, acted as a motivating force to villagers. The task of community building, however, remained the major challenge. The next section examines how this was undertaken.

The AGY and Beyond: The Building of a 'Community'

Hivre Bazar was one of the 300 villages first selected under the AGY. Pre-AGY work in the village provided the basis on which Hivre Bazar could stake a claim to be considered a potential AGY village. The fact that Popatrao Pawar was personally close to Anna Hazare helped as well. The work under the AGY began in 1994–95 and tapered off by 2003. The bulk of the physical work was done before 2000. Popatrao's main task was to persuade villagers about the physical works that were needed in the village to increase the availability of water. The gram sabha was used to discuss the details of the project. The first step that they took was to form an NGO (Yashwant Krishi Gram and Panlot Vikas Sanstha [YKGPVS]—literally Yashwant Agriculture, Village and Watershed Development Trust) as they did not want an outsider organisation to implement the programme.[11] In other words, the NGO comprised local villagers only and in fact had a number of members common with the gram panchayat (which also doubled up as a watershed committee). Sub-committees were formed in each of the watersheds. The NGO employed some technical staff to make plans and prepare technical reports and also recruited volunteers to do the organisational work. The plans were then discussed in the gram sabha and opinions sought from people. If anybody's land was to be affected by planned new structures or if any land was needed for some structure then that villager was requested to make a sacrifice.

Unlike in many other watershed programmes launched in the late 1990s (Karnataka Watershed Development Project or KAWAD; Andhra

Pradesh Rural Livelihoods Projects or APRLP), in the AGY there was no provision for physical works on private agricultural lands, though there was provision for the construction of *nala* (stream) bunds and check dams. All the soil and water conservation works in Hivre Bazar were implemented on common lands or on private wastelands/grass-lands.[12] Continuous contour trenches (CCTs) were dug on the hill slopes to arrest the erosion of soil, harvest water and encourage the growth of grass. The CCTs were dug mainly on panchayat lands and on some private lands also.[13] A number of water-harvesting structures like check dams, percolation tanks and loose boulder structures were also built in the village. Plantations on forest lands and roadsides were also part of the programme.

When the AGY was launched in Maharashtra, there were some pre-conditions set for selection of villages as mentioned earlier. Most important were the four bandis made famous by the Ralegaon Siddhi experience. The four bandis were *kurhad* bandi (ban on the felling of trees), *charai* bandi (ban on free grazing), *nasbandi* (family planning) and *nashabandi* (ban on liquor). People also had to agree to a certain amount of shramdaan, except for the landless who were exempt from it.

The significance of these five ideals needs to be understood keeping in mind the situation in Hivre Bazar at that time (the late 1980s). The felling of trees and open grazing were common amongst both rich and poor households. The surrounding hillocks, according to many local people, had a barren look, soil erosion was prominent and groundwater levels were very low. In addition to this, fodder and fuelwood shortage was common in the village. The ban on liquor, as discussed in detail earlier, had both a historical and moral dimension to it. People tended to look at wage labour for relief work as an opportunity to collude with authorities and to receive wages without doing commensurate work. Shramdaan was meant to tackle this problem, harness a collective spirit and express a commitment to the interest of the village. Collectively, these measures were meant to promote community feeling in the village.

There are other bans in the village which were not part of the AGY but were added later. These are bans on the use of bore wells for irrigation, growing sugarcane and banana and selling one's land to outsiders.[14] These measures illustrate that issues of long-term sustainability (especially in terms of water use) were very much central to the vision of watershed development in Hivre Bazar. Another novel, and in a way bold, social measure that has been adopted is that anyone marrying in the village needs to undergo the ELISA test (for AIDS).[15]

The bandis were not mere proclamations but it did take effort and mobilisation to maintain them. The bandis were instruments of community building; more and more people had to identify with the common purpose that lay behind the bandis. But it was not always a smooth affair. For example, take the case of the ban on liquor. Till almost 1997, Popatrao ran the campaign on a low key. Later that year, members of the Gopal community, traditionally involved in brewing liquor, were first requested to end their business (some people also mentioned that they were forcibly evicted). They were offered loans to start new businesses and some of them bought cows and buffaloes. A few of them left the village. Apart from exerting 'moral' pressure, at times the village youth also resorted to physical violence, probably a spin off from having Ralegaon Siddhi as a role model. Despite the use of aggressive tactics, the attempt to stop the liquor trade in Hivre Bazar won support from many, especially women.

Another occasion where 'community formation' faced hurdles was with the selection of a police *patil*[16] for the village in 1995. Factional divisions emerged and the 'old' atmosphere returned (Warghade, 2003). Popatrao withdrew and refused to come to the village for almost six months. He was finally persuaded by the youth that they would not let such a thing happen again and he returned.

The wider social challenge that Popatrao faced was that of inter-clan (*bhavki*) rivalry. Although Hivre Bazar is dominated by the Maratha-Kunbi caste, there has historically been a rivalry between the Thanges and Pawars, both belonging to the Maratha-Kunbi caste. Although Thanges are numerically a majority in the village, most of the big landowners are Pawars. The sarpanch, a Pawar himself, has made it a point to shy away from clan allegiances. Next we illustrate in more detail how the agenda for development has addressed the needs of the landed castes as a whole; suffice it to say that such an agenda has been well received by the Thanges, the Pawars and other landed castes and has acted as a means to keep inter-clan tensions out of the limelight. Moreover, the sarpanch has made it a point to constantly engage with all communities in an informal way. The fact that he himself has married a Thange woman from the same village has not hurt matters. One should also not forget that his attempts at social reforms (education, anti-liquor drive) had enough support across caste lines. He is no longer identified as a 'Pawar' by the villagers. In a way one can say that through the combined effect of development, social reforms and personal behaviour, Popatrao has been able to help the village community transcend clan, caste and other rivalries and the interests they represent and project the image of a unified community to the outside world (this is not to say that all the internal contradictions have been resolved).

Multiple strategies were, therefore, used to garner support for a new approach towards development in Hivre Bazar. Although, as we shall illustrate in more detail later, many of the measures taken originated from the sarpanch and a few others, the fact that a better life was promised to people meant that overall the people in Hivre Bazar were willing to support such initiatives. Developments over the last decade have bolstered collective support for the sarpanch and also given the village an identity of its own. From a fragmented, degenerated, divided village it has attained an identity—a collective identity that is important in the context of CBNRM. This is not to say that there are no internal asymmetries or power relations or no exploitation. But today the villagers, cutting across class, caste and gender, see themselves as part of this new identity. They feel proud that Hivre Bazar is considered an 'adarsh gaon'—an ideal village—in Maharashtra and the fact that thousands of people including officials and politicians visit the village. Moreover, 'their son'—Popatrao—attended the World Water Conference in Japan! Not only is Hivre Bazar now a household name in rural development circles, à la Ralegaon Siddhi, but the 'brand name' Hivre Bazar has become sellable. Hence, the willingness to support the sarpanch has not been only because of promises of economic benefits, but also due to the label that goes with Hivre Bazar. The next section explores whether Hivre Bazar's 'ideal' village status has been warranted or not.

Livelihood Enhancement:
Changing Trajectories

As highlighted earlier, a number of watershed-based activities have been undertaken in Hivre Bazar including CCTs, building of water-harvesting structures and social forestry. These have been supplemented with some restrictions that include a ban on the use of bore wells for irrigation and a ban on water-intensive cash crops like sugarcane and banana. Our transect walk across the watershed revealed a good forest cover especially in the upper reaches of the catchment and a significant grass cover on many of the hill slopes. Moreover, unlike in many afforestation projects, in Hivre Bazar a number of indigenous varieties of trees such as *babool*, tamarind, *shisham*, and *khair* have been planted along with fast-growing exotic varieties like gliricidia, *subabul*, eucalyptus and Australian babul.[17] Many of the water-harvesting structures had considerable water in them even in December (though it should be noted that the village witnessed an above normal rainfall in 2004).

For villagers locally, the more important question, however, is to what extent water augmentation activities (WAA) have benefited them or in some cases hindered them in their quest for meeting their livelihood needs. As highlighted earlier, the main aim of watershed development in Hivre Bazar has been to increase water availability (both for drinking and irrigation purposes) and to improve agricultural productivity—the various rules and regulations have been aimed at meeting these priorities.

According to everyone we spoke to, WAA have resulted in increased water availability. The parrot-like fashion in which people made this claim was substantiated upon by more detailed stories as to how this was true. The most common claim was that various measures adopted to impound water have helped recharge groundwater and that this was noticeable in terms of the water levels of the wells. Lalit Pawar, who has about 10 ha of land and four wells, says he pumps water for about 6–8 hours in a day (when there is electricity). As a result, the water level goes down by about 5 ft, but it takes only a couple of hours for the water level to come back to the pre-pumping level, something he says was never the case before the AGY.

The very fact that there has been a more than two-fold increase in the number of wells in the last 10 years illustrates not only that water availability has increased, but that it is due to watershed activities (digging an open well would cost between Rs 50,000 and 100,000). Also, discussions with villagers from different hamlets of the village confirmed that earlier throughout the village most of the wells in the village used to hold water only till December–January. Now there is water all through the year. The area under protective irrigation has also increased more than three times. According to the NGO, the area irrigated for summer crops has increased from 7 ha to 72 ha. In a year of normal rainfall, like in 2004, there is enough water in the wells to irrigate not only the kharif bajra, but also the rabi jowar and some summer vegetable crops. People are now growing two and sometimes three crops in a year of normal rainfall (see Table 2.1). Even in un-irrigated land, the improvement in soil moisture level has helped to increase productivity. Moreover, the crop basket is considerably more diverse than in the past with people growing cash crops such as potatoes, onions, fruits (grapes and pomegranates)[18] and flowers and wheat.

TABLE 2.1
Cropping Intensity in Hivre Bazar

Land Use	1996–97	1998–99	2002–03
Gross cropped area (ha)	821	1,007	1,125
Net area cropped (ha)	723	730	748
Area cropped more than once (ha)	99	276	377
Cropping intensity	1.14	1.38	1.50

Source: Talathi (village accountant) records.

Wage employment opportunities in the village have also increased. The increased intensity of agriculture and multiple cropping seasons has meant that the demand for labour has increased, although wage rates have increased only slightly from Rs 30 to Rs 40 for women and Rs 50 to Rs 70 for men since WAA started. Perhaps the most significant development is the fact that people have to migrate less than in the past as many households are now growing a second crop. Also, a lot of other activities, like land improvement by individual farmers, construction of buildings, laying of pipelines, etc., have emerged.

As drought was a regular phenomenon between 2000–01 and 2002–03, most farmers, even those with large holdings, had to depend on EGS work. A number of works like construction of the panchayat building, guest house and village roads were started during the drought period. It should also be noted that unlike in other villages where horticulture has been encouraged and hence the demand for labour is low, in Hivre Bazar crops requiring more labour input continue to be grown. Also, the amount of person-hours spent in collecting water, fodder and fuelwood (both in normal rainfall years and drought years), was much higher in the pre-WAA period than at present. The extra hours available now can be spent on earning extra cash through wage labour.

Another major component of the local economy is livestock rearing. Charai bandi introduced in the early 1990s was aimed at reviving the vegetative cover on the hill slopes of the watershed. The restrictions on grazing during the initial treatment period were rotational and some private lands (located on hill slopes) have remained accessible for grazing (see also Box 2.3). Also, people can cut and bring the grass from the common lands (forest land for example).

The other major intervention in terms of the livestock economy has been stall feeding. Stall feeding and cows of better breeds (jersey) have been introduced in the village. This is a normal practice in most of the watershed projects. Stall feeding is encouraged to protect the newly-planted trees. This helps in the process of natural regeneration of vegetation and prevents soil erosion due to open grazing. Stall feeding has also helped increase the production of organic manure. As a result of WAA animal husbandry has become one of the important livelihood options for a lot of people in the village. A number of people have opted for rearing buffaloes as its milk fetches a high price, but the village dairy accepts only cow's milk. These developments are clearly linked to the fact that fodder availability[19] has increased (due to better productivity) and incomes have improved.

Concerted efforts have in fact been made to promote Hivre Bazar's dairy industry as a means to improve the livelihood of all villagers. Loans have been given to many farmers. As a result, the number of milch animals in the village has increased from 19 in pre-WAA days to 476 now. Milk production in the village has also witnessed a more than 20-fold increase from 140 to 3,000 litres per day (according to information provided by YKGPVS). Cross-bred cows (the most popular is Australian jersey) are better suited to stall-feeding and produce more milk than the local variety. While most people sell their milk to the village dairy cooperative, a number of people sell it at Ahmednagar market as they get a better price there.

In addition to the direct benefits to agriculture, a number of other benefits need to be noted. The most important of these is drinking water, an issue that is often ignored in most watershed efforts (Joy et al., 2004). Special care was taken in Hivre Bazar to see that drinking water needs were not compromised: the conscious decision to ban bore wells for irrigation was a means to ensure drinking and domestic water to all households in the village throughout the year. Stories are often told with pride by villagers of how only their village did not have to depend on imported water after three consecutive drought years. Even the neighbouring village of Pimpalgaon Wagha, which was considered a 'success' story of the Indo-German watershed programme, had to depend on water tankers. The government has also constructed 12 handpumps at various locations within the village intended mostly for those who do not have private sources of water. In fact, in Hivre Bazar, it has been decided that if the rainfall is below normal, first preference is to be given to drinking water, other domestic needs and water for cattle. Therefore, in the years of drought, people are allowed to irrigate only half an acre of their field even if they have water in their own wells.

Importantly, watershed activities in Hivre Bazar have not undermined the importance of other developmental activities. The 'leadership' within Hivre Bazar seems aware of the possible limitations of watershed development as a sustainable livelihoods strategy and hence has tried to find other ('watershed plus activities') ways in which households can supplement their income.[20] One major activity in Hivre Bazar has been self-help groups (SHGs). The SHGs have had two main functions: (*i*) to generate savings (that could be used later to start some gainful employment); and (*ii*) to improve the overall well-being of women. In Hivre Bazar, three SHGs were started at the outset, but the only one that seems to be functioning at all is the below poverty line (BPL) group. Members of this group have availed of bank loans that they have used mostly to buy goats. It is unclear, however, to what extent even the BPL group is sustainable. Many of the women complained that they were not able to make their contributions last year during the drought. Moreover, the long-term vision of SHGs acting as a platform through which women can diversify their livelihood strategies is currently not being realised. While the panchayat has experimented with other activities such as *papad* making (for which it provided training), the women of the BPL SHG say that they do not have time for such activities due to adequate agriculture-related employment.

The limited success of SHGs in Hivre Bazar, however, may have more to do with the specific circumstances in Hivre Bazar than with the limitations of SHGs per se. The fact that agriculture has become more prosperous has also meant that women have more work and hence less time for other activities. In such a scenario, the role of SHGs might have to be different. Also, the SHGs (non-BPL) that have become more or less dysfunctional have become so because of individual households defaulting for reasons other than the lack of ability to make monthly contributions.

Household assets have also increased significantly. The number of tractors have increased from none to around 20–25 (we did not do a survey, but this figure is based on individual interviews that we conducted).[21] Similarly, the presence of motorcycles, television sets and other assets has increased in the village.[22] People have money to reinvest in agriculture and do so in terms of buying pump-sets, digging wells, laying pipelines for irrigation and levelling lands. The use of fertiliser (both organic and chemical) has increased from nine to 163 tonnes per year in the village (according to information provided by YKGPVS). Similarly, people are investing in higher priced high-yielding variety (HYV) seeds. There is also a spate of construction of new houses, especially in the farmlands. While earlier the reason for shifting of houses to farmlands was the infighting in the village, now people are shifting because agricultural activity has increased and because travelling from the main village to the farmlands (a distance of 0–3 km) is difficult.

Equity and Livelihood Enhancement: Who Got How Much?

Implicit in the approach taken in Hivre Bazar are two main assumptions: (*i*) the overall improvement in the status of agriculture will benefit all in the village and (*ii*) inequalities are best addressed in non-redistributive ways. In that sense, clearly the major beneficiaries of watershed interventions are the landed as more water means more irrigated land and likely higher yields. Having said that, the examples illustrated next also suggest that attempts have been made to address the concerns of smaller farmers through informal mechanisms like those related to sharing water. This has at least been partly possible because social barriers have been reduced due to the 'collective' feeling of being a 'citizen' of Hivre Bazar.

As mentioned earlier, a significant transformation of agriculture has taken place in Hivre Bazar. Increased water availability has not only meant more water, better productivity and crop diversification, but has also resulted in a more intense form of commercialised agriculture. These changes are clearly linked to the question of land ownership, that is, those with more land benefit more and are able to avail of the benefits more as well. For example, the ability to construct a well or to take the risk of diversifying the cropping basket is more likely to be borne by better off farmers. Location is, however, also a factor. Although water is available in the entire village, there is a distinct variation between the different sub-watersheds and also within sub-watersheds. Lands in the central and northern parts of the village are the best lands with the highest productivity. The silt content in the soil in this sub-watershed also seems to be more than in other areas in the village. Moreover, wells located near *nalas* or near water-harvesting structures (nala bunds, percolation tanks and storage *bandharas*) have more water in them and retain it for a longer time. Hence, the extent to which class (seen largely in terms of operational holdings) is important given the changes in agriculture locally, must be understood keeping in mind these locational factors as well.

The transformation of agriculture in Hivre Bazar, however, has not been on traditional capitalist lines. For one, as mentioned earlier, there is a rule within the village that land should not be sold to people residing outside the village and hence an extensive market for land has not emerged. Second, access to water, though determined by size of operational holdings and location has been mediated by a few measures: the sharing of wells (with no market for water) and the transport of water through pipes across large distances. Binod Thange who lives in Pawar Basti is a case in point. He had about 3.2 ha of land that was originally not irrigated. In 1986, he dug a well near the percolation tank and laid a pipeline of 4,000 ft (more than a kilometre long) and started irrigating his 3.2 ha of land. Bansi Thange has built up his assets because of the higher returns to irrigated agriculture. He had a flourishing dairy even before the AGY started and was financially in a position to invest in pipelines and a tractor. He also used his profits to buy land (about 2 ha) in the early 1990s. In some cases, the land of poor farmers is advantageously situated in terms of water availability, but they do not have the wherewithal to dig wells. Sharing arrangements between rich and poor farmers have emerged in such situations (see Box 2.1 for an illustrative case).

Box 2.1
An Equitable Arrangement of Water Sharing

Raju Thange comes from a very poor background. He is a primary schoolteacher belonging to the Maratha-Kunbi caste and has about 2.8 ha of land. This land was not very productive prior to the WAA as it was located on the slopes in the upper portion of the watershed where the quality of soil cover is not good. He also does not have a well. His father is an ex-serviceman and after retirement undertook mostly labour work. Rajendra, a large farmer on the adjoining land, however, agreed to invest money to dig a well on Raju's land, as it was favourably situated in terms of groundwater availability, and install a motor and pipeline. He arrived at a water sharing agreement with Raju. Rajendra who invested in the well, has rights to the water for four days, while Raju, the owner of the land, has rights to the water for two days. Now out of the 2.8 ha, 2 ha are under fully irrigated crops and the remaining area is kept for grazing. Access to the well water has made all the difference for Raju. He has invested the savings from his salary as a teacher, levelled the land and brought parts of it under citrus cultivation.

The question of the landless is equally important. At one level, WAA have not benefited the landless at all in terms of access to water. Unlike in experiments such as the Pani Panchayat or Sukhomajri where the landless too were entitled to water, this has not been the case in Hivre Bazar. However, the landless have indirectly benefited, due partly to the increased prosperity of the landed and hence increased requirements for labour and partly because labour itself is scarce (see Box 2.2). Wage employment opportunities have also increased for the rural poor. As pointed out earlier, the increase in intensity of agriculture and cropping seasons has led to an increased demand for labour and a lot of other non-farm activities, like land improvement by individual farmers, construction of buildings and laying of pipelines.

Box 2.2
Employment Opportunities for the Landless

Ramu Dhotre is landless and belongs to the Vadar community. (The Vadars are traditionally involved in construction work, breaking stones, etc. There are two Vadar families in the village.) His father was a contractor in Mumbai. He had to come back to the village because of personal problems—his mother fell ill and he spent all the family's savings on her treatment. Ramu also works as a nurse in a private nursing home in Ahmednagar. His father goes for labour work, especially earth work like excavation of foundations for buildings, pipelines, etc. He believes that employment opportunities in the village have increased. Rather than working as a daily labourer he prefers taking up contracts for specific jobs. Ramu also took a loan of Rs 10,000 from Allahabad Bank and bought five goats. He has already repaid the loan and every six months he earns about Rs 2,000–3,000 from selling the goats. He further added that there is no social discrimination based on caste in the village. He says, 'Marathas also come for our family functions.... Though there are only two families of the Vadar Samaj we have not faced any problems from anybody in the village.' These families also take private wastelands on rent for grazing—this year they have rented some land at the rate of Rs 100 per year for grazing from Anjabai Gunjal.

Sharecropping has also become increasingly prominent in Hivre Bazar.[23] The most common sharecropping arrangement in Hivre Bazar is one in which both the landowner and the sharecropper share the costs and profits in a ratio of 1:1. The total cost does not include the costs of irrigating the fields (the cost of electricity charges for pumping water, the cost of digging a well and the cost of buying a pump set) and land revenue which are borne by the landowner. Due to labour scarcity, the bargaining power of the sharecropper has improved. In many cases, due to this shortage the landowner prepares the field and then does the sowing. From here on till the harvesting the sharecroppers take charge of the operations, though the cost is shared by both. The sharecropper basically takes care of arranging the labour input during the most labour intensive phase. The profit in this case is shared in a ratio of 2:1 between the landowner and the sharecropper. Moreover, due to better water availability, risks have decreased. Many sharecroppers we spoke to said that a sharecropping arrangement was a better option than wage employment (see Box 2.3).

Box 2.3
Innovative Sharecropper

Bhiku Girhe is a small farmer (0.2 ha) and a former liquor trader. He sharecrops for Rameshwar Pawar (one of the biggest landowners in the village). At the time of our visit to the village he had entered into a sharecropping arrangement with the landlord to cultivate potato on a little less than 1 ha of land. He believes that this arrangement suits him as he will earn much more this way than he can through daily labour. Moreover, it involves minimum risk and investment. He deals with the landlord on more equal terms than he would as a daily labourer. He expects to earn a good amount from just one crop of potato. Bhiku also mentioned that he did not stop his liquor business because he was forced to. He did so in the interest of the village. He also donated some of his land to the village for the construction of the village school. He was later compensated for this. He was given another piece of land where he built a house with the help of a government scheme.

The possible negative impact of charai bandi is expected to be felt mostly by the landless, small and marginal farmers who do not have alternative grazing lands. However, this has been overcome because charaibandi initiatives have been devised with this in mind. The restrictions on grazing during the initial treatment period have been rotational and hence some lands have remained accessible for grazing. Moreover, people have been allowed to cut and collect grass from common lands. Yet, many poorer households (especially the landless) prefer to graze their cattle on lands close to their homes or in the fields in which they work because they lose less time that way.

Stall feeding could also be an issue for the poor. The poorer households possess small ruminants and they do not find the option of stall feeding acceptable. Nonetheless, the number of small ruminants in the village does not seem to have declined over time, as is generally witnessed in other places where grazing bans are imposed. Overall, as pointed out earlier, animal husbandry has become an important livelihood option for a lot of people in this village, even for the poor. A number of landless and small farmers too have invested in cattle in order to supplement their income from agricultural labour. Loans were specifically provided to the landless to start new businesses and most of them bought milch animals. Members of the BPL SHG also bought animals from the loans they took. Also, that the 'leadership' within Hivre Bazar has not undermined the importance of other developmental activities has helped poor households supplement their sources of income with other options. During the off-season, the EGS plays a major role in providing employment opportunities especially to those below the poverty line though records on how much employment has been provided are not readily available. Since part of the payment is in kind (that is, grain is given in lieu of payment), resource-poor farmers can meet part of their food requirements in the lean season.[24] Other associated benefits in Hivre Bazar are improved education and health facilities.

Thus, although watershed interventions in Hivre Bazar, like in most watershed experiments, have been biased in favour of the landed, this has not meant that the concerns of the marginal farmers and the landless have not been addressed. We have already illustrated how attempts have been made to address the concerns of marginal and small farmers through water sharing arrangements given the fact that WAA largely benefits the landed. A different example of this in terms of the livestock economy is that of cattle camps. The years 2001–02, 2002–03 and 2003–04 were drought years and farmers were unable to produce more than one crop on average. Partly as a result of this there was also a fodder scarcity and the government had to run a cattle camp. The usual practice is that the government pays money to the institution organising the camp—gram panchayat, village cooperative society, youth club, sugar factory, etc.—on the basis of the number of animals in the camp. This money includes the cost of labour charges for bringing the fodder, setting up the camp and other maintenance charges. Unlike other villages, Hivre Bazar managed the cattle camp in such a way that it saved money: villagers provided shramdaan and also used their own tractors to transport the fodder. The money saved was then used for developing the village milk cooperative. The village leadership is already planning a milk processing unit within the village. As these initiatives are aimed at all, they do not exclude benefits to marginal and smaller farmers.

Having said this, there are specific biases against the poor that have remained unaddressed. For example, kurhad bandi seems to have impacted the poor more. The practice of planting trees on farm bunds and then lopping off branches for fuelwood is quite common in Hivre Bazar, but one needs land to do that. Moreover, it is mostly the well off farmers who have shifted partly or fully to liquid petroleum gas (LPG). The poor either make use of cowdung cakes, the production of which has increased because of enhanced dairy activity, or broken twigs from fields and common lands. Moreover, the poor use a number of alternate sources of fuel, but very often of very low quality, like sawdust (bought from the Ahmednagar market), stalks of bajra/jowar and leaf litter. Some of the landless who work either as daily labourers or sharecroppers in fields of others are allowed by the latter to collect broken twigs from the fields in the off-season. The low numbers of landless in the village is partly responsible for the fact that this ban has not been challenged. No effort has been made to compensate the landless who, unlike the landed, do not have any private source of fuel supply.[25]

The impression one gets from Hivre Bazar, therefore, is that the overall standard of living has improved significantly. Almost all households we spoke to, across land holding size, mentioned that their incomes have increased. A pamphlet of the YKGPVS claims that the WAA have been responsible for the reduction of the number of BPL families. It also claims that the per capita income has increased from Rs 832 before WAA to Rs 11,893 after WAA, an eight-fold increase. The change in the income status is evident from the new assets people now have. Claims made by the YKGPVS, however, about the reduction of people living below the poverty line between 1992 (168) and 2002 (11) need to be seen in light of the recent reduction of BPL families all over Maharashtra because of changes in government BPL norms. Moreover, the landowning class has benefited more because of the increased agricultural productivity, production of cash crops and increase in cropping intensity. The poor have benefited mostly from increased availability of work and other collective improvements such as infrastructural facilities.

Social transformation is an important component of social equity and cannot be ignored. Interviews with Dalits in the village revealed that there is very little overt discrimination or exploitation in the village on the basis of caste. People from different castes eat and celebrate together and interact with each other freely. A common cremation ground has been made where a dense plantation of trees has also come up. Though the village has only two Muslim households, a mosque and a Muslim burial ground are also being constructed adjacent to the Hindu cremation ground. All this has come about because of a politically conscious and benevolent leadership. Also because of the shortage of labourers and sharecroppers in the village, the labourers' position vis-à-vis the landowners' has seen a drastic change. The workload has decreased, work timings have become flexible and the latter's behaviour has also improved. There is a general realisation that every single individual's needs and aspirations have to be met at least partially if the village has to sustain the positives of the WAA.

A Sustainable Experiment?

We have so far looked at the overall impact of watershed development across different groups and locations and studied to what extent (if at all) these interventions had a transforming effect on social relations within the village. There are, however, 'bigger' questions about the Hivre Bazar experience that also require attention. Foremost amongst these is the question of sustainability, a concern that has both an ecological and social component. The main achievement of watershed development in Hivre Bazar is increased availability of water which then has had a positive impact on irrigation, agricultural productivity and domestic water supply. In many villages, including the neighbouring village of Pimpalgaon Wagha, increased water supply has also resulted in a burgeoning number of bore wells, a pumping race, increased water-intensive cash crop production and, consequently, over-exploitation of water. This has not been the case in Hivre Bazar because of the ban on bore wells and water-guzzling cash crops. During our complete coverage by foot of the village, it was evident that this ban has not been violated anywhere inside the village. One danger that potentially exists is that a proliferation of wells in the neighbouring village of Daithane Gunjal might have some impact on water availability of those households which have their wells located near the village boundary (though none of the villagers reported such a thing happening in the past).

Though the decision in Hivre Bazar to ban bore wells for irrigation is often seen as a sustainability measure, it also had equity implications.[26] Invariably farmers from all locations of the village and size holdings told us that in addition to protecting water levels, the ban on bore wells gave a chance to poorer farmers to utilise the water as well. The investments for bore wells and submersible pump sets are much higher than that for shallow wells and hence bore wells are by and large concentrated in the hands of the big farmers who can make the necessary investments. Wherever bore wells have been allowed, by and large the open shallow wells have gone dry. Thus, by banning bore wells, water has become available in a more dispersed manner within the village and can be utilised through open shallow wells. Coupled with this, the ban on water-intensive crops like sugarcane and banana has ensured that water is used in a more dispersed manner throughout the watershed thus raising the overall productivity of the watershed as a whole. It has not created 'eco-system islands' where small pockets of high input (including water) agriculture co-exist with large tracts of low input, dry land and low productivity agriculture.

Apart from the institutional arrangements that have been worked out, another factor which seems to have had an impact is the bio-physical setting itself. The three micro-watersheds in Hivre Bazar form the upper portion of the larger milli-watershed (or sub-basin) that contains them. Moreover, the major Hivre Bazar watershed is a classical watershed in the sense that it is properly bounded from all three sides and has one common exit point. For this reason, watershed development activities could be planned more scientifically adhering to some of the cardinal principles of watershed development (like ridge to valley approach). Enlisting the participation of other villages to comprehensively treat the entire watershed has, therefore, not been necessary. The only thing that remains to be seen is whether the emergence of bore wells in the neighbouring village could have an impact on farmers who have land near the boundary. Another important physical feature is that the per household watershed area (PHWA) is about 5 ha which is quite high for Maharashtra. Generally, it is around 2.5 ha or 3 ha. A higher per household availability of watershed area makes a difference to resource regeneration and resource availability especially the quantum of water that one can harvest. For example, sometimes even with very small interventions the situation can improve in much shorter time spans and also very dramatically. Probably, this higher availability of PHWA has something to do with the thumb rule of water balance that Popatrao has worked out. According to him, 'if we get 100 mm of rainfall our drinking water and some rabi crops will be assured. If we get another 100 mm of rainfall we will be assured of rabi and some summer crop, and if we get another 100 mm of rainfall we will be assured a third crop for most households in the village.'[27]

However, there is a caveat to this. There are people in the village who do not have wells and who do not have access to irrigation water. On the other hand, there are larger farmers who have as many as six wells, all with pumping devices.[28] Popatrao has, in other words, consciously chosen the soft option of not talking about equitable access to water, water sharing and minimum assurance. Hivre Bazar, moreover, unlike Ralegaon Siddhi has made no effort at constructing collective wells as yet. Another important factor to be noted is that a village which has about 140 households has about 340 wells, almost all of them with pumping devices and most of them were brought after the watershed development. Thus, though the ban on bore wells has helped to improve the water regime in Hivre Bazar, the proliferation of dug wells does have implications for sustainability in terms of total water use, investments made and energy used for pumping.

One other possible ecological threat to Hivre Bazar's watershed experiment is the nature of agriculture. Although there are bans on tube wells and cash crops, intensive agriculture has become the order of the day. We have already highlighted the manner in which wells have proliferated, pipes are strewn across fields to irrigate land throughout the watershed and cropping patterns have intensified. The use of mechanised implements has also increased in the village as has the use of HYV/improved seeds and fertilisers. Mechanisation has come largely in the use of tractors for levelling of fields, transportation of goods, ploughing and other agricultural uses, and use of electric pumps for irrigation in contrast to the use of animal power (including in *mots*[29] for drawing water from the wells) in the past. Although mechanisation has helped, more so given the shortage of labour, the trajectory of future developments has become more uncertain. Villagers in Hivre Bazar, however, are quite sure that they are concerned about the fertility of their land and are convinced that they have put in place rules and restrictions that will prevent serious ecological problems of soil erosion and over-exploitation of water from arising. They stated that while they continue to use chemical fertilisers, the use of organic fertiliser has increased because of increased dairying activity.

It is the non-ecological dimensions of sustainability, however, that deserve more attention in Hivre Bazar. As highlighted earlier, there is a realisation amongst Popatrao and others that watershed development alone might not be able to address the long-term livelihood needs of the people or at least cater for year-round employment particularly in periods of stress. The year 2003–04, for example, was a drought year and farmers were at best able to produce one crop. One manner in which the village has been able to respond to such threats is by saving money through government schemes. Most government schemes have a people's contribution component and a subsidy component. Through an increased amount of shramdaan, the village is able to save a significant component of the subsidy, thus building up reserves. An example of this is savings of the order of Rs 1.7 million which can be used to provide employment in periods of stress.

At an aggregate level, moreover, the village seems to be self-sufficient in most of its basic needs—food, drinking and domestic water, fodder and fuel. However, although the resilience of the village to shocks of drought has increased, the experience of the past year shows that the village still needs outside intervention to tide over continuous droughts for two to three years. Food for work programmes and cattle camps have been a regular feature in the village.

Another important point is that Hivre Bazar is now strongly linked with the market. The increased agricultural and dairy production of the village finds its market in the nearby town of Ahmednagar (a number of bigger farmers informed us that they have tried to tap the markets of cities as far as Bangalore). Also, for most agricultural inputs people are dependent on Ahmednagar. At present, people's relationship with the market appears to be quite problematic. Farmers generally sell their produce immediately after harvest as there are no storage facilities. Prices are often at their lowest then. A case in point was the sale of jowar at half of last year's market price. Onions were sold at Rs 30–50 per quintal due to the glut in the market. Input costs for agriculture, on the other hand, are increasing because of the use of HYV seeds and chemical fertilisers. The market price of these inputs is more during the sowing period and farmers have little option but to buy at these higher prices. But in the case of milk, the farmers get an assured price throughout the year from the village milk cooperative. Though this price is lower than the price in the local market, they do not have to worry about or pay for transportation.

There are preliminary efforts on in Hivre Bazar to deal with the vagaries of the market by packaging a Hivre Bazar brand. Already collective efforts are being made to market agricultural produce from the village in the regulated wholesale vegetable market in Ahmednagar. This is not entirely a post-watershed development, but the scale of operations has certainly increased. Added to this is the fact that now many more farmers are marketing their produce. An ex-army officer has bought a van and does much of the transporting although individual households often go on their own as well. However, there is a lack of clarity with regard to what would constitute the Hivre Bazar brand. At the moment Popatrao and others are thinking of organic vegetables and fruits. They also want to set up a chilling plant for milk and then market Hivre Bazar milk in packets to urban centres. There has been some talk of agro-industry being the future of Hivre Bazar, but there is little clarity on this.

While these are all interesting possibilities, they are difficult to implement as Hivre Bazar is basically a one-village initiative and so far as we could make out there does not seem to be an attempt to link this initiative with any larger grouping or experiment in marketing. This would require an immense amount of effort, planning and state support to show results. Moreover, it is pertinent here to raise the question—to what extent can these smaller experiments or success stories succeed (especially in matters related to market linkages) keeping in mind that not much is happening at the macro level. Also, in the absence of any lateral linkage with other such experiments (partly because of the lack of success of such experiments), there is not much one village can achieve on its own, with or without government support. Even if it achieves some degree of success, the moot point is can this model be replicated successfully?

The question of supra-local institutions is, therefore, important in the wider context of sustainability as well. Supra-local institutions have become quite central in the CBNRM discourse as often long-term concerns with regard to ecological sustainability and even economic sustainability are located within spatial limits that go beyond the village. Hivre Bazar's locational advantage (it is set right at the foothills) has protected it from the vagaries of the behaviour of those in the surrounding villages. In that sense, the immediate need for federations of organisations undertaking similar watershed work does not arise. Yet, despite the strategic location of Hivre Bazar, attempts have been made over the last few years to promote similar watershed activities in neighbouring villages. Yashwant Vikas Sanstha has started work in eight villages (implementing DPAP programmes). This cluster approach (as Popatrao calls it) is focused on similar watershed activities as well as sanitation. The main aim, according to the sarpanch, is to share the experience of Hivre Bazar with others and help in providing ideas and logistical support. However, Popatrao believes that any external agency, including the local NGO, cannot achieve the same level of work successfully as that which can be achieved by the villagers themselves. He is very clear that leadership has to emerge from within other villages.

It is too early to say whether or not the NGO's involvement will have an impact. A cluster approach might in the long run be important in terms of carving out a niche for the production of particular fruits and vegetables that bear the trademark of the area. Economies of scale could be an important factor particularly in terms of storage and marketing facilities though the NGO has not addressed these concerns yet. The trade-off could, however, be that Hivre Bazar may lose the advantage that it seems to have created in Ahmednagar.

The dimension of sustainability that is perhaps the most important in terms of the 'achievements' of Hivre Bazar is that which has been hinted at earlier, namely, social engineering. Popatrao is convinced that the biggest change in the village is the change of attitude. There are two dimensions to this change of attitude: (*i*) promulgation of certain social norms and (*ii*) the creation of a 'collective identity'. It is these factors, rather than the actual physical works or institutional rules with regard to watershed management that Popatrao feels have led to Hivre Bazar's success. The ban on alcohol, he feels, resulted in the change of the village's 'self' image as well as its image in the outside world.

This 'change of attitude' needs more critical scrutiny. At one level, the ban on alcohol is enforced strictly in the village. It was enforced quite forcefully at the outset. But there are cases even in the present of people going to Jakhangoan, a neighbouring village, and drinking there. In fact, people in Jakhangaon joked with us that their sales had gone up. Nonetheless, the ban has been welcomed especially by women and it certainly has had an impact in terms of people's willingness and actual participation in the watershed efforts in Hivre Bazar. In that sense, the longevity of this measure is central to sustaining the Hivre Bazar experiment both in terms of sustaining people's commitment individually and collectively to watershed development.

The 'collective' commitment to watershed development has, however, also had its down side. As we have illustrated in detail earlier, the main thrust of watershed development has been on improving the productivity of agriculture. The inherent bias, therefore, in favour of the landed has meant that questions of landlessness have not been tackled directly. In fact, Popatrao is clear that the main aim is to maintain the collective spirit of the village (community formation) and that raising 'redistributive' concerns might go against that. As only approximately 10 per cent of the village derive their incomes more from non-land based activities, the sarpanch feels that it would be detrimental to focus on questions of land redistribution—rather other measures such as increased employment opportunities will continue to be the focus.

There is, in other words, a possible trade-off between sustainability and equity concerns in Hivre Bazar. The desire to retain cooperation within the village and the positive image of the village in the outside world (including the bureaucracy) seems to be a major component when decisions are made in Hivre Bazar. This is not to say that particular measures aimed at sustainable use of resources such as the ban on bore wells do not have a positive impact on marginal and small farmers who cannot afford to dig bore wells, but rather that the desire to retain the 'unity' of the village prohibits embarking on potentially realisable redistributive measures because they might upset large landowners. In that sense, as we have suggested earlier, the social sustainability of the Hivre Bazar experiment is very much privileged in the planning process.

An equally important point perhaps is how not addressing the concerns of the poor directly could affect the watershed experiment. Whether or not weaker sections (socially or economically) will continue to support the initiatives in the village depends on how they perceive of the initiatives taken. The absence of the most important asset, land, might turn out to be crucial in determining the involvement of the poor in the long run. Though not articulated publicly, there is a feeling amongst the Dalit community that if plantations were not taken up on common lands, then that land could have been redistributed to the landless. This took place in the adjoining village of Jakhangaon where the landless families got about 1–2 ha of land in their names. In Hivre Bazar no land has been redistributed under the Land Ceiling Act, though there are people who have more land than the ceiling law prescribes. At present these have not become points of conflict because of the incremental/trickle-down benefits that the marginal and landless have gained and the fact that many exclusionary social norms have been discarded. One should also not forget that since marginal and landless farmers (and Dalits) constitute a small percentage of the population, they can be ignored.

Unravelling 'Participation'
in Hivre Bazar

Right from the outset, the initiatives within Hivre Bazar have been located within a discourse that privileged the 'community' and the need for this community to be in charge of its own destiny. The AGY was an opportunity for the village not so much to benefit from a government programme only, but also to shape the collective future of the village. The formation of the NGO within the village comprising villagers only and the fact that the gram sabha partly by design but mostly out of choice is considered the most important body in the village reflects in different ways the importance of the village as a collective unit.

The privileging of the village community in the context of participation has consequences in terms of both the intra-village dynamics of participation and in terms of how the state interacts with this community. The first impression that one gets in Hivre Bazar is that Popatrao is the central figure, an impression, in fact, that one carries to the village itself. When one first arrives in Hivre Bazar, one is informed almost immediately whether Popatrao is available in the village or not. The aura that Popatrao carries with him is based on the belief, at least by many, that he is responsible for the transformation of the village, a belief shared by the outside world and acknowledged through a series of awards he has received.

Although Popatrao has without doubt been central to developments in Hivre Bazar, the dynamics of participation are quite complex. Most of the planning was and is done by Popatrao in consultation with some influential and knowledgeable people of the village, including the members of the gram panchayat,[30] the body which actually is formally responsible for watershed activities.[31] The 'smooth' functioning of the panchayat needs to be seen in the context of Popatrao's acumen in social and political engineering. As suggested earlier, his attempts at building a community by distancing himself from intra-village rivalries, both personality and caste based, is one main explanatory factor in terms of why the panchayat has remained largely unified. It is also important to note that Popatrao took specific steps to accommodate influential figures and possible opponents in positions of power within the gram panchayat, cooperative societies and various committees.[32] However, he has seen to it that they do not acquire unlimited power. Moreover, he has helped 'needy' people from within the Dalit community by arranging for jobs and providing financial help.[33]

Another means by which Popatrao has bolstered his support is by privileging the gram sabha. All ideas are taken to the gram sabha where they are debated and discussed with the village community (adults). These meetings are well attended. Although most people (including most women) remain largely silent during the meetings, most people we spoke to endorsed the ideas tabled by Popatrao saying that what he does is best for the village.

This might give the appearance that there is hardly any participation in any of the programmes and that the Hivre Bazar experiment is basically a 'one man show'. However, at one level what the Hivre Bazar case illustrates is the fact that typologies of participation are themselves problematic in nature. For example, the fact that most ideas have emerged from a small number of individuals suggests that most people have not participated. Nonetheless, most people attend gram sabha meetings and though they might not speak much, they are very much aware of and endorse past programmes and future plans. This is the case because informal modes of communication are much more important than formal ones in Hivre Bazar. Informal discussions take place regularly in the village square, temples, agricultural fields and the panchayat building. It would be a fallacy to underestimate the influence these informal discussions and debates have on the final decisions taken in the gram sabha meetings.

But it is also true that these informal channels and networks are balanced and tilted in favour of the landed and rich of the village and that the interests of the landless seem to have been bypassed. What happens in such circumstances is that the dominant view/opinion is accepted by most of the villagers as the best possible in the given circumstances. The danger is, therefore, that the views of the minority, in this case the landless in particular, are not adequately addressed and that their silence is based more on the fact that what is there today in Hivre Bazar is much better than what was there in the past.

A deliberate attempt has also been made in Hivre Bazar to keep away from party politics,[34] the assumption being that party politics leads to internal fighting and erosion of the collective spirit of the village. Our intent here is not to judge whether such a claim is correct or not, but to highlight the fact that the absence of party politics has been central to the Hivre Bazar experience. As highlighted earlier, collective well-being has been at the centre of watershed experiments in Hivre Bazar and this has been counterposed to the divisive nature of panchayat politics. This feeling was strengthened in 1995 when some people in the village who wanted to replace Popatrao at the helm of things, and who had connections with district-level leaders of political parties, attempted to discredit him. Popatrao actually resigned as sarpanch and did not enter the main village (he lives in one of the bastis) but eventually returned when his supporters convinced him to return. Popatrao's contention is that panchayat politics are not meant to be party based. His concern is that politics should not divide the community. The danger, however, exists that certain views are not heard at all as they are considered 'divisive'. Nandrao Thange, for example, was a member of the panchayat, but opted out because he felt that his views were not listened to adequately. Moreover, in such a context where disagreements appear to become divisive even special interests such as those of the landless get marginalised.

Hence, it would seem that participation in Hivre Bazar has clearly defined contours. While the silence in meetings in terms of speaking out is not necessarily a sign of non-participation, it is perhaps indicative of a certain hegemonic practice of participation that privileges the 'collective well-being' of the village over 'group claims' of particular sections of society. Moreover, people in general, and the landed especially, have put their faith in the decisions made by the leadership. Marginal farmers, the landed and Dalits view developments in Hivre Bazar mostly in comparison to a past that was worse. In that sense, participation maps well with the overall thrust of developments in Hivre Bazar where the making of the 'community' has taken precedence. But development itself often tends to destroy the context that brought it about. If commodification, increasing dependence on the market, branding and such developments begin to take over, and if sharper differentiation results, the present notion of community might become more problematic.

The forging of the collective community within Hivre Bazar has also been welcomed by the state. The fact that Hivre Bazar became a 'success story' in a short duration has meant that bureaucrats will be more willing to allocate money to the village, as it is an easy option for meeting targets.[35] This may seem to be in line with Baviskar's observations in the context of the Rajiv Gandhi Watershed Mission initiatives in Jhabua that villages likely to be success stories are often the target of government spending (Baviskar, 2002). However, though government officers may be attracted to Hivre Bazar as an easy option, unlike the Jhabua case where soft targets were turned into and held up as success stories, Hivre Bazar has earned that position. No government officer was prepared to touch Hivre Bazar in the pre-1989 days.

The dynamics of state–village interaction in Hivre Bazar, however, is no longer that of the state simply dictating what and how things should be done. While the agriculture department, minor irrigation department and the department of social forestry of the government of Maharashtra were involved in implementing the project in Hivre Bazar, their involvement was only in offering technical inputs and sanctioning of plans and budgets prepared by the NGO staff. Moreover, the line departments have worked in tandem with the panchayat and not independent of it as they continue to do today. In that sense, Hivre Bazar is a case of the local community playing an important role in decentralised development and the state supplementing that role.

Conclusion

The 'success' of Hivre Bazar offers important clues as to what constitutes success and how one arrives at it. More so than the other case studies, the Hivre Bazar experience has been very much about building a collective pride and showcasing it to the world beyond the village boundaries. This pride no doubt is built on real achievements, namely, a successful watershed intervention that has resulted in better water availability, increased agricultural production and a number of other watershed plus benefits. Unlike in most watershed experiments, moreover, concerns of ecological sustainability have been translated into institutional measures aimed at long-term sustainable use of natural resources. Equally important, the village's achievements are not due to an outside NGO's intervention, but are a result of the work of the village leadership who continue the good work through the gram panchayat.

But Hivre Bazar also highlights the generic problems that might exist when the collective community is privileged. Though attempts have been made to address the concerns of marginalised sections, namely, the marginal farmers, the landless and the Dalits, these attempts have shied away from addressing the major issue of land inequality. While this silence might not hamper further developments in Hivre Bazar, it does suggest possible limits to the vision itself, what local communities and NGOs can or cannot do and possible trade-offs between different concerns such as equity and sustainability.

Hivre Bazar also illustrates the possible limits to watershed-based rural development. Despite the substantive success of WAA in Hivre Bazar, during acute drought periods villagers have had to depend significantly on the EGS. While this is not a critique of community-based development, it does suggest that CBNRM (at least in semi-arid areas) needs to be understood and practised keeping in mind the need for other types of 'development' interventions.

Notes

1. For a fuller account of the 1972 drought in Maharashtra, see Brahme (1983).
2. It is worth noting that water lifted from the Kukadi Canal through the Krishna Pani Puravatha Society supplemented water from local watershed development, a fact often missing from the bulk of the literature available on Ralegaon Siddhi (see Joy et al., 2004).
3. To get selected to this programme, a village had to have some physical characteristics, namely, 'it should be located in a drought prone area, scarcity of water should be the first major problem in the village, total irrigated area in the village should be less than 30 per cent and the population in the village should be less than 4,000' (Hazare et al., 1996).
4. The acceptance of the four bandis and shramdaan has to be endorsed by the gram sabha. In addition to this, 70 per cent of the village has to agree to participate in the AGY for the application to be processed.

5. Yevle basti (south of the road), Phuphata basti, Pawar basti, Gohad basti and Padir basti.

6. The history of Hivre Bazar and the arrival of Popatrao Pawar narrated in this and the next section is important not necessarily because it represents the 'true' factual account of what happened, but because it is this account that is implanted in the 'collective' memory of the village and which forms the basis of future developments in the village. In other words, the history narrated here is important more in terms of understanding the community-based watershed initiative that emerged in the village than for its actual veracity.

7. The increasing degeneration of the situation in the village seems to be only part of the reason for this shift away from the main village settlement or *gaothan*. This phenomenon is also not very specific to Hivre Bazar alone. From the late 1960s and early 1970s, with easy credit facilities available, people went in for wells in a big way. The increased availability of water resulted in intensification of agriculture in most parts of Maharashtra, especially south Maharashtra. With increased agricultural activities, many households preferred to stay close to their lands and eventually built houses and shifted there. This resulted in the development of bastis or hamlets and very often these bastis had a clannish look as people from the same clan or sub-caste or bhavki stayed in the same basti.

8. Panchayat *samiti* is a local government body at the tehsil or taluka level in India. It is a link between gram panchayat and the district administration.

9. An *akhara* (wrestling pit) has also been subsequently built adjacent to the temple. It is significant that Popatrao has consciously tried to revive activities that can strengthen the sense of a village community. The youth of the village are encouraged to have clean habits and engage in sports. This has also helped the youth get employment in the army.

10. Maharashtra was the first state in India to pass a legislation in the 1970s that made the state responsible for providing employment to people during periods of drought.

11. Popatrao believes that any programme can be successful only when the local people are directly involved in the implementation. He argues that leadership has to come from the village and that outside organisations cannot provide this leadership. He also implicitly argues that the gram sabha should be involved as the local leadership will eventually take up the leadership of the gram sabha too.

12. The same rules or bans applied to private wastelands/grasslands that were treated under the AGY.

13. Note that this was the only case where private land was treated. But even here this was done on condition that common property principles would be followed on private treated lands as well. Also, for the larger good of the watershed, private lands on the hill slopes had to be treated.

14. The decision to not allow the sale of lands to outsiders was based on experiences elsewhere. It had been witnessed in many places that after successful water augmentation activities (WAA), outsiders, especially the resourceful ones, buy lands from the villagers at a high price and the benefits of WAA are then enjoyed by them rather than the people who worked hard for it. This decision has especially helped the smaller and marginal farmers who sell off their land in the first few

years post-WAA when the benefits gained from land treatment measures are not fully realised. Individuals cannot resist selling their land because of the sudden rise in land prices. Though this decision appears to be in favour of the rich and big landlords of the village, who are least likely to sell off their lands and most likely to buy, the smaller and marginal farmers believe that this decision was taken in consultation with them as they were well aware of the benefits they were going to realise soon and they feel it has benefited them the most. People who had left the village have returned to give more time to agriculture. No transfer of land to outsiders has occurred after this decision was taken.

15. There is an interesting story behind this. Some time back, the village school conducted an essay writing competition on village development issues and one girl from the village wrote that although their village was an 'adarsh' village, this could change if a person from outside who got married to a resident of Hivre Bazar had AIDS. Sarpanch Popatrao was touched by this and realised that AIDS was in fact a real issue for the young generation. Today it is compulsory for both men and women in the village to undergo the ELISA test for AIDS before marriage.

16. A police patil is a village official who, though not formally part of the police force, performs some police functions and liaises with the police/outpost/station that has jurisdiction over the village.

17. Kurhad bandi, though not monitored, seems to have been followed by the people strictly. Even trees on the farm bunds are not logged. A *chowkidar* (watchman) was appointed by the forest department to keep strict vigil. In the initial years there were some cases where people of neighbouring villages were caught cutting trees in the newly planted forests.

18. On the face of things, fruits are not that common in Hivre Bazar, but the YKGPVS pamphlet claims that the area under horticulture has seen an increase from 7.1 ha to 54 ha. Some farmers had started the cultivation of grapes, but had to stop during the drought of 2000–01 to 2002–03 as they did not have water. Drip irrigation is now being experimented with and the leadership is encouraging people to start using it for some crops.

19. A much more reliable supply of fodder from the village has been the main reason behind the upswing in the dairy economy. The treatment of hill slopes and other lands has not only led to soil conservation and water harvesting, but it has also increased fodder growth. This coupled with the ban on open grazing has been a major impetus. The gram sabha has made rules for the sustainable harvesting of fodder from the panchayat lands. On the payment of Rs 100 (the fodder collection fee is Rs 30 for the poor of the village) anybody can cut and bring as much fodder from the grasslands during the period designated by the gram sabha on condition that it is transported only by headload. One headload per household per day is allowed. This rule is operational even now.

20. The total money allocated to Hivre Bazar through the Adarsh Gaon Yojana Trust till April 2003 was about Rs 7.4 million. But the total fund committed to the village for implementation of various components under the AGY was around Rs 4.7 million. The AGY was a three-year project and the implementation of the main programme ended in 1997. This means that an extra amount of about

Rs 2.7 million were spent in Hivre Bazar in what may be termed as 'watershed plus' measures.

21. Warghade (2003) gives the number of tractors post-WAA as 16.

22. Warghade mentions that the number of motorised two-wheelers has increased from five in the pre-WAA period to 230 in the post-WAA period (Warghade, 2003).

23. We should keep in mind that at present sharecropping rooted in feudal relations does not form a significant part of the agrarian relations in Hivre Bazar and for that matter for most part of Maharashtra especially southern Maharashtra where the ryotwari system was prevalent (also because of the implementation of the tenancy act) and presently consists mostly of independent landowning peasants. Also the type of sharecropping practiced in Hivre Bazar and for that matter for most part of Maharashtra is pretty different from that of West Bengal or Orissa both in terms of scale and the way sharecropping arrangements are made.

24. About 125 people worked on the EGS sites last year for about four to five months, as people needed work and also foodgrains because of the continuous drought for the past three years. Each labourer gets 5 kg of foodgrain and cash according to the quantum of work done. The approach road to the temple is being constructed under the EGS. About 50 labourers came from outside the village.

25. For example, in Ralegaon Siddhi, an effort was made to set up a community biogas plant and the Dalit families were given connections from it.

26. It should be noted that the impact of this 'sustainability' measure on equity seems more of an unintended consequence.

27. Personal communication with Popatrao Pawar (11 December 2004).

28. Since lands of individual farmers are dispersed throughout the village, farmers have to dig wells near each of their landholdings.

29. *Mots* are indigenous water drawing mechanisms which use animal power to draw water from open wells.

30. Gram panchayat meetings in Hivre Bazar are open to all.

31. It should also be noted here that the AGY was a government programme and the outline/design of the same was made in the drawing board of government offices. Therefore, people had a limited role in the planning process, especially in the decision to adopt the five rules/bandis.

32. Nandrao Thange, for instance, who still remains a bitter critic of Popatrao, was made the *up-sarpanch* during Popatrao's first term as sarpanch. Likewise, Popatrao sees to it that people from all castes and sub-watersheds are given representation in the gram panchayat. But it should be noted that most of the members (mostly women) hardly have any role to play and that they are clueless as to what is expected from a gram panchayat member. Nandrao Thange believes that Popatrao only selects people who will not oppose him while he completely ignores his critics. He believes that there is hardly any collective decision-making process; Popatrao, along with his acolytes, takes all the decisions. Though Nandrao Thange was the only villager who openly gave a different opinion, there were others, like Ulhas Samble (an ex-sarpanch), who was also at one point of time opposed to Popatrao's leadership. Samble too was a member of the gram panchayat from 1991–2001.

33. Gopal Gaikwad, for instance, is a landless Dalit belonging to the Matang caste. He was appointed gram panchayat *sipahi* or soldier (with a salary of Rs 500). He also doubles up as worker in the village dairy (with a salary of Rs 700). His other source of livelihood is in kind, namely grains that he gets from each household as a *pujari* (priest; Matangs have traditionally been the pujari of the local goddess Mumbadevi). He believes that after WAA he receives more grains from each household. His brother sharecrops 2 ha of the sarpanch's land.

34. Discussions with Popatrao indicate that he was approached by one of the mainstream parties to stand for the recent assembly elections in Maharashtra. Apparently, he refused the offer because he did not want to identify himself (and the village) with a particular political party. He feels that without identifying with any political party he is playing a much larger role and he wants to keep his broad appeal intact.

35. Compare this with the pre-1989 days in Hivre Bazar when government officers feared implementing any programme because of the attitude of the villagers and infighting in the village. The villagers say that bureaucrats and government functionaries favour Hivre Bazar because there is no bickering and infighting in the village on how/what/where to implement any scheme. This is advantageous to the bureaucrats as they have to spend money within the financial year and show the success of the programme to their bosses.

3

Utthan's Work in Nathugadh, Gujarat

Introduction

Utthan literally means uplift. Since 1981, Utthan and another NGO named Mahiti have been known largely for their grassroots work in the Bhal region of Saurashtra, Gujarat that has focused on a woman and community-based approach to solving the problem of drinking water scarcity. Since 1994, Mahiti has concentrated on the Bhal area and Utthan has looked outwards to other geographical areas as well as other spheres of work. We selected Nathugadh, a village in Bhavnagar district in Saurashtra, for our case study. Utthan worked in the village first on drinking water and sanitation issues and later took up and completed a watershed project. What primarily attracted us to this village was the community (village) initiative at setting up and managing a piped water supply system. We were interested in how the process of community formation had shaped Utthan's work on CBNRM.

The chapter is divided as follows: the next section provides a brief regional-historical context to Saurashtra and Bhavnagar district where Nathugadh is situated. The two sections that follow provide a brief background of Utthan's organisational history and of its work in the water sector. Later, a summary of the drinking water and sanitation programme and the watershed programme activity are provided. The rest of the case study is devoted to discussing some of the salient issues related to Utthan's intervention and the manner in which it has shaped and been shaped by the particularities of Nathugadh's 'community'.

The Regional Context: Saurashtra and Bhavnagar District

Saurashtra is the roughly triangular peninsular portion of Gujarat that lies between the Rann of Kachchh and the Gulf of Khambat (earlier called the Gulf of Cambay). The thick neck of this peninsular region joins Saurashtra to the Kathiawar region of Gujarat, abutting it on the north. Agro-climatically, Saurashtra may be divided into two zones: north and south Saurashtra. North Saurashtra is classified as a semi-arid region. It has low rainfall (average rainfall of 537 mm) and has medium to light soils that are calacareous. South Saurashtra has a higher rainfall (average rainfall of 844 mm) and better soils. Bhavnagar district falls in the north-east part of north Saurashtra (Directorate of Agriculture, Government of Gujarat, quoted in GEC, 2006a).

Before Independence, Saurashtra and Kathiawar between them comprised literally hundreds of small princely states and principalities, with the largest of them being the princely state of Junagarh. In 1947, after Independence, 217 of these princely states were merged to form the independent province of Saurashtra, which was later merged into the bilingual state of Bombay in 1956. Subsequently, the bilingual state of Bombay was divided into the two linguistic states of Maharashtra and Gujarat, and Saurashtra is now a part of Gujarat.

The erstwhile princely states were ruled by Rajput princes. The Rajputs are also known as Darbars in this area, with the obvious connotation of royal couriers since they were tied by kinship to the princes in the area and also formed the dominant feudal landowning class.[1] Their immediate tenants were the Kanbi Patels; the more prosperous among them are now favoured by the name of Patidars. The land reforms of the late 1950s and the early 1960s benefited the Kanbi Patels the most and the title to a very substantial portion of the land passed to them. They now form the main landed section in rural Saurashtra. Besides the Scheduled Castes (SCs), who formed the bottom of the social heap and were restricted to the most menial and onerous tasks, the most important intermediate caste is that of the Koli Patels. They too claim a peasant status and elsewhere they do own land, especially in tracts contiguous with the tribal belts, but in north Saurashtra they are mainly landless or marginal holders. They provide the bulk of the labour that the Kanbi Patels employ. During the colonial period, the Kanbis (sedentary cultivators) were elevated to the category of landowners called Patidars[2] (Shah and Rutten, 2002) whereas the Kolis were essentially shifting cultivators. Through the change in the land tenure system during the colonial period, Kanbis encroached upon land belonging to and cultivated by the Kolis and tribals (Gidwani, 2001; Rutten and Patel, 2002). Since the early- to mid-nineteenth century, the Kanbis—who eventually got categorised as Patels—steadily ascended in terms of economic and political power. *Patel* was originally a title given to a village officer in charge of tax collection and law and order. The title was adopted by all members of the Kanbi community (Gidwani, 2001).

What is interesting is the structure of landholding that has emerged. In at least five districts of Saurashtra, holdings greater than 2 hectares (ha), according to government records, form more than 50 per cent of the total holdings, while for most of the rest of Gujarat they form 21–50 per cent (see Table 3.1). This implies that rather than a small rich peasantry, a broader class of rich peasants have emerged as the dominant section. It also means a greater likelihood of a much stronger large farmer ethos dominating the farming community.

During the 1990s, Saurashtra, especially its coastal portions, attracted a growing number of industries and emerged as an increasingly attractive industrial destination in Gujarat. For example, one study (Hirway, 1998) points out that during 1991–96, Saurashtra accounted

TABLE 3.1
The Number and Area of Operational Holders (in percentage)
in the Districts of Saurashtra (non-SC/ST)

| District | Marginal Holders | | Small Holders | | Other Holders | |
	Number	Area	Number	Area	Number	Area
All-Gujarat	26.40	5.37	27.86	15.21	45.74	79.42
Bhavnagar	17.43	4.42	34.62	18.38	47.95	77.21
Rajkot	13.67	3.36	33.57	17.25	52.76	79.39
Amreli	14.58	3.42	31.25	16.02	54.17	81.62
Jamnagar	10.63	2.26	31.41	15.05	57.96	82.69
Suredranagar	8.34	1.42	22.11	7.82	69.54	90.75

Source: Agricultural Census, 1995–96 (quoted in GEC, 2006b).

for 21 per cent of the investment in, and 16 per cent of the employment generated by, medium and large-scale industries in Gujarat as against the corresponding figures of 2 per cent and 8 per cent for the period 1983–90. However, the study also points out that the Saurashtra region is primarily characterised by a high degree of environmental degradation: highly erratic and less than 600 mm of average annual rainfall; shrinking of forest area to only about 5 per cent; overdrawal of ground water and salinity ingress. In fact, 80 per cent of the blocks are declared drought prone or desert area, 95 per cent of its villages are declared as 'no source villages' and there has been a significant decline in the net sown area, the cropping intensity and the number of milch animals— all between the early 1980s and 1990s (Hirway, 1998).

Nathugadh, the village selected for the case study is situated in Ghogha tehsil of Bhavnagar district. Bhavnagar district forms the northernmost and easternmost part of Saurashtra, and Ghogha tehsil forms part of the eastern coastal part of the district. The immediate coastal strip of the Gulf of Khambat in Bhavnagar district has saline soils and is plagued by the problem of salinity ingress as well (Anonymous, 1995b). This is the area known as the Bhal area and this is where Utthan started its work and where it has its strongest base.

The proportion of forest area is very small, a little more than 3 per cent of the reporting area, while the cropped area accounts for almost 65 per cent of the reporting area (see Table 3.2). The main food crops grown are *bajra* (pearl millet), *jowar* (sorghum), wheat and, to a much smaller extent, rice. The main cash crops are groundnut and cotton and in some areas mangoes and coconuts.

TABLE 3.2
Land Use Pattern: Bhavnagar District (1991)

Land Use Category	Area '00 ha	Reporting Area (%)
Geographical area	11,155	–
Reporting area	9,789	100.0
Forest	315	3.2
Area under non-agriculture use	724	7.4
Barren and unculturable land	1,017	10.4
Permanent pasture and other grazing land	709	7.2
Land under miscellaneous trees and crops	0	0.0
Culturable wasteland	307	3.1
Other fallow	12	0.1
Current fallow	417	4.3
Net area sown	6,288	64.2
Area sown more than once	206	2.1
Total cropped areap	6,494	66.3

Source: Census of India, 1991.

Bhavnagar was founded in 1723 by Bhavsingh Gohil of the Gohil dynasty when he decided to shift the capital from Sihore. The Gohils are said to have ruled in Sihor as well as in Ghogha and Umrakla since the sixteenth century and many feudatories in the region including the one for Nathugadh are Gohil Rajputs. Bhavnagar boasts of being the first princely state in Saurashtra and the third in the country after Baroda and Hyderabad to have its own railways. Bhavnagar district has about 40 per cent urban population and the city of Bhavnagar is the biggest urban area in the district. Cotton textiles and food products are the major industries in the city. In the Alang shipyard, the district boasts of the largest ship-breaking shipyard in Asia. Bhavnagar also has a flourishing diamond polishing industry. Like most of Gujarat, especially coastal and central Gujarat, it has a large non-resident Indian (NRI) population and it is not uncommon to see NRI-initiated and sponsored or assisted village improvement activities and organisations in the countryside.

Groundwater is the only source for whatever little irrigation there is in the district; only 1,900 ha are irrigated by canals. Tube wells have not been very successful in the district, and irrigation is mainly dependent on dug wells.[3] There is a general feeling all over Saurashtra that groundwater use has exceeded recharge. Hirway (1998) notes the high degree of commercialisation of agriculture along with the tendency to use groundwater without adequate recharge. The data for 1997 published by the Gujarat Water Supply and Sewerage Board (GWSSB) show that both Ghogha tehsil and Bhavnagar district as a whole fall in the white zone though more likely it should be classified in the grey zone.[4]

This then is a brief profile of the region to which Nathugadh belongs. We shall now briefly discuss the background of Utthan, the NGO that has worked in Nathugadh.

Utthan: A New Endeavour[5]

Utthan's emergence is closely linked to that of another organisation called Mahiti and their joint birth itself is intertwined with another organisation called the Ahmedabad Study Action Group (ASAG) set up in the early 1970s. The ASAG was a group formed predominantly of architects and its initial focus was on low-cost housing. Members of ASAG took up a block-level planning (BLP) exercise, then being promoted by the Planning Commission, as one of its many activities. While the idea initially was simply to understand better how block plans were formulated, in the course of the work they became acutely aware of the gap between the government's and the people's perceptions of the planning process. As against a target-oriented government approach, ASAG was committed to area and people-oriented planning and development and had prepared a BLP along those lines. A spate of skirmishes with the government soon made it clear that the government would not implement the BLP as it stood. The BLP group in ASAG decided to go ahead and take up the implementation of the BLP and, given ASAG's focus on low-cost housing, decided to break away from ASAG precipitating a crisis within the organisation. There were also differences in approach. Amongst other things, many of the original founders, with a more architecture-oriented background, wanted ASAG to limit itself to a planning exercise, but the members who had joined recently (among them was Nafisa Barot, the present executive trustee of Utthan) wanted to move on to implementing the BLP. Eventually a group of members separated from ASAG in 1981 and founded two organisations named Mahiti (literally information) and Utthan. Though Utthan (formally Utthan Development Action Planning Team) was a registered society and public trust, Mahiti was a more informal group. However, up to 1994, they operated like Siamese twins and one always implied the other. Mahiti, Utthan and the joint appellation Mahiti–Utthan were used interchangeably.

Mahiti–Utthan was founded based on the idea that lack of information amongst the local people was the main reason for their non-participation in the planning process. The organisation's motto was to supply appropriate information to people to empower them. However, the members of Mahiti–Utthan soon realised that there were drawbacks in this approach. Even after supplying relevant information to people, little progress was achieved in terms of people's participation in the planning process. A vibrant debate erupted amongst Mahiti–Utthan members with regard to understanding the difference between 'outsiders' (urban activists who played a catalysing/facilitating role) and 'insiders' (villagers who played an active role in the organisation and villagers more generally). How to make villagers independent of outsiders was a point of fierce debate. While the differences and inter-linkages between outsiders and villagers were recognised and debated, what was also realised was the fact that *mahiti* (the provision of information) alone was not enough to initiate action both from the people and the government.

As a result of these debates in Mahiti–Utthan, a two-way process was started. An inward-looking approach was aimed at creating democratic organisational space that did not make villagers dependent upon outsiders while at the same time an attempt was made to remain committed to certain normative values such as non-discriminatory treatment to members of the organisation and the villagers more generally. Intense debates and conflicts within the organisation and with villagers emerged in the process of debating basic normative values and social practices such as untouchability. At the same time, outward-looking concerns were debated such as how to influence policy at the local, national and even international level.

There were two kinds of changes that were taking place that affected the organisational structure that were evolving. There was a steady outflow of the original initiators of Mahiti–Utthan as one person after another left for personal or professional reasons. At the same time, their work was finding a response, especially among the women in the Bhal area of Saurashtra who faced very particular and difficult problems because of salinity and the harsh conditions in the area. Mahiti–Utthan was spreading and taking root in many more villages. The solution that evolved was simultaneously to look for more professional 'outsider' leadership at the top and look for more participative structures and greater participation of 'insiders' in the whole effort. This was reflected in the changes that took place around 1985. An intensive interaction that Mahiti–Utthan had with Professional Assistance for Development Action (PRADAN), a large professional NGO, led to the latter deputing Tejinder Singh Bhogal as a professional programme coordinator. He took charge of the main programmes. At the same time, a new crop of local workers and activists was also inducted; among them Devoobehn, a Dalit militant activist who now heads Mahiti.

In 1994, Mahiti and Utthan became two separate entities. Recognising its formal priority, Utthan, as its website puts it, 'withdrew from Bhal—after having helped Mahiti to build its organisational capability to continue with the development efforts in the region'. Mahiti, headed by Devoobehn, has taken over work in the core area where Mahiti–Utthan's work originated and took shape. Utthan now looks to work in the rest of Gujarat and sees its present work as comprising three programmes: (a) Tribal Area Development Programme (TAD), which was started in 1995 in Limkheda taluka, Dahod district, (b) Coastal Area Development Programme, Gohilvad, initiated in 1992 in Bhavnagar taluka, Bhavnagar district and (c) Coastal Area Development Programme, Kathiawad (started in 1996) in Rajula taluka, Amreli district. Today, according to them, their work covers 108 villages in seven blocks of four districts of Gujarat.

Utthan's Work: From Drinking Water and Sanitation to Watershed Development[6]

Though Utthan–Mahiti worked on many issues, it is most well known for its work aimed at resolving the drinking water crises in drought-prone areas of Gujarat, especially in the Bhal area, with women leading the struggle. A large part of Saurashtra and Kutch faces severe drinking water shortage during the summer months. Utthan–Mahiti's efforts to resolve the drinking water crisis involved a three-part strategy. First and foremost, it aimed to find a technical solution to preserving local water sources. The Bhal region, as mentioned earlier, is a low rainfall coastal area with problems of salinity ingress and saline soils. Hence, harvested and conserved rainfall in dug out ponds or even in wells turns saline very soon.

According to Utthan activists, the emphasis on drinking water as well as the direction in solving the technical constraints have both emerged from the gender-sensitive and people-centred approach of Utthan. Drinking water was identified as a priority issue by women, whereas men prioritised employment generation. Moreover, in the process of undertaking participatory situational analysis to identify alternatives, the idea of harvesting rainwater in a lined pond emerged.

Utthan–Mahiti struggled for years to find a proper technical solution to this problem. The most viable remedy was spreading a plastic sheet at the bottom of the dugout pond to prevent salinity ingress. A model was finally developed after several experiments spanning almost a decade (21 lined ponds were developed from 1986 to 1994) and involved fixing technical parameters to deal with issues such as slope, the desirable type of dugout pond, the type of construction material necessary for the sides of the dugout pond, the type of side slopes, the location of such ponds and the positions of the inlet and outlets.

However, while finding technical solutions was one problem, convincing both local communities and policy-makers was quite another. Utthan–Mahiti worked uncompromisingly with self-reliance as its principle. It focused on women as the main agency in bringing about change in the rural areas. The decades of the 1980s and 1990s was the time when the Gujarat government was trying its best to provide drinking water to several thousand villages through long pipelines or through tankers. These sources were not only unreliable, but also made local communities dependent on outsiders. Villages with which Utthan–Mahiti worked closely were situated at the tail end of these supply lines and suffered the most. Inter- and intra-village fights were common. And problems of water shortage in the summer months remained. Fighting vested interests both at the village level and beyond needed courage and commitment. Utthan–Mahiti largely organised women in the village. These women regularly fought with village landlords from the Darbar caste who were deeply ingrained with feudal values and practices rooted in feudal privileges. Patriarchal resistance at the family level was also common. Darbar landlords had a nexus with local politicians and contractors—both groups also having vested economic interest in the pipeline and tanker projects. After a decade of intermittent, but sustained struggle between local women and the landlords/contractors/politicians combine, Utthan–Mahiti finally won the battle when several villages collectively recognised the importance of being self-reliant in acquiring drinking water.

However, the third major challenge is still to be fully overcome. The struggle to convince policy-makers of the importance of local sources and village self-reliance for water requirements is still going on. Finding funds for plastic covered dugout ponds has proved very taxing for Utthan–Mahiti because authorities tend to think in terms of external water being brought into the area rather than strengthening local sources. After a protracted dialogue with the Water Supply Board and the World Bank in the 1980s and 1990s, Utthan–Mahiti finally got funds to construct eight dugout ponds. The ideas of self-reliance and decentralised water management have since become almost a movement due to the relentless efforts of Utthan–Mahiti.

After 1994, Utthan has carried the message of Utthan–Mahiti's earlier work outside the Bhal area. Several institutional reforms have been initiated in the state and also in civil society that remain committed to community-based, decentralised water management. In 1997–98, a pilot project involving communities in planning for water and sanitation began in the Ghogha district of the state. The experiment covered 82 villages, one of which was Nathugadh. The partnership of communities, NGOs and the state government for the Ghogha project led to the formation of the Water and Sanitation Management Organisation (WASMO) in 2002, an initiative supported by the Dutch government. The WASMO implemented community-based water and sanitation schemes in hundreds of villages in the drought-prone region of Gujarat. Recently, two civil society-based networks have also been initiated and Utthan is a part of both: Swajaldhara is a national network committed to decentralised water and sanitation management and Pravah is a movement of self-reliance that claims to cover 1,200 villages in 24 out of the 25 districts of Gujarat.

One of the consequences of the division of responsibilities that took place in 1994 has been a virtual restructuring of Utthan along increasingly professional lines. It was our impression during our discussions with the staff of Utthan that the bulk of the personnel are now young professionals or semi-professionals with a social work orientation rather than representatives of local struggles. Now, after analysing the problems of the different geo-climatic regions of coastal and tribal areas and in consultation with the members of various village-level institutions and community-based organisations, Utthan identifies the following five areas of work: (*i*) NRM, (*ii*) gender equity and empowerment, (*iii*) health and sanitation, (*iv*) networking and advocacy and (*v*) disaster management. While the concern for community participation remains strong, increasing professionalisation is also apparent and that may bring its own problems.

While Utthan has always been involved in water and soil conservation work, it became seriously involved in watershed projects from about 1997–98 onwards. By 2001–02, it had taken up and completed 29 watershed projects, 20 of them under the National Watershed Development Programme (NWDP) and nine under the Integrated Wasteland Development Programme (IWDP). Nathugadh belongs to this cycle of watershed development projects that Utthan took up. Subsequently, it has been given the responsibility for implementing 23 more watershed projects under the NWDP and another five watershed projects under the Indo-German Watershed Programme of NABARD (National Bank for Agriculture and Rural Development).

Selection of Nathugadh

When we began our discussions with Utthan about the selection of a village(s) for the case study we were drawn in two directions. The first direction was to concentrate on the oldest cluster of villages where Utthan–Mahiti had spent the longest time and from where Utthan had practically withdrawn now. Here the community and the leaders/ activists who had evolved from the movement have taken over the work with the help of Mahiti. We also made a few initial visits to one or two of the peripheral villages in the Bhal area and also had a very detailed discussion with Devooben. However, it soon became apparent that given the time of the year, it would not be possible to intensively study those villages because of heavy rains that usually resulted in the area being cut off from the mainland. We were, therefore, advised to select a village(s) from outside the Bhal area.

The other direction that we were pulled in was to take up a village(s) where Utthan had taken up not only issues of drinking water and sanitation, but had also gone on to take up watershed development which called for greater and more comprehensive involvement with NRM concerns. Ideally, of course, it would have been best to select both, but constraints of time and other support dictated the choice of one or the other. The rains had already made a decision for us. One consequence of this has, however, to be noted. Our discussion with Devooben made us realise that Utthan–Mahiti's work in the Bhal area, especially in the oldest cluster, represented a community involvement in the form of a vigorous movement from below spearheaded by women, whereas, the later expansion of Utthan's work represented a phase when the organisation had become a much more professional body. To the extent that this may have a bearing on the issues discussed, it should be noted that some of it may not apply to the older cluster.[7] However, it is also to be noted that after Utthan's present phase of expansion from 1994, the older cluster now forms a small part of the total work that Utthan–Mahiti has implemented.

We selected Nathugadh for our case study from the number of villages in which Utthan is carrying on its watershed development work for a number of reasons. First, Utthan has carried out both its main programmes, the drinking water and sanitation programme as well as the watershed programme in this village. For the case study, we also needed a village where the watershed programme had been completed. Since Nathugadh belonged to the first cycle of watershed programmes that Utthan had taken up, for all practical purposes the watershed development project has been completed. However, the most important reason that tilted the balance in favour of Nathugadh was the information that the villagers had come together and set up a piped delivery system for drinking water within the village on their own and were running it as well for the last few years, an indication of considerable community initiative.

Nathugadh: A Profile

Nathugadh is a small village in Ghogha tehsil[8] of Bhavnagar district, Saurashtra (see Figure 3.1). It is situated 24 km away from Ghogha, the taluka headquarters and nearest city, and a few kilometres away from the larger village of Vavdi. Vavdi is a marketplace and also Nathugadh's parent village. According to the oral accounts given by villagers, Nathugadh was settled sometime in the late nineteenth century when a number of presumably Kanbi Patel tenants from Vavdi requested Nathu Singh, the then ruling Darbar of the village to allow them to form a settlement away from Vavdi because the distance of their fields from Vavdi made life very difficult for them. He assented and the group formed a settlement named after him—Nathugadh— in the Saurashtra tradition of Bhavnagar of naming villages after lords who initiated them or permitted their settlement and construction.

By virtue of this fission, almost all of the permanent households in the village are Kanbi Patels. In 1991, according to the Census, there were 152 households in Nathugadh with a population of 889. This increased to 156 households and a population of 1,070 respectively by the year 1998 when Utthan carried out a participatory rural appraisal (PRA) in the village. Of the 156 households, 142 are Kanbi Patels, seven are Koli Patels, one is SC, one is Brahmin and five belong to miscellaneous castes. It is important to note the complete absence of the Darbars, the traditionally dominant feudal landlords in Saurashtra. All the Koli Patel households are landless and the caste divide between the Kanbi Patels and the Koli Patels is very sharp.

Thus, it would seem that Nathugadh is apparently a single-caste village of Kanbi Patels, barring seven settled families of Koli Patels. However, this is the case only if we consider permanent resident households. There are at least 50 and possibly more Koli Patel families who

FIGURE 3.1
Location of Gogha Taluka

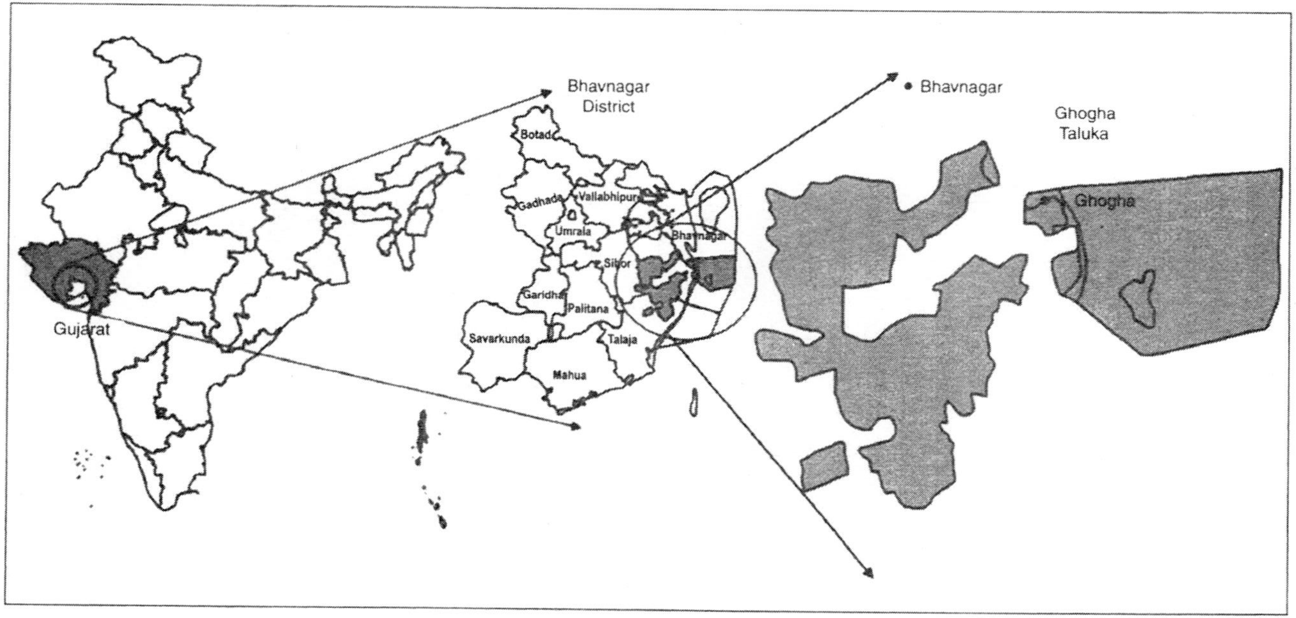

are sharecroppers/yearly labourers called *bhagiyas*, as opposed to the daily labourers who are called *dadhiyas*. They live in the village practically all the year round (at least 10 months in the year), but in makeshift dwellings in the *wadis* (fields). Many of them have been staying this way for decades. Yet, they are barely acknowledged as part of the village; even the seven Koli households who have built or rented houses in the main village are grudgingly acknowledged by the villagers. It is only because of the cultural mechanisms that ignore the year-round presence of the Koli sharecroppers in the village that the village presents itself as a 'single-caste' village.

Like most Saurashtra villages, Nathugadh has no forest land and very little common land. The land use pattern seems to be quite stable: the 1991 Census land use figures are very similar to those collected by Utthan in their 1998 PRA. About 20 per cent of the geographical area is recorded as irrigated (by wells)—a high percentage by Saurashtra standards—65–70 per cent is unirrigated, 4–5.5 per cent are cultivable fallows and 7–10 per cent is uncultivable land. Most of the land is privately owned (little less than 500 ha) and only about 70 ha is common or panchayat land. This is mostly pasture and wasteland (see Table 3.3).

The landholding pattern is also similar to the overall Saurashtra pattern, showing a sizeable section of large holders. The landholding pattern according to the *talathi* (revenue official at the village level) records, as recorded by Utthan in 1998, shows that 42 per cent of the landholders have an average holding of more than 4 ha, thus forming a substantial large holding peasantry. There are 35 agricultural

TABLE 3.3
Land Use Pattern in Nathugadh

Type of Land Use	1991 Census		Utthan PRA 1998	
	Area	Per cent	Area	Per cent
Irrigated	113.31	19.98	110.11	19.08
Unirrigated	377.69	66.59	394.00	68.26
Cultivable fallows	22.63	3.99	30.68	5.32
Forest	—	—	—	—
Uncultivated	53.58	9.45	42.44	7.35
Total	567.21		577.23	

Source: District Census Handbook, 1991 and Utthan PRA, 1998.

labourers in the village, but these figures exclude most of the bhagiyas (see Table 3.4).

The cropping pattern is typical of commercialised agriculture. Bajra and maize are the staple cereal crops and are grown mostly in the kharif season. *Mung* (*Vigna radiate* or golden gram) and *matki* (moth bean) are the major pulses (see Table 3.5). Jowar is mainly grown for fodder along with other fodder grasses. Wheat, groundnut, cotton and onion along with vegetables are the major cash crops. Cotton and vegetables are grown in both kharif and rabi seasons whereas groundnut is a kharif crop and onions and wheat are rabi crops. Cash crops account for more than 70 per cent of the gross cropped area. Our discussions highlighted that even a substantial portion of the cereal crops like

TABLE 3.4
Landholding Pattern in Nathugadh

Landholding Class	Number	Per Cent	Area (ha)	Per Cent	Average Holding (ha)
Small farmers	61	28.37	87.08	17.68	1.43
Marginal farmers	28	13.02	17.76	3.61	0.63
Big farmers	91	42.33	387.68	78.71	4.26
Agricultural labourers	35	16.28	–	–	–
Total	215		492.52		2.29

Source: Panchayat records, talathi (village accountant) books, 1998.

TABLE 3.5
Estimated Crop Pattern in Nathugadh

Type of Crops	Crop Names	Area (as per cent of Gross Cropped Area)
Pulses	Mung, matki	6
Cereals	Wheat, bajri	30
Oilseeds	Groundnut	30
Fodder	Jowar, maize, fodder grasses	7
Vegetables	Guar, chavli, cabbage, gourds	3
Cash crops	Cotton, onion	24
Total		100

Source: Utthan PRA, 1998 and gram sabha elders' discussions.
Note: Although the table suggests that only cotton and onion are cash crops, many of the other crops are also sold commercially. This accounts for our contention that 70 per cent of crops are cash crops.

bajra, matki and mung were sold; hence, the subsistence economy formed a very small portion of the agricultural economy. It was reported that area sown to cotton was declining, but has picked up in the last three years, especially after the availability of 'loose' packed Bt cotton seeds.

According to the PRA, there are about 250 bullocks and about 300 she buffaloes in the village. So far as draught power is concerned there are more than 10 tractors in the village and they are hired in by many families for the sake of timeliness and speed. Dairying is also an important activity. The main grazing practice is free grazing in the 70 ha of common land supplemented by fodder grown on private land. Nathugadh farmers have not switched over fully to chemical agriculture and reported substantial use of organic manure for all crops though in combination with chemical fertilisers.[9] Crop productivities reported were good, but not very high.[10]

All irrigation is from wells and the situation remains so even after the watershed development project. The village reportedly has about 300 functioning open wells. Interestingly, there are only a few functioning bore wells in the village, unlike other areas in Saurashtra where bore wells are common. But this is not the result of any regulation or conscious choice. Apparently, according to the villagers, there is a solid black rock at a depth of about 100 ft and attempts at drilling deeper, even up to more than 300 ft have not yielded more water. In fact, five deep bore wells that were drilled had to be abandoned later, though many of the wells do have horizontal bores. In some sense this is an advantageous position, in that water use is directly linked to annual recharge and does not draw from accumulated storage in a big way.

A substantial number of people have migrated from the village. Two to three people seem to have migrated from every household to nearby cities and to places as far away as Surat and Ahmedabad. According to the villagers, an exodus of sorts took place in the 1990s. Most of these migrants have been absorbed in the diamond cutting and polishing industry and to some extent in the shipbuilding industry. According to the Utthan PRA reports, more than 400 persons from the village are currently employed in the diamond industry. Interestingly, however, there are no NRIs in the village.

The village has an active Swadhyay movement.[11] Hour-long weekly Swadhyay meetings take place on Tuesdays. The meeting starts at 9 p.m. by when almost all routine chores are done and everyone can attend. No liquor is allowed inside the village and no meat is eaten. Apart from prayers, discourses and *bhajans* (religious songs), many issues are discussed at these meetings, including those pertaining to community action for drinking water, well recharge or other activities. At these meetings, a member of the Swadhyay movement may conduct the proceedings or a cassette or written message might be sent by the movement and this forms the basis of discussions.

Overview of the Drinking Water and Sanitation Programmes

Community action for drinking water had begun even before Utthan entered the picture. The first identifiable action seems to have been the building of a common reservoir, a *gaon talav*, as part of a Nirmal Jal programme initiated by Swadhyay. Farmers from three villages had come together to perform *shramdaan* (voluntary labour) for the gaon talav. This and another private talav are near a large well in the common land (known as *jalum* by the villagers) of the village (the gaon talav is also on common land). Since then a number of elements have contributed to the present drinking water and sanitation system in the village.

Ghogha Regional Water Supply and Sanitation Programme

The first drinking water intervention was the water supply scheme set up by the Gujarat Water Supply and Sewerage Board (GWSSB) under the Netherlands-aided project unit of the Ghogha Regional Water Supply and Sanitation Programme (GRWSSP). At a total cost of about Rs 1.2 million, this programme supported the construction of a pump house, sumps, an overhead (buffer) tank, rising mains up to the panchayat overhead tank as well as a bathing place or wash facility (a *snaanghar*), which has not been used since its construction, four public latrines, a cattle trough for drinking water for the cattle and a couple of stand posts.

The GWSSB scheme presupposes supply from the Mahi River pipeline. The Mahi pipeline passes very near the village (it passes by the side of the main road, which is about 2 km from the village habitation). As part of the Mahi water supply scheme, a pipe has been laid up to the GWSSB sump. However, water from the Mahi pipeline has yet to be delivered to the village and the village is engaged in a running argument with downstream villages on the issue. The village elders are demanding 200,000 litres per day from the Mahi pipeline.

In the absence of water supply from the Mahi pipeline, water from the jalum is brought into the sump by a 10 HP pump. From there it goes to the overhead tank of GWSSB and from there to the overhead tank opposite the panchayat, from where piped supply serves households or village stand posts.

Soak Pits

Utthan also took up a supplementary programme with the Netherlands-aided programme concentrating on awareness building, roof water harvesting and soak pits. This last part of the programme has been a remarkable success and almost 80 per cent of the houses are reported to have built and used soak pits. Waste water and sewage do not run down the village lanes and the village environment is generally very clean. As a result of constant advocacy by Utthan, *paani samitis* (water committees) and other partner organisations in the area, WASMO has contributed about Rs 150 per soak pit or approximately half the excavation cost of the soak pit. In practice, a proper soak pit costs much more, but it is to the credit of Utthan and the community that many farmers were ready to spend the required money to construct a proper soak pit and maintain it. Utthan has helped women's groups and paani samitis set up youth teams to build and manage soak pits on a large scale in a short period of time.

Roof Water-Harvesting Tanks

The roof water-harvesting (RWH) tank programme has not made much headway. About 10 RWH tanks were built with assistance from the programme and many more were to be built. Contributions (Rs 100 per household) were taken from many households. However, this was just around the time of the Kutch earthquake and after the earthquake all the grants for RWH tanks were withdrawn and remobilised for earthquake relief. In fact, the story goes that a cheque for RWH tanks was received on 25 January 2001 but could not be deposited the next day since it was Republic Day. The Kutch earthquake hit on the morning of Republic Day (26 January). On 27 January, when they went to deposit the cheque, the villagers found that all the grants had been withdrawn! Presently, because of the good groundwater situation in Nathugadh, especially after the implementation of the watershed and WASMO programmes, government support for RWH has dwindled.

Although a number of RWH tanks have emerged in Nathugadh, either supported by Utthan or undertaken individually by households, most of the RWH tanks are being used mainly as storage tanks for tap water and are being filled from the common water supply. There are also differing views on the effectiveness and potability of harvested roof water. Many people feel that the harvested roof water is not safe for drinking. A minority feels that so long as roof water and tap water are not mixed they both remain safe, but mixing them renders them unsafe. Be that as it may, the result is that practically no one harvests roof water and the tanks have been turned into buffer storages for domestic water supply.

House-to-House Supply of Piped Water

Normally, domestic water supply schemes in rural areas stop at the panchayat overhead tank or with a few stand posts. In Nathugadh, however, there is a piped delivery system that provides a tap for every house. Besides for an initial grant of Rs 100,000 from the local Member of the Legislative Assembly (MLA) scheme, for laying pipelines, the system is entirely designed, implemented and operated by the villagers. Keshubhai, a villager, has been the impetus behind the success. Keshubhai spent his working life in the cities. On returning to the village after retirement, he masterminded the water supply scheme and put his heart and mind into making it a reality.

A pani samiti, a committee of five persons, was formed at the gram sabha. The funds collected from the villagers are entrusted to the samiti which deposits the money in a designated bank account that is run with three signatories; it includes Keshubhai, the person who initiated the idea and looks after most of the routine tasks involved in the implementation of the scheme. Every household had to pay Rs 1,000 for a connection. For the operation and maintenance, Rs 500 per household is collected. When that money runs out, another round of collection of Rs 500 per household is made. The first Rs 500 per household that was collected was enough to see the scheme through almost two years, that is, the annual charge worked out to approximately Rs 250 per household.

Almost every household in the village, barring 10–15 households, have taken a tap connection. There have been small rumblings of discontent. Most of the work of monitoring the scheme and the maintenance tasks are done on an honorary basis by Keshubhai, though the actual expenses involved or labour performed is paid for. There is a tendency on the part of many households to criticise harshly, impute motives and make accusations—all of which have created stress. Second, there is the issue of flouting of rules/norms. For example, one of the rules is that the samiti should be informed and prior permission obtained if anyone is to use water in large quantities and that a charge has to be negotiated if necessary. One person has been drawing water for construction work without informing the samiti. He maintains that the water belongs to everybody and that the samiti has no right to impose its will. At present, no charges have been levied on him and the arrangement is still holding together.[12]

However, more serious is the complaint voiced by the Koli Patel families. They found that the charge for a tap connection was too much and opted for a stand post connection for which they claim that they had to pay Rs 400 and that too annually.[13] As will be discussed in more detail later in the chapter, many of the Koli Patel families live in wadis (fields), especially during the cropping seasons. Koli Patel families also pointed out that the standard charges have not taken into consideration the fact that there is substantial difference in the amount of water consumed by the Kanbi Patels, landowning and cattle-owning families, and Koli Patel families who do not own any cattle. In fact, cattle consume a significant amount of water especially during the summer months when all other sources dry up. There is truth to their claim. If we consider the Rs 400 charge for a stand post connection as a one-time collection, then those drawing water from stand posts are being charged 80 per cent of the tap charge; this is somewhat on the higher side. In fact, a migrant family that was staying in the main settlement area decided to go and live in the wadi since it could not afford to pay the stand post charge which it would have to pay if it stayed in the village.

Overview of the Watershed Development Programme

The watershed development programme for Nathugadh was taken up under the IWDP implemented under the rural development department guidelines. Utthan was the project implementation agency (PIA). As is customary, Utthan began with a gram sabha, in which they explained the project objectives and requirements and then conducted a PRA. The PRA was a fairly detailed exercise.[14]

The PRA also involved marking out possible sites for check dams and *nala* (stream) plugs. The more promising sites were studied and cost estimates were drawn up. A project proposal was prepared and finalised after discussing each of the issues. At a subsequent gram sabha, the watershed committee was selected. It comprised 20 persons, including three women. The secretary was drawn from the sole Brahmin family. He and two other members received honoraria from Utthan. The others worked on a voluntary basis.[15]

The watershed development programme consisted of many components grouped as follows:

(a) Soil and water conservation measures: well recharge, check dams and nala plugs.

(*b*) Agricultural development: land levelling, bunding, farm ponds, crop demonstration plots.

(*c*) Environmental improvement: kitchen gardens, plantations and pasture improvement.

(*d*) Livestock improvement: gobar gas plant, vaccinations, fodder cutters and animal clinics.

The check dams and nala plugs accounted for the major portion (almost 60 per cent) of the project cost. Land levelling as an item was practically dropped.

Check Dams

An issue of utmost concern was how the check dam sites were decided upon and how arrangements were made to construct them. The check dam sites seem to have been selected through a two-way process. First, the 10–15 village elders who participated in the transect walks during the first gram sabha and the subsequent PRA identified possible sites. The criteria for this initial selection of check dam sites were described by the farmers as follows: (*i*) they should be favourably located, in that sufficient water should be expected to collect there (a purely technical reason), (*ii*) the water that collects should stay there for a longish period, (*iii*) they should be on common property land as far as possible and (*iv*) they should cause as little submergence as possible. The next step involved preparing estimates and identifying beneficiary (user) groups. From then on people were left to themselves to come to an agreement.

For each check dam, the foundation work was supposed to be completed by the beneficiaries. This was some sort of a test, related to people's readiness to bear the cost and their acceptance of the check dam and its potential benefits. A tentative list of farmers whose wells were likely to benefit due to rising water tables after the check dam was built was prepared. Each group then shared the foundation cost of its check dam. On this score it is clear that the farmers seem to have accepted the potential benefits of the check dams. The farmer's enthusiasm for this project can be gauged from the fact that instead of the eight check dams and nala plugs planned, 10 were constructed.

But it was not a smooth affair. There are at least two instances where some of the beneficiaries refused to comply and were brought round only with great difficulty. Reportedly, there were also instances of free riding where some of the beneficiaries refused to contribute knowing fully well that the others would nevertheless go ahead and build the check dam, at times paying the share of those who did not contribute.

Most of the check dams have withstood the heavy rains of the last two years, but a few have not. A special case is that of a check dam that was built on part of a stream that flowed between two villages. The farmers on the other bank belong to a different village and hence are not part of the watershed development effort. They were duly consulted and the concerned farmer had agreed to the check dam being built. However, he discovered last year that the overflow was scouring part of his land so one day he breached the check dam. There were at least two other incidents of check dams being breached.

Well Recharge

The programme had developed a common well recharge package that consisted of guiding run-off and letting it into a well through a filter chamber, commonly called a *kundi*. Some changes to accommodate people's views have also been made in this design. In one case, a farmer had a different design in mind which comprised two kundis instead of the standard one-kundi design that Uthhan was working with. Though the extra cost was not allowed, the farmer was allowed to build the kundis according to the design he had in mind. Our visit to the well recharge sites, however, brought out the fact that there is no proper upkeep of the well recharge structures and that they are in danger of getting clogged and silted.

Other Measures

Apart from check dams and well recharge measures, a watershed programme entails other measures. But in Nathugadh there is a singular lack of a ridge to valley approach, largely because the village community was primarily interested in check dams. Land levelling as a programme was dropped by the government because they found that it was quite expensive and would benefit only a few people in the village. Some amount of bunding was done and some farm ponds were also dug, but these were isolated cases. Often these measures were undertaken on the lands of those people who had not benefited from the check dams. The kitchen gardens did not materialise properly be-cause the women's self-help groups (SHGs) did not really take off.[16] Saplings were distributed under the plantation programme but the survival rate was very poor and practically nothing of it remains. The tree component of the pasture development programme likewise did not show good performance. The different items under the livestock programme were dutifully carried out, but these have not left any lasting impact on the livestock situation in the village.

Community Formation: Exclusions and Inclusions

The Well-to-Do Kanbi Patel Community

The Nathugadh experience raises many issues that are important in assessing the role and extent of CBNRM initiatives. They involve notions of community, of being community based and what relation and impact they have on issues related to equity and sustainability. We start with trying to delimit the notion of 'community' that under-pins 'community action' in Nathugadh. How Nathugadh came to be-come, and view itself, as a single-caste village is a process that is rooted in history. As mentioned earlier, Nathugadh was settled when a number of Kanbi Patel tenants of Vavdi split off and formed a settlement away from Vavdi, close to their fields with the permission of Nathu Singh, the then ruling Darbar in Vavdi.

The settlement, in those times, was not free from the menace of the Darbars. Villagers graphically described to us how the common village courtyard was adjacent to the Darbar's granary, how everyone had to bring all their produce to the courtyard, how strict the vigilance was, how even a woman carrying water had to show that the vessel really contained water (and not smuggled grain), how the produce had to be apportioned into the lord's share and their own (3:2 according to some and 2:1 according to others) and how the tenant himself had to carry the lord's share and deposit it in his granary.

It was finally the abolition of tenancy in 1956 that gave the Kanbi Patel tenants their freedom and turned them into independent farmers. Everyone in the village remembers this as the *barpat* transaction, whereby they had to pay the government 12 times the land revenue in order to be free of the Darbar's tenancy. How they came to collect the cash required in those cash-strapped times are stories in themselves. The most daring act according to the stories is one about a tenant who approached the Darbar himself for a loan promising to pay it off at the harvest! Probably taken aback, we will never know, the Darbar agreed.

Free of the yoke of the Darbar, free even of the Darbar's presence in the village, the Nathugadh farmers came to acquire substantial chunks of land. All the families report their original holdings as several hundred bighas, now whittled down to smaller portions because of sub-division. Nevertheless, when we consider that these were joint families and migration was common, operational holdings of households would still be large as attested to by the landholding pattern, where even today more than half of the landholders (individual holders, not households) have more than 4 ha on an average. Historically, therefore, the community acquired an identity not only of a Kanbi Patel community, but also a largely well-to-do or rich farmer community. Both the caste and class dimensions are important and nowhere is this more apparent than in their relationship with their largely Koli Patel bhagiyas—nominally sharecroppers, but in fact yearly labourers.

The Institution of Bhagiyas: Making Community Invisible

The term for the daily casual labourers is dadhiyas (which literally means those who survive on daily earnings); the bhagiyas by contrast are contracted for a year's labour and they are paid a share of the produce. The standard share for a bhagiya is one-fourth of the net produce, though for the last few years, bhagiyas get one-third of the net produce for the area sown to cotton. Unlike a true sharecropper tenant, who is in possession of the production process and the management of the land, the bhagiya has no control over the production process and its management and does not share in any of the costs. His role is simply to supply the necessary labour and do so under the control of the owner. There are no illusions on both sides that the bhagiya is but a labourer who is being paid in the form of a share.

Most if not all the bhagiyas in Nathugadh are Koli Patels and with one exception all of them are landless. The number of bhagiyas in the village, that is, the number of bhagiya families employed in the village, is reportedly anywhere between 60 and 100 or even more. Except for the seven families who have either rented or built houses in the main village area, others stay in the wadis in makeshift huts or pump houses or in the corner of cattle sheds. Even those who come from adjacent villages stay in the wadi during the cultivation season and return to their villages for a few months after the season is over; only a few travel every day from adjacent villages.

The interesting thing is that many of these bhagiyas have been coming to the village for the last 25 years. Moreover, they stay in the village almost the whole year. They go back to their villages for a few months in the summer. Though they are not responsible for the ploughing, which is done by the owner, they have to be there at the time of sowing, immediately after the first good rains and have to stay back in the wadis till the last harvest of the summer season. They spend practically their whole time in this village (although not in the village settlement proper but in the fields), take part in all the productive activity in the village (down to the smallest tasks), accompany the owner on every marketing trip and yet neither the owners nor the bhagiyas themselves see the bhagiyas as part of the village community!

This exclusion is automatic and happens without thinking, it is not in that sense deliberate; but that it is not deliberate is all the more significant. The implicit construction of a normative model of the village community as composed of settled dwellers with a *pucca* or permanent house in the village habitation area automatically excludes the bhagiyas. At least in Nathugadh, the bhagiyas are outsiders on two counts, both by virtue of having originated in another village as well as belonging to a different, hierarchically 'inferior' caste. These notions run through the discourse, surfacing momentarily in chance remarks: like the remark that all in the village being from one caste there is no problem in cooperation; or that being all from one caste there is no danger of bad habits being formed—the implication is that the 'other' castes are non-cooperative or have bad habits.

A similar exclusion operates with respect to watershed activity. Watershed activity is an activity that is relevant primarily to landowners. No wonder that exclusion from watershed activity and its planning is accepted with equanimity by the bhagiyas, especially the men. They acknowledge that Utthan did call them to the meetings, but they themselves refrained from attending them. This was partly because they would lose valuable time—they did not have as much spare time as would have been needed. They also did not attend the meetings because they did not see themselves as being involved. This was especially true for the bhagiyas, as distinct from the landless *casual* labourers, who could benefit from the extra employment generated by the watershed activity or from other activity aimed at the landless, though there was little of such activity in Nathugadh. As much as others excluded them, they also excluded themselves. The process of their exclusion thus ran two ways. Kanbi Patel villagers did not consider them as insiders and the Koli Patel themselves did not ask for participation since they too did not see themselves as insiders nor did they anticipate any benefits for themselves.

They are doubly excluded in another sense as well. They have insider status and are entitled to participation in 'their' villages, but are not present there to realise this potential; whereas in the villages they are present they do not have an insider status and are not entitled to participation. In this respect, the situation of the bhagiyas is somewhat like the situation of the nomadic (and erstwhile so-called criminal) tribes. However, there are two important differences: first, the bhagiyas often do have a specified place where they are present, whereas the nomadic tribes are truly without a place to call their own; and second, the situation of the nomadic tribes originates in their caste/traditional way of life whereas for the bhagiyas it originates in their specific class position.

The important thing that the Utthan experience highlights is that sincere but simple efforts to increase participation of the bhagiyas, which do not engage with and challenge the underlying normative and institutional boundaries that define inclusion and exclusion of different groups in different activities, is unlikely to bring about true participation.

The Story of the Dissolved SHG

The story of the dissolved women's SHG in the village and the disputes around it throw more light on the caste and class identities and conflicts that operate between the Kanbi Patel landowners and the Koli bhagiyas. The SHG in Nathugadh comprised both Kanbi and Koli women. Initially they contributed Rs 20 per month and later Rs 25. They accumulated savings of a little more than Rs 28,000 in three years, when internal strains led to the SHG being dissolved. There were many factors that operated. The main reason, at least the most visible one, and the one which was cited by Utthan activists and had been in turn cited to them by the SHG leaders, was that one woman who had taken a sizeable loan, did not repay it in time. There were also charges of favouritism against the leading woman activist, a strong-willed person who finally decided to dissolve the SHG rather than continue in an atmosphere of mistrust.

Utthan activists recognised their failure and also gleaned a few learnings from the experience. They concluded that they had put too diverse a group together and that they had given undue importance to savings over empowerment.[17] They also realised that many of the women were well off and that their sons had good jobs outside the village and hence did not really need micro-credit. They wrote off this experience, and rightly so, as a failed attempt but not one without its learnings. Utthan activists say that what they have learnt from Nathugadh has been very important in other villages. They now insist on the watershed committees having a membership of which 50 per cent are women, in giving representation to landless and other vulnerable sections and facilitating their participation so that they will truly represent the voices of those communities.

But there was another sub-text to this failure that had not been communicated to the Utthan activists. This was uncovered in our discussions with the Koli women, who in many ways stood in greater need of micro-credit, but complained that they were denied it. According to the Koli women, there was not a single Koli woman to whom credit was extended. The Koli women were given clear-cut reasons for it. They were told that they were not given loans because Koli women do not return loans. At the same time, it was also said that unlike Kanbi women who had cash in hand only at harvest time, Koli women had cash in hand all the time because they also hired out their labour and so they did not really need credit.

This complaint and this course of argument had not been put forward in meetings with Utthan activists. The Koli women found themselves too inhibited to discuss these issues in these terms with the staff of Utthan and this itself was a problem. In this light, there is a need to discuss the learning from the experience again. For example, the learning that Utthan activists have identified is the need for more homogeneously-composed SHGs. Faced with such division, this implies that it would be better to form separate SHGs for the Koli bhagiya women and the Kanbi women from landowning families. In all probability this would also 'work' in the sense that the SHGs would run more efficiently, both groups of women would have access to credit, etc. However, in some ways this is a way of getting around the original problem that caused the breakdown of the joint SHG as opposed to a way of truly resolving it. The prejudices that caused the breakdown have not been eliminated, but simply separated physically and socially. There is a danger here of a kind of ghettoisation, of sealing rather than healing divisions within communities. The need to discard stereotypes and form a wider notion of community still remains. How this has to be done may neither be clear nor very easy, but recognising the problem is an essential component of any solution.

Livelihood Issues

Since drinking water and sanitation have been the focus of a separate programme that predated the watershed project, it is best to start with a consideration of those needs. The work on sanitation is the most impressive. As noted earlier, the village is clean and there is no waste water flowing in the streets. All repairs on the soak pits and their maintenance are the responsibility of individual households and everyone maintains them well.

The drinking water scenario is not as straightforward. Almost all of the RWH tanks are practically defunct as RWH tanks; they mainly function as buffer storages for the normal supply. There is a dispute about using harvested roof water, about how good it is, how long it stays potable and how far it can be mixed with other sources of water.

The piped water scheme has been running well. Except for the last one or two months in summer, when the jalum (small tank) along with some tanker refilling is required, the piped water scheme takes care of the main drinking water needs of the village. The schedules are well maintained and the distribution is generally fair. There is some dissatisfaction about the rates, especially by the Koli Patel households as noted earlier. But on the whole the scheme has been running well.

The farmers were very reticent about any kind of quantitative estimate of the benefits they derived from check dams, well recharge and soil and water conservation activities. Most farmers responded by saying that it varied: from farmer to farmer according to the location of his land in respect to the check dam, and from year to year according to the rainfall. Most of them also said that in good years the watershed work proved to be beneficial, but did not help much in bad years. Moreover, in bad years (as in good years) there was no collective regulation or understanding of any sort with regard to water use and priorities of water use. In fact, there is an almost total lack of any collective control over any form of productive activity, a feature that we shall shortly discuss.

The bhagiyas, however, were much more forthcoming in an assessment of the benefits of watershed work, and in some ways were more closely associated with it too. They acknowledged that there had been a definite increase in irrigated area and/or irrigation and in yields, anywhere between a 10–25 per cent increase in both irrigation and yields. The total benefit would then be of the order of 25 per cent. They also acknowledged that as sharecroppers, they too had correspondingly benefited from the increase. In fact, recent renegotiation of sharecropping rates for cotton has seen an increase in the bhagiyas' share from 25 to 33 per cent. It is difficult to assess as to whether it is watershed development or the tremendous surge of cotton cultivation in Gujarat after the illegal introduction of genetically-modified BT cotton since 2000 that is responsible for the changed scenario, but it is likely to be result of both. Bt seeds[18] cotton cultivation in Gujarat has increased more than four times (for further discussion on this, see Shah, 2005). However, the benefits to the bhagiyas are essentially trickle-down benefits rather than direct benefits and their participation remains practically negligible.

As far as fodder and fuel are concerned, no major change was reported either by the landowners or the bhagiyas. This was expected since the tree component of the pasture development programme on common land as well as the plantation programme on private land had failed.

In conclusion we would say that two things stand out from a consideration of livelihood issues. While the Utthan team tried to conceive and plan the programme as one of different livelihood options around watershed development, the community's view of the programme was one of a water augmentation programme. The community identified the programme as such and, it is our contention, participated in the programme based on this belief. As a consequence, all other elements of the programme did not receive as much acceptance and participation as the check dams, well recharge and other water augmentation programmes received, resulting in the failure of these other programmes.

Similarly, Utthan understands the drinking water programme as one aimed at creating self-reliance in drinking water through maximum utilisation of local water resources. However, here too, even while the community supports Utthan's work, the operative framework that underlies the community's overall actions, as reflected in the minds of their leaders, appears different. One of the reasons for this could be that people think more in term of overall reliability rather than self-reliance. If this were so, people would concur with strengthening local resources, but within a framework that would try to acquire access to as many different sources as possible. In such a scenario, people may try to develop local water-harvesting ponds, build RWH tanks and simultaneously lobby for water supply by tankers and/or water from the Mahi pipeline. Leaders devote as much time in lobbying for Mahi water as they do for getting access to WASMO's funds and programmes or funds for the jalum and RWH tanks. When access to external water or a large dedicated source is not on the horizon, the community does indeed participate in strengthening local water resources, but as soon as other possibilities emerge, a mix emerges in which local resources may receive less priority and eventually even suffer neglect. Signs of this process are evident in Nathugadh as also elsewhere. The Mahi pipeline has yet not arrived in the village. If it does, what impact it would leave on the state of local water harvesting is difficult to predict. We think that augmenting and harvesting local resources may suffer a decline unless the community as a whole not only supports Utthan's work, but also Utthan's objective and framework of self-reliance.

Equity and Sustainability Issues

Discussing equity issues in Nathugadh was very revealing. If we pointed out an inequity—for example, if we pointed out that different people would benefit to different degrees based on how far they were from the check dam and some might not benefit at all, the typical response was to agree, but have nothing to say about rectifying it in any manner. The question of rectification did not arise because that was how it was; if someone owned property in a certain place and another owned it in a more favourable place, the latter would derive the additional benefit. It slowly became clear to us that inequity was recognised, but it was accepted rather than challenged. We were expecting this kind of acceptance in respect of inequities rooted in hereditary relations and accumulations. However, we were not expecting the same attitude to extend to those inequities and asymmetries that arose from the watershed development process itself which led to unequal sharing of its benefits. That is why we were surprised to find that spatial asymmetries and consequent unequal sharing—for example, those related to upstream and downstream as well as those related to distance from stream in the case of benefit from recharge—were similarly recognised and accepted. The most that could be said was that there were attempts to compensate for the lack of benefit by offering some other benefit—for example, those upstream were offered silt from the check dam for their fields.

So far as social and economic inequalities were concerned, the issue of the bhagiyas had already been rendered invisible as part of the process of watershed development by their exclusion from the community referenced by the watershed development process. By virtue of their being bhagiyas, this exclusion also extended to the seven bhagiya families who were now residents of the village. What about the other landless and marginal sections? The only other landless family in the village was the Brahmin family. The head of the Brahmin family, as mentioned earlier, was made secretary of the watershed project. Another marginal farmer was made a paid office bearer of the watershed project and worked as a general factotum. And, of course, there was ample employment provided during the watershed development project phase; in fact, many outside labourers were also inducted for that work. There was however no effort at any conscious equity measure for reducing inequities.

What appeared to be the minimum common perception of the principle of equity within the community may be formulated as: everyone gets something. This was most explicitly put forward by a woman who had land upstream of one of the check dams and whose only benefit from the dam was that of silt. As she said, 'Someone gets silt, someone gets well recharge and someone gets water for cattle. Everyone generally gets something', a point made by others also. Even the bhagiyas benefited from increased productivity.

'Everyone Gets Something': Potential of Changing the Discourse

What is interesting here is that the principle 'everyone gets something' is perceived as an acceptable principle of equity. What is interesting about it is the potential it has of leading us into a discourse that could sharpen the sense of equity—a potential that was not exploited in Nathugadh. One could then begin by saying, did everyone get something? Let us have a hard look. Who got something and who did not? Going further one could then ask, is there not a minimum that everyone should get, or does any benefit however small become acceptable?

What is important here is to treat the principle as the beginning of a new area of exploration. Since this is not a principle being 'imported' from outside, but so often expressed within the community, it could form a more acceptable starting point for discussion of equity issues within the community. Such an exercise would possibly bring us to some kind of concept of a minimum acceptable benefit, qualitative as well as quantitative, that should be assured for everyone. This is because though there is no explicit mention of a minimum threshold here, the very principle that everyone must get something implies a tacit understanding that there is some kind of a minimum threshold benefit that everyone should get. In this sense, equity is seen as an assurance of a minimum benefit. Acceptability then has a reference to this minimum. In practice, the details of this principle will have to be worked out (or contested) during the complex negotiations that take place around site selection, participation in the foundation cost and the building of the physical structures.

However, since minimum assurance is only an implicit notion embedded in this kind of consensus, it needs to be brought out and made explicit, especially because there is a sense of an entitlement that underlies it that it is important to make explicit. In this respect, it resembles the traditional concept of entitlement to social security. However, it also needs to be remembered that traditional concepts never aimed at full security, but were much more oriented towards avoiding destitution and mitigating misery while the more modern notions look at security nets more in terms of aids to eliminating poverty.

Obviously, this still falls short of the more proactive and radical conceptions of equity that are possible, especially those that seek to rectify the inequities built into the social and economic structures of society. That can happen only when organisations begin to consider what the minimum entitlement/assurance should be: should it not incorporate the rectification of historical inequities? What is important is that the implicit principle that may be very widely acceptable offers us something to build upon. Making the principle explicit and questioning it for its many meanings allows us to start from a common perception and expand its scope by making it an object of collective discussion—and contestation. This is the kind of role that organisations like Utthan are expected to play; something that they have done effectively in respect of drinking water in their earlier work, but have not been able to do effectively in dealing with watershed development in Nathugadh.

Gender Issues

Women's issues and women's participation have always been a strong point of Utthan–Mahiti. Women have led the struggle and shaped its outcome. Women have not been very active in the whole process of watershed development planning and implementation. A few women in Nathugadh have also been prominent; for example, the leader of the SHG while it was in existence. Similarly, another woman, who looked after the inventory of the stores and supplies, has played an important role in the watershed development effort.

However, the general level of participation of women has been low. Women have very little say in the running of the drinking water schemes. The women did not participate in the first crucial gram sabha, mainly because the SHG was being formed at the time. And subsequently, important matters related to women's concerns have been discussed in the SHG separately. But a small number of women did begin to attend the gram sabha meetings regularly in the later phases; three women members attended the watershed committee meetings as well. Many of the women also attended training courses in SHG activities as well as participated in those meant for watershed activity. What remains missing is a linkage between the SHG programme and the watershed programme. The SHG leader described the strong patriarchal norms that prevented women's participation in social life and also how a few of them had now learnt to stand up for themselves and to exercise their right to participate. After the SHG closed down there was of course no formal organisational forum for interaction between the women and between the women and Utthan.

There seem to be two possible reasons for the lack of participation of women and the comparatively minor role they have played in this village as compared to women in Mahiti's drinking water and sanitation projects. One of them has to do with the nature of watershed development activity as opposed to drinking water and sanitation activity. The other relates to the separation of Utthan and Mahiti. Drinking water and sanitation has been in some sense associated with women's traditional role as homemakers. It is well documented that women supply the major portion of the labour that goes into finding and fetching drinking water and water for other domestic uses. It is easier for patriarchy to cede leadership to women in this sphere since it does not violate the lines that define the patriarchal division of labour. However, it is different with watershed development. Production in fields and agricultural plots is an exclusively male sphere. It is highly unlikely that patriarchy will cede leadership to women in this sphere or even allow for much participation.[19]

Second, in the Bhal villages, and with Devooben, the sense of a strong women's movement was evident and the associated discourse was also militant and self-confident. However, in Nathugadh, the approach and the discourse, in general, had less the character of a movement and more that of professional social work. There is no denying that what is reflected in the Bhal area is the result of sustained work over two decades or more whereas the work in Nathugadh is only a few years old. But there is a distinct change in approach and sensibility that also perhaps needs to be recognised. Can work in Nathugadh produce the same kind of results given the more projectised mode of work?

Sustainability: To Regulate or Not

During our discussions, there was an acknowledgement—a cautious and sometimes grudging acknowledgement from the Kanbi Patels, but a much more emphatic acknowledgement from the Koli Patels who worked their lands—that there was a significant increase in water availability, its assurance and benefits from irrigation after the watershed development programme. However, at the same time, there were complaints that this had not resulted in alleviating drinking water stress in the summer months and also that in bad years, watershed development work did not help much.

During our first discussion outside the village temple, in which more than 20 farmers participated, everyone agreed that summer stress was still a problem, that things were still very difficult in bad years and that they needed Mahi water or water from a big talav or a river if their problems were to be solved. Were they not using more water than was available? Yes, it was acknowledged, but then their needs too had grown. It was also quite clearly acknowledged that much of this was because of changes in lifestyle, which were now irreversible.

In other private discussions, we raised the issue of regulating water use. And there was a strong reluctance to collective regulation. The analogy utilised was that of someone making more money and someone less. If farmer A was using more water and making more money, can we tell him not to make more money?

What is worth noting here first of all is that water, especially groundwater, is treated clearly as a private resource. Second, the sense of property is quite strong and it is not within the domain of peers to say how one should make a living or a profit from one's property. (There is a similar lack of regulation in respect of free grazing as well.) What is interesting, and this is common to many of the dark zones in other parts of Gujarat, is the acknowledgement of the collective impact of individual action of overwithdrawal juxtaposed with the reluctance to acknowledge collective regulation by peers as a solution. What is the way out? As one of our respondents put it, 'Maybe, if the state puts restrictions and makes it illegal, then there may be some impact'. What is interesting here is that this is an acknowledgement that at least as things stand today, these farmers see state regulation rather than community regulation as more likely to succeed in matters related to groundwater use.

Decision-Making: Who Decides and How?

Most of the decisions involving the entire gram sabha were taken at the first two meetings of the gram sabha: the first in which the PRA was held and the second in which the committee was formed. Thereafter, the gram sabha seems to have become more of a body to which accounts were rendered (this happened roughly once every six months) rather than a deliberative body in which important decisions were taken. This former function, one that ensures transparency, is no less important than the latter; however, they do remain distinct functions.

It seems that after the first two meetings of the gram sabha, the role of the gram sabha reduced and focus shifted directly to stakeholders and their role in constructing and maintaining different physical structures and interventions that were planned. Discussions and deliberations happened at two levels. First, discussions took place to determine and enumerate likely suitable sites. This involved a group of elders as part of the process of the first meeting (the PRA) of the gram sabha. This was later narrowed down by the technical people in the watershed development team. In that sense, the context may be seen as being provided by the decisions made by the watershed development team. However, it should also be noted that, overall, there was a lot of common ground between the criteria that the farmers had suggested and those that the Utthan team utilised. Second, the 'gainers' and 'losers' thrashed things out amongst themselves. There have been instances of 'voting with one's feet' in instances in which farmers did not oppose the structure but refused to pay and the other beneficiaries made up their share; there have also been instances in which those who stood to lose have not allowed some structures to be built. Outcomes have been to a large extent determined by the strength of bonds between farmers who are located near particular structures.

On the whole, we may say that decisions were taken in a transparent manner so far as the community was concerned and that unlike in some of the government-run projects, there was good scope for direct stakeholders to negotiate and influence the process. Nathugadh certainly serves as an example of CBNRM, and even though it is difficult to see it as an effort originating in the community and being initiated by it, the community is quite active in shaping the final result; in fact, even constraining it in many ways.

Conclusion

As a CBNRM initiative, Nathugadh is one of the better examples of Utthan's work. Utthan gained an entry into the village in the course of its drinking water and sanitation work in the WASMO phase. The village is remarkably clean and the drinking water scheme has been well organised and is functioning well despite cases of bigger farmers drawing extra water without intimation or payment. The RWH tanks have not always harvested roof water, but have served well as buffer storages.

As a watershed effort too, so far as water harvesting goes, Nathugadh shows better performance than most. Most of the structures are standing and the few that were breached have been repaired. Even the one breached by the farmer on the opposite banks was in the process of being renegotiated and repaired. Though soil conservation and a ridge to valley approach could have been given more attention, water availability has increased significantly, leading to increased irrigation and better productivity.

Even more importantly, Utthan has been successful in involving the community and maintaining transparency in its activities. It has been able to resolve the disputes around the structures and interventions it planned by allowing sufficient space for stakeholder negotiation and engagement to take place. Moreover, it has been prepared to take suggestions on board. It has faced two very difficult situations and dealt with them well. The first related to money collected for *tankas*. The second issue was with regard to a whisper campaign about fund disbursement. In the former case, the contributions for RWH tanks taken from each household during the WASMO period were converted to a contribution towards common building that was taken up as an entry-point activity. The decision had been taken in the committee and not everyone was aware of it. However, the books and accounts were clear and rectification could be done easily. In the latter case, Utthan got the deputy collector to visit Nathugadh and explain the reasons for the delay and the committee offered to open its books to anyone who wanted to scrutinise them. In both these instances it was transparency and the readiness to face public scrutiny that saw them through. This has been its greatest success.

It also needs to be noted that Nathugadh was one of the first watershed development projects that Utthan took up after it decided to expand its activities in a systematic manner from drinking water and sanitation issues and move outside the Bhal area. It was evident in our discussions with Utthan staff, both at the local centre in Bhavnagar as well as in its headquarters in Ahmedabad, that it was carrying many of its experiences as learnings into its other watershed development projects. As Nafisa Barot expressed, those who join Utthan also come from the same society that they are hoping to change and often are imbued with the same prejudices and notions that surround them. This can only change over time and through a process of learning. Organisations need to allow sufficient space and time to engage with such concerns.

This self-critical awareness of the need for organisations to allow sufficient space and time for people to learn and to engage with their own ideas has been documented elsewhere.[20] Moreover, this is a process that is not a one-time process but a continuing one and it is all the more so when an organisation is expanding into new geographical areas as well as new areas of work. While the organisational centre may comprise time-tried and tested personnel, the outward fringes constantly absorb new persons who come with pre-conceived notions and in need of space and time for engagement and learning.

Nevertheless, if we look at the Nathugadh case from an equity and sustainability (environmental sustainability) point of view, watershed development has fallen short. It has not enhanced equity, except marginally and in a 'trickle-down' manner. In the absence of a clearly enunciated normative framework, it could perhaps be argued that it was never an objective of watershed development per se. It is nevertheless useful for two reasons to point out the silences. First, it shows that even with an organisation like Utthan, committed to participation and with a long history of such work behind it, the socio-economic context does shape and constrain what and how much can be achieved in this respect. The community asserts itself and constrains efforts that attempt to go beyond its bounds and it needs a more than routine effort to engage with and transform the notion of community that has become dominant and exercises such strong influence. Second, it also highlights areas that need to be proactively tackled if watershed development is not to work mainly as a water augmentation measure for the landed. Otherwise, NGO efforts are likely in the final analysis to remain somewhat more efficient systems of delivery than state programme but primarily beneficial to the dominant elites.

Notes

1. The broad description of the social relations described here is based on the following studies: (Gidwani, 2001; Patel, 1957; Rutten and Patel, 2002; Shah and Rutten, 2002) as well as conversations with Utthan activists and the farmers of Nathugadh.

2. The term literally means those who have formal ownership right over a piece of land.

3. Figures provided by the agriculture department show the total (net) area irrigated in 1999–2000 in the district was 135,600 ha of which only 1,900 ha was irrigated by canals and the rest by dug wells (quoted in GEC, 2006c).

4. Both the tehsil and the district should be classified in the grey zone for two reasons. First, as the utilisation figures for both the tehsil and the district are close to 70 per cent, they actually do fall in the grey zone category. Second, the assumption that 80 per cent of groundwater recharge is utilisable is highly questionable.

5. This section is largely based on Vohra (1990), information available at Utthan's website (www.utthangujarat.org) and our talks with Nafisa Barot, Devooben and activists of Utthan and Mahiti.

6. This section is based on the following sources: Barot and Chatterjee (2004), Bhatt and Iyer (1994) and Wilkinson (1990) and discussions with Nafisa Barot and Devooben.

7. Many of the dilemmas and challenges that Utthan–Mahiti faced in the initial work around issues of drinking water, especially as they related to divisions within the village, have not been as central to their interventions in watershed villages.

8. The peculiar shape of Ghogha tehsil is a consequence of the disjoint and chequered territories that the different princely states in Saurashtra controlled.

9. For most crops, 13 cartloads per ha of manure are used. In the case of cotton, it is approximately 20 cartloads. Urea and Di-ammonium Phosphate (DAP) are the main fertilisers. Around 125 kg of both DAP and urea are used per ha for most crops other than cotton for which 250 kg of both DAP and urea are used per ha.

10. The average productivity for wheat and bajra is 3 quintal per ha. In the case of cotton, the productivity is 8 quintal per ha and for onions 19 quintal per ha.

11. Swadhyay, literally self-study, is a movement started by the late Pandurangshastri Athavale. It is a spiritual movement that concentrates on self-study, the Gita and Krishna bhakti. But it also stresses that the study must be of practical help and improve life. Swadhyay encourages a particular kind of self-help and the movement has been quite active in the Saurashtra region. The contention is that Swadhyay has been responsible for thousands, maybe millions, of wells being recharged as well as a number of conservation works being taken up.

12. Utthan activists informed us that subsequently the issue was brought up in the gram sabha and other meetings and that there is strong support for not allowing such things in the future. According to the activists, since this was the first case they did not want the issue to be discussed further; moreover the person involved also realised that what he was doing was wrong.

13. There is some doubt whether the charge is annual or a one-time collection paralleling the Rs 500 collected from tap owners.
14. A lot of the information presented earlier is based on the PRA information.
15. Subsequent discussions revealed that two members owned three wells each (wells, we were told, were more important and easier to identify than land), 11 owned two wells each, three owned one well each, three did not own wells and one (the Brahmin secretary) did not own any well or land.
16. The failure of the women's SHGs is discussed separately later because it sheds some light on the Koli and Kanbi Patel relations in the village as well.
17. Utthan activists point out that there are a number of women leaders from Nathugadh who have been involved in various women's issues such as violence against women, health, gender discrimination, etc. It is expected that the women's groups will be rebuilt with these activities as a starting point with the savings emphasis following as and when the need is felt. They feel that the time and space they allow for a community to come to terms with the issues of inequity (gender) and empowerment separates their programme from that of the government.
18. Farmers did not always recognise the seeds they cultivated as genetically modified. However, the information on the back of the seed box and/or the dramatically reduced pest attacks reported confirmed to us that the cotton grown in this village was genetically modified.
19. There are isolated examples like the Deccan Development Society (DDS) which works with *sangams* or voluntary associations of poor village women, mostly Dalit agricultural labourers in about 75 villages of the Medak district of Andhra Pradesh (Reddy et al., 2006).
20. The insider–outside dynamic, the pulls between rootedness, proficiency and local prejudice and the space needed to overcome them are among the problems discussed in Vohra (1990).

4

Community-based Natural Resource Management in Gopalpura, Rajasthan

Introduction

It is widely believed that state intervention since the colonial period has resulted in widespread degradation of environmental resources and marginalisation of local communities. This view is often based on the premise that alienation of communally-held knowledge—often termed as traditional or local knowledge—and communally-owned natural resources went hand in hand (Alvares, 1991). Several studies have contended that a rich heritage of traditional water management structures went to disrepair and correspondingly traditional knowledge was disregarded as a result of the state's apathy towards anything not modern. It is widely argued that this disregard for communal and cultural heritage has resulted in widespread degradation of environmental resources and deprivation of livelihood options for rural communities (Agarwal and Narain, 1997).

In this chapter, we focus on Tarun Bharat Sangh's (TBS) rejuvenation of traditional water-harvesting structures—*bandh*s, *johad*s, *medhbandi*s and *anicut*s[1]—which is aimed at conserving water and improving livelihood options for rural communities in arid and semi-arid parts of Rajasthan. Like several other semi-governmental organisations and NGOs in current times, TBS also believes that rural communities and not the state should have complete control over natural resources. It thinks that an answer to ecological degradation and rural impoverishment lies in reviving traditional community institutions, their knowledge and their lost heritage. It is thus committed to organising communities to conserve and rejuvenate traditional water-harvesting structures.

This chapter tries to capture the processes of community formation around the revival of 'traditional' systems of NRM. As discussed in Chapter 1, communities are rarely idyllic, harmonious, homogenous and static in nature (Agrawal and Gibson, 1999; Manor, 1999; Mosse, 2003a). It is, therefore, necessary to examine TBS' work of reviving traditional water-harvesting structures as a process, paying attention especially to issues of caste, class and gender in a certain historical and cultural context. In other words, this chapter intends to show that the political and cultural aspects of community formation and transformation are central to the manner in which community-based development takes place (Mosse, 1999).

Given the staggering scale of TBS' work in Rajasthan, the focus is on the NGO's work in one village. After getting a bird's eye view of TBS' work in Alwar district, we decided to select the village of Gopalpura for a detailed case study for two reasons. First, TBS started its work in Gopalpura almost two decades ago. A long association with the village thus has mutually shaped TBS' ideology and Gopalpura's community.[2] What we also illustrate is that TBS is just one among many actors influencing the process of change in Gopalpura. In this chapter we pay more explicit attention to show how NGO-driven CBNRM is part of a wider process of development intervention and hence always needs to be understood not purely in terms of success or failure. The second reason to select Gopalpura for an intensive study is more practical, pertaining to division of (academic) labour. Two other villages where TBS has been working for a long time—

Bhaonta-Kolyala and Mandalwas—have been relatively well re-searched.[3] Hoping to tap the insight generated by other researchers in these villages, we decided to focus on Gopalpura where at the moment no other academic attention is directed.

The chapter will discuss issues such as sustainable use of natural resources and enhancement of livelihood options as a part of the dis-cussion on the process and dynamics of community formation in a specific social context. Sustainability, equity and enhancement of livelihood options have not been, therefore, separately dealt with in the chapter, but have been interwoven with the question of community formation and dynamics.

Tarun Bharat Sangh and Its Work

As mentioned earlier, TBS focuses on reviving the tradition of rain-water harvesting in the semi-arid regions of Rajasthan. It also believes that traditional informal village institutions (*gram sabha*) can better manage these structures than government departments and hence has decided to revive, and work with, these village institutions. As a result, it has avoided involving any state agency, including the panchayati raj institutions at the village level, in its work. Further, it has stayed clear of state-sponsored programmes and has preferred to approach donor agencies for financial support to carry out its programmes.

The scale of TBS' work is staggering. As an organisation it has been active since 1975; however, its significant work in Rajasthan started in 1985 (Jamal et al., 2002). By the end of the 1990s, it had expanded its work to 10 districts of Rajasthan (Kumar and Kandpal, 2003). It is claimed that by 2005 TBS had worked with rural communities of 700 villages and had built thousands of water-harvesting structures in various parts of Rajasthan (McCully, 2002). Evaluation reports, how-ever, indicate that TBS built 2,264 water-harvesting structures be-tween 1984 and 2000 (Jamal et al., 2002). In terms of the number of villages covered and structures built, the scale of TBS' work could, therefore, be potentially compared to that of state agencies. Further, claims are also made that the organisation has helped in the conservation of forests in the region. But as we will see in this chapter, this activity has not always been the result of TBS' interventions.

In line with its objective/vision of encouraging community control and management over natural resources, TBS has employed strategies that it believes takes it closer in understanding the needs of people and implementing its programmes efficiently. First, it employs local youths to liaison with the villagers and facilitate the mobilisation and organisation of the village community. The TBS solicits community labour for the construction of water harvesting structures, and this it considers an important tool of mobilisation. The villagers or the group of individuals benefiting from any structure have to contribute one-third (earlier it was one-fourth) of the cost of any structure. Third, it has used various social and cultural techniques to spread awareness about the need to locally harvest water and its sustainable use. It has taken the help of religious leaders and has used religious and cultural symbols to enthuse villagers to organise and manage their own local resources. Fourth, as highlighted earlier, it has avoided engaging with any state programme or agency. Its strategy is to involve the gram sabha in its work, while avoiding engaging with the formal village-level institutions created under the process of political devolution.

Gopalpura

Gopalpura is one of the many villages situated in the narrow valleys between the ridges of the Aravallis in Thanagazi tehsil of Alwar district (see Figure 4.1). Until 50 years back the hills of the Aravallis were densely forested, but now these have a barren look with few patches of bushes and grasses. In Gopalpura, between the agricultural fields and the hills on the eastern side, there is a narrow area with a plantation of *palash* (flame of the forest) trees. This is also the *gauchar* (grazing) land. On the western side of the village is a small hillock which acts as a boundary between Gopalpura and Bhikampura village. This hillock only supports shrubs and thorny bushes (a local variety of bush, *anushtha*, which has white flowers, is used as firewood). A number of streams originating from the eastern hill meet the stream passing

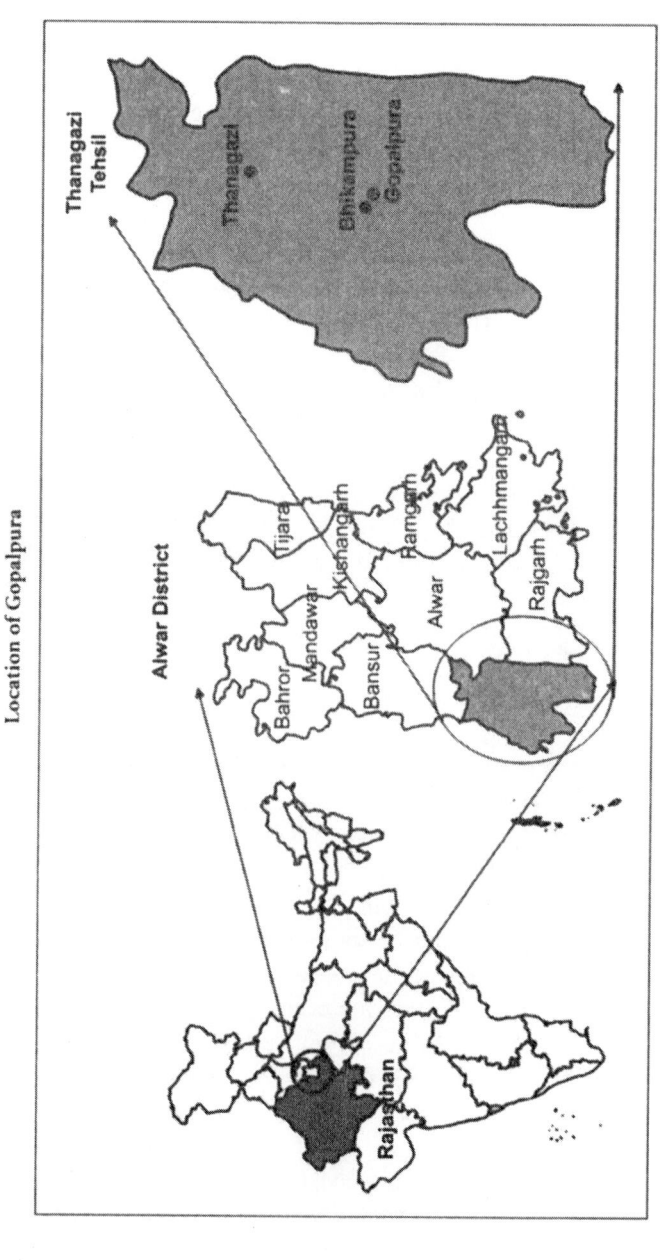

FIGURE 4.1
Location of Gopalpura

Thanagazi
Tehsil

Thanagazi

Bhikampura

Gopalpura

Alwar District

Tijara

Kishangarh

Ramgarh

Lachhmangarh

Mandawar

Bahror

Bansur

Alwar

Rajgarh

Rajasthan

through the middle of the village. The main/middle stream is now visible only in some patches as in a number of places it is covered by a thick silt deposit. This stream, after joining the various other hill streams, is later known as Sarsa *Nadi* (river). About 52 villages fall under the catchment of the Sarsa, with Gopalpura situated in the upper end. Babul and *neem* trees are found near the village settlement and the stream.

Due to the sparse vegetation on the hills, sunlight and heat are reflected back to the valley, making it exceptionally hot even in the beginning of summer. The flat landscape of cultivable land is populated by a beautiful jungle of palash trees covered by vermilion-coloured flowers. Stories about wild animals routinely roaming around the village are common.

Gopalpura has 97 households and a population of 627. The village is situated close to a bigger village called Bhikampura and can be approached by a dirt road from the motorable road passing by Bhikampura.

Samaj Seva (Social Service) in Gopalpura

Changing people's thinking is projected by TBS as one of the main aims of its work. This crusade towards change, as per the story repeatedly told by the staff of TBS and also by the villagers of Gopalpura, started when an elderly farmer (an ex-Lambardar) of Gopalpura village, popularly known as Mangu Baba,[4] confronted Rajendra Singh, the founder of TBS, about his ideas regarding social service almost two decades ago. By then Singh along with his three friends had been staying in neighbouring Bhikampura for a couple of months and were teaching the children of the village and educated people about health issues. As the story goes, one evening while talking to Singh on a *charpai* (wooden cot) in the front yard of Mangu Baba's ramshackle *haveli* (pucca residence), where the men of the village regularly assembled to smoke the *hukka* (water pipe) and talk, Mangu Baba engaged Singh on the issue of social service. Mangu Baba, on hearing Singh's desire to change village society by social service, suggested that Singh should organise repair work on the dilapidated johads. Singh took Mangu Baba's advice seriously and soon started a collaboration with other villagers to repair the old johads. The village then was suffering from a second consecutive year of drought and there was a serious shortage of water. As a result of this livelihood options were severely circumscribed. Singh was then talked into devoting himself to raising funds to organise the rejuvenation of traditional water-harvesting structures. This is, to a large extent, considered as the pioneering event in the birth of TBS in Rajasthan. In what follows, we try to situate TBS' work in the context of other developments in Gopalpura.

'Nayi Sarkarne Sabko Ek-Nek Kar Diya' (The New Government Made Everybody Equal)

Gopalpura has people from four castes: 18 households of Balais classified as SCs, 56 households of Meenas classified as STs, six households of Brahmins and 17 households of Banjaras classified as Other Backward Castes (OBCs).[5] While the Balai and Meena castes (and the *karmkandi* Brahmins too) have been in the village for more than three generations (roughly 150 years), the Banjara families are a recent addition. They have been allotted land from the common land of the village and are administratively considered as part of the village although culturally and socially they are not. Their habitation is 1 km away from the main village habitation. The conflict around the settlement of the Banjara families will be elaborated upon later in this section and the rest of the chapter.

The Meenas of the village belong traditionally to a landowning caste known as zamindar Meena.[6] The descendents of three sons of Mangu Baba's great-grandfather now largely inhabit the village. All the Meena caste members thus are related by kinship of three to four generations barring a couple of families. As history goes, Mangu Baba's great-grandfather bought the village for Rs 200 during the British period. Purchase of the village meant that Mangu Baba's great-grandfather was expected to pay Rs 300 as revenue to the British every year on behalf of the whole village and in return was entitled to collect Rs 13 per *bigha*[7] of revenue from all landowners of the village. According to Mangu Baba, at that time only Meena caste members were allowed to own land; the Balais worked on the lands of Meena farmers as tenants. While Mangu Baba nostalgically narrates the lost glory of the past, he also complains that everything changed after 1947. He describes the period commencing from 1947 as the rule of Congress in contrast to the earlier rule of the British. 'Nayi sarkarne sabko ek-nek kar diya', is how Mangu Baba disapprovingly describes the whole process of change in the post-1947 period. This is so because it was then that the zamindari system was abolished.

However, the actual change in the landowning and tenancy pattern in the village did not occur until 1975. Two rounds of land reforms in the decades of the 1950s and 1960s failed to have any significant impact on the landowning structure of the village. It was during Indira Gandhi's rule in the Emergency period that 10 Balai families of the village were allotted land from the communal village land called savai chak.[8]

The Balai tenants becoming landowners did not pass without a ruffle. Enraged by the fact that their dependents would now be landowners, which would mean losing their services, the Meena farmers objected to the allotments made to the Balais and also violently attacked them. As an elderly Balai farmer recalls, a mob of 100–120 Meena farmers (many of whom had gathered together from the surrounding villages) attacked 10 Balai families with long sticks soon after the allotment was declared. A timely phone call to the police in the nearby village and the intervention of the army stationed at Alwar prevented large-scale killings although several people from both sides sustained serious injuries as a result of the violent scuffle. Land measurement and subsequent procedures for allotment to Balai families were then carried out in the presence of the army.[9] In the immediate aftermath of the violent event, the adult Meena men of the village ran into the surrounding hills and stayed there for a few days to avoid arrest by the police. Several men were also rounded up and put in prison by the police. Later, a case was filed against the Meena men of the village, which was eventually settled outside court almost a decade later with the consent of both the parties who were tired of following court procedures.

Meena farmers still hold a grudge against the Balais because of land allotment to them. Balais traditionally worked as tenants on Meena land in return for one-fourth of the produce. The old arrangements were much more frugal in terms of payment to the tenants. Now that the Balais have been made landowners, traditional practices that were largely exploitative to the Balais are under threat. In the aftermath of the allotment of land and also the violent scuffle, the Balais of the village stopped working on Meena lands. And the social relations between the Meenas and Balais have also been severely strained. As per the old custom, at the time of death of Meena people, Balais were expected to remain present and perform some of the rituals concerning the dead body. The Balais also cleared the village of the remains of dead animals. As a mark of resistance, the Balais have now stopped performing all these rituals. Other customs pertaining to the social restriction on sharing of food and water continue. As per custom, Meenas and Balais do not exchange food and water and Balai men are not allowed to sit across Meena men and smoke from the same hukka. Four Balai families have now started living outside the main village habitation; their settlement, located close to the village, has been named Amlika ki Dhani.

In the second round of allotment in 1989, encroachments made on common lands were regularised. According to the rules, allotments are made only in case there are no objections from anyone. Most of the land for which the Balais had filed applications was reported by other villagers to be common land. For example, an application for the allotment of a piece of land cultivated by a Balai family was objected to by the Meenas because a communal bandh existed on that land. The Balai family had no means to counter the claim of the Meenas. This land now remains uncultivated. This communal bandh was earlier called Balaiwala Bandh, but the Meenas insist on calling it Bavdiwala Bandh[10] as the Meenas do not want a Balai name to be associated with the bandh. In the second round of allotments, 20 Meena families ended up getting some land, but the Balais got no lands.

Another caveat pertains to the economic context of feudal practices in present times. Rajasthan's economic pyramid is considered by Narain and Mathur (1990) flat at the apex owing to the absence of both landless labourers and also absentee landlords. Although their argument is based on old NSS data that shows a very low number of landless labourers, the same can be largely found true even in the current context (Narain and Mathur, 1990). Even large landholders in Rajasthan do not exercise a lot of economic and political clout owing to much lower productivity of their predominantly dry landholdings.

Economic disparities, therefore, between the dominated and dominant of agrarian society are not highly pronounced as in the case of the wet/irrigated regions with higher land productivity.

Things changed further in the post-Independence period due to state policy. Mangu Baba's aphorism about the new government making everybody equal, in fact, aptly describes the process of change. On the one hand, while the Balais in the village acquired land and thus improved their economic and social positions as a result of state policy, on the other hand, the Meenas lost their economic supremacy due to the state's agricultural policy that favoured irrigated and cash-crop agriculture.

In Gopalpura, farmers from all castes fall in the category of small and marginal farmers. Out of roughly 500 bighas of cultivable land, 40 bighas belong to the Balais, 30 bighas to the Brahmins and the rest to the Meenas. Mangu Baba's family owns the largest amount of land in the village, that is, 40 bighas. As that is divided among seven brothers, each brother's share is 5–6 bighas, which is the average landholding in the village. Furthermore, dry agriculture has become less and less self-sustaining due to larger forces of political economy of agrarian change. As dry land agriculture has not been self-sustaining additional income from other sources is needed to continue investment in agriculture. At least 25–30 men from the village, from all castes, migrate every year mainly to different parts of Gujarat to work as manual labourers and earn an additional income. After the sowing of the rabi crop during Diwali, men usually migrate and return before Holi in March. Many of them work in iron factories in Ahmedabad and as manual labourers in Kandla Port. A few Balai men spend almost the entire year outside the village. Those who have cattle have partial relief. Nevertheless, all farmers in the village—some more and some less—face insecure livelihood options and in that sense remain relatively on the margins of society.

Such a scenario gives an impression to outsiders (including NGOs like TBS) that the village does not have internal differentiation or inequality (Singh, 1995). Moreover, insiders like Mangu Baba make a similar point. However, what did not fit in Mangu Baba's dictum were gender relations and the relations with the Banjaras—a new addition in the village. Gender relations are not equal. Patriarchal and feudal norms heavily control women's lives in the village. Married and adult women are not only supposed to cover their heads in front of the older men of the village, but are also not allowed to express their opinion in front of them in public. They are practically not allowed to participate and speak in the public space. The ritual observance of cultural practices was not the only mark of inequality. Women in Gopalpura perform a large part of the agricultural and household work. Most of the adult women were found in the fields attending to the harvest of wheat during our visit and hence we could barely talk to the women at length. However, we had no problem finding and talking to men in the village. In fact, machines and women's labour seem to have replaced the labour previously provided by the Balai tenants for agricultural tasks. It is not as if men do not work, but women do a disproportionately higher load of agricultural and household work compared to men.

The relationship of villagers, both Meena and Balai, with the Banjaras has been openly antagonistic. Swami Agnivesh in the early 1990s had led a *bandhuwa mukti andolan* (free bonded labour movement) in Rajasthan. The movement was aimed at stopping the practice of bonded labour in the marble mining industry and rehabilitating the freed labourers. Several Banjara families thus freed were rehabilitated in different parts of the state. As per the villagers' account, TBS had applied for an allotment of 60 bighas of savai chak land after planting trees on it between 1990 and 1992. The idea was to convert barren common village land into a communal forest. At the time when the application was made, the government was in search of appropriate pieces of lands to rehabilitate freed Banjara families. This piece of land was immediately allotted to six Banjara families with no regard to the objections put forward by the villagers and TBS. A meeting was called in the village, which was attended even by the district collector. However, the order was not reversed. Initially six Banjara families were rehabilitated and later another 18. There are 18 Banjara families in the village as per the census data (Census, 2001b). However, informal accounts of the villagers put the actual number of Banjara families residing in the village in the range of 30–40. The piece of land allotted to the Banjaras is close to the cultivable land of the village. The Banjara settlement thus is physically not part of the village. The antagonistic relationship between the villagers, the Balais and Meenas, on the one hand, and the Banjaras, on the other, has many facets. All villagers openly expressed their prejudices against the Banjaras. Their eating and hygiene practices, their language and their culture were all adversely commented upon. The Banjaras, on the other hand, have also developed antagonistic relationships with the villagers. They not only have not been allowing the repair and reconstruction of the Bavdiwala Bandh (as reported by the Meena farmers), located close to their settlement, but have often allowed their cattle to graze in the standing crops of the villagers. The villagers think that the Banjaras are capable of committing murders and hence strong opposition to their actions at times is discouraged. Interestingly, not all exchanges between the villagers and the Banjaras is antagonistic. In the midst of open hostility, the two groups also do business. The villagers sell their non-milking cows or buffaloes to the Banjaras, who eventually sell them to slaughter houses in Delhi and in some other towns. The villagers also regularly buy salt from the Banjaras. However, apart from the occasional business transaction, there exists no cordiality between the villagers and the Banjaras.

This is the socio-cultural and historical background in which TBS' intervention needs to be situated. Its efforts at rejuvenating traditional water-harvesting structures through community organisations in order to enhance water availability and livelihood options has occurred in a context where social relations are hierarchical, antagonistic and even violent. The consequences of this are delineated in detail in the following sections.

Agricultural and Water Management Practices

At least a century (if not more) before TBS arrived in Gopalpura, three water-harvesting structures existed in the village: the Mevala, Nava and Ullala bandhs. The impounded water in these bandhs soaks the land throughout the monsoon season as a result of which ground-water recharge happens. Sometime around the end of October, the impounded water is released by opening the overflow structures provided in the embankment. In current times, wheat is sown in the land after the release of the impounded water. There are several advantages to this practice apart from the widely acknowledged fact that impounding rainwater recharges groundwater. Obstructing the stream when it flows torrentially during the monsoon reduces soil erosion, renews the land by depositing fertile silt and thus reduces the need for fertilisers. The yield of wheat on this land is roughly double that of wheat grown on land irrigated only with well water. Silt deposition in the nearby uncultivable lands also helps bring these lands under cultivation. Land in the submergence area of the bandh (from now on bandh area) is the main irrigated land in the village. Part of the land outside the impounded area of the bandhs is also irrigated with well water and the rest is cultivated with rainwater.

Each of these three bandhs belongs to the descendents of the three sons of Mangu Baba's great-grandfather. Mevala Bandh was first taken up for reconstruction by TBS in the year 1985. The bandh was repaired and the height increased. In subsequent years, both the Nava Bandh and the Ullala Bandh were also reconstructed. A series of water-harvesting structures (18 in all) eventually came up in the village over a span of two decades with financial support not only from TBS, but also from other state and donor-aided programmes for drought relief and from the People's Action for Watershed Development Initiative (PAWDI).[11] The location of these structures within Gopalpura are schematically indicated in Figure 4.2 and a detailed list and description of these structures is given in Table 4.1.

Agriculture in Gopalpura consists largely of dry crops, but the importance of irrigated wheat has increased since the reconstruction of the bandhs. Dry crops such as barley, chickpea, *sarson* (mustard), gram, *jowar* (sorghum), *bajra* (pearl millet), *kala jeera* (black cumin), tobacco (in very small quantity) and maize are grown on mainly rain-fed land occasionally supplied with some irrigation depending upon the availability of water in the wells. Wheat was barely grown in the bandh area before the reconstruction, but is now the main crop. Wheat has now emerged not only as the main crop in the bandh area, but also in the lands irrigated by wells. However, there is a significant difference in the yield of wheat in both the lands. The yield is 15 quintals per bigha when the crop is watered four times and 20 quintals per bigha when it is watered

FIGURE 4.2
Gopalpura Village: Location of Water-harvesting Structures

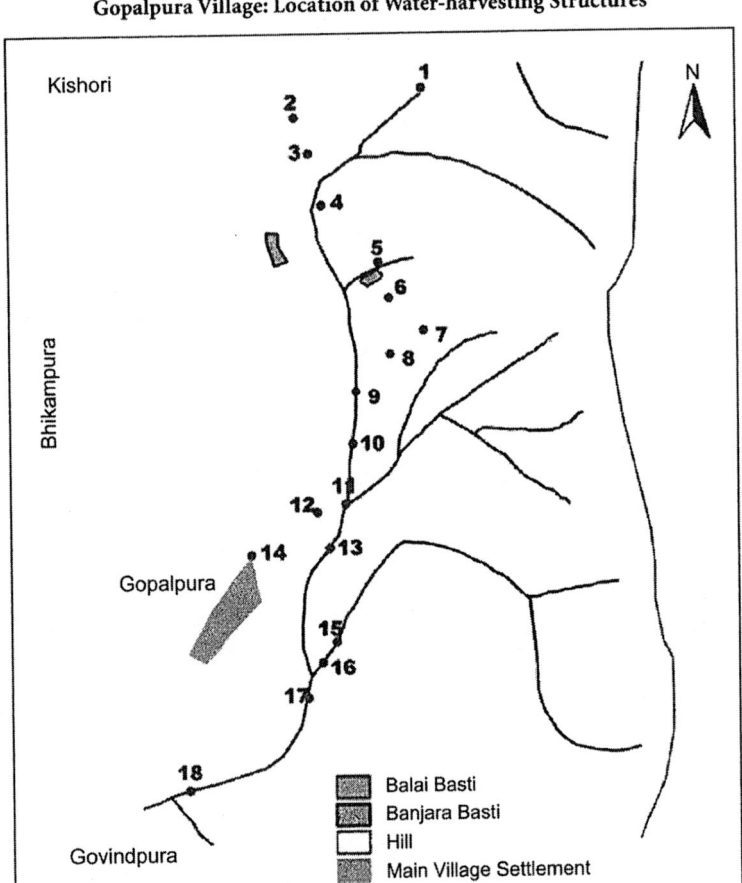

five times. The bandh area, especially the land in the Mevala Bandh,
produces 40 *man* (1 man = 40 kg). Wheat grown in the bandh area
too is irrigated with some water from wells.

Wheat has also substantially replaced the cultivation of barley in
the last two decades and, correspondingly, even the food habits of the
villagers have changed. Wheat *roti* (Indian bread) was made only on
special occasions earlier, but now it has become the staple diet. A more

TABLE 4.1
Water-harvesting Structures in Gopalpura

Name and Type of the Structure	Constructed with Financial Support from	Current Status of the Structure	Specialty
Kanhaiyalal ka anicut or dolibhat	PAWDI	Standing	It is constructed on the land of Kanhaiyalal Meena, who has constructed his house near this land (on the northern limits of village).
Kadkhayawala Bandh (public)	TBS	Partially broken. The people of Gopalpura are not taking interest in its repair because they think that it their land in any way.	Water impounded in this bandh flows towards a neighbouring village called Kishori. The land behind the bandh belongs to the descendents of Sukaram and Nainuram (not related to the main family of Meenas). It is alleged that lands in Gopalpura have not benefited from this bandh. The villagers think that Kishori village benefits from this bandh.
Badriwala Bandh (private)	TBS	Standing	The land belongs to one of Mangu Baba's sons.
1. Bavdiwala or Balaiwala Bandh (public)	First constructed with TBS' help. An anicut (or waste weir) was built under PAWDI assistance on the right flank of the bandh.	The bandh is in disrepair now, but the anicut is in good condition.	The bandh is not allowed to be repaired because the storage land of this bandh is disputed. The Balais of the village were cultivating this land before they were allotted lands by the government in another location. The Meenas now argue that since the Balais have already been allotted land by the government, they have no right over this land
2. Rampal ka anicut built under PAWDI			

(Table 4.1 continued)

(Table 4.1 continued)

Name and Type of the Structure	Constructed with Financial Support from	Current Status of the Structure	Specialty
Radhvali or Balaiwali Johdi (public)	Drought relief (2001–02, 2003–04)	Standing	as this was given to them by the villagers for their sustenance in the first place. The Meenas also contend that the Banjaras do not allow its repair. The villagers like to call this bandh Bavdiwala (because of the presence of a very old bavdi or step well in its vicinity) and not Balaiwala.

Provides water to cattle. |
Mahendra Balai ki medh (private)	PAWDI	Standing	A field embankment constructed on Mahendra Balai's land.
Ramjilal Balai ki medh (private)	PAWDI, 1995–96	Standing	Field embankment on Ramjilal Balai's land.
Badri Balai ki medh (private)	PAWDI, 1995–96	Standing	Field embankment on Badari Balai's land.
Shravanlal ka anicut (private)	PAWDI	Standing	This anicut has helped bring new land under cultivation and this land has accrued to Shravanlal.
Mewala Bandh (private)	This was an old bandh with low paals (embankments). Tarun Bharat Sangh helped	Standing	Roughly 22 bighas benefit due to impoundment. All the land of this bandh belongs to the Thakarsi branch. It has increased by roughly 20 bighas of

	in its repair and increased the height in 1986. It was damaged in 1988 and reconstructed subsequently by those who have land in this bandh with help from the TBS. A waste weir was also built subsequently.		land as a result of silt deposition. In all 40 bighas are sown with wheat after the impounded water is let out. When the paals were low, only 2 bighas were submerged, but now the entire area of 40 bighas are submerged.
Nava Bandh (private)	First built with the TBS' help in 1987 or 1988. It was breached in the following monsoon season and repaired. Badri Prasad ka anicut was built with PAWDI assistance on this bandh.	Standing	The land in the bandh submergence area belongs to descendents of Mangu Baba's grandfather (Gobanda branch). Earlier only 12 bighas of land were there in the bandh submergence area, but new land has been formed and now about 22 bighas are under cultivation. Only one rabi crop (wheat) is grown in this area.
Badi or Chabutrawali Johadi (public)	Was constructed from the drought relief fund of the panchayat in 2003. Tarun Bharat Sangh has also contributed in the past to its repairs.	In working condition	Provides drinking water to village cattle.
Ullala Bandh (private)	Constructed with the TBS' help in 1987. Was breached in the following monsoon season.	Breached. Not repaired.	The land in this bandh belongs to the Parsaram branch of the family. Four brothers having land in Ullala Bandh are in conflict over its repair and hence it has not been repaired.

(Table 4.1 continued)

(*Table 4.1 continued*)

Name and Type of the Structure	Constructed with Financial Support from	Current Status of the Structure	Specialty
Choti Johadi	TBS	Broken and no trace of it remains.	The land in this johad has been encroached upon by members of the Parsaram branch of the family. Water impounded in this johad used to recharge the wells in the village settlement.
Mohar pal ka anicut (private)	PAWDI	Standing	Does not impound too much water, but has helped prevent erosion of land near the hill streams and built up a silt cover on these lands.
Rewad ka anicut (private)	PAWDI	Standing	Does not impound too much water, but has helped prevent erosion of land near the hill streams and built up a silt cover on these lands.
Girdhari ka nala (private)	PAWDI	Standing	Diversion of a stream with masonry walls to prevent erosion.
Goryala Bandh (private)	TBS in 1990–91	Standing. Silt deposited behind this bandh has almost reached the height of the waste weir. Will need desilting and/or increase in height of bandh or waste weir.	Land belongs to several people from the village, mainly from the Meena caste.

detailed description of land types and also land use change between 1971 and 1997 is presented in Table 4.2.

There is an obvious problem in the figures cited in Table 4.2 as the figures do not add up to the total land in the village and the gains/ losses do not match. Despite this, Table 4.2 can be used to show the changes in land use. There is a huge addition to *barani* land. This gain has been from the *charagah* and *banjar pratham*. This means that there is now more cultivable land.[12]

There are two types of land that have been regularised during the allotment rounds. The first type of land is encroached upon from various forms of common land, mainly from banjar and charagah

TABLE 4.2
Land Use Change in Gopalpura

Category of Land	1971 (in bigha-biswa)	1997 (in bigha-biswa)
Chahi Pratham	29-12	29-12
Chahi Dwitiya	16-12	16-12
Chahi Swayam	29-1	29-1
Barani Jau Swayam	1-13	1-13
Barani Swayam	0-0	13-0 (+13-0)
Barani Dwayam	120-17	298-18 (+178-1)
Banjar Beed	0-5	0-5
Banjar Pratham	227-3	117-2 (−110-1)
Gair Mumkin	4-7	0-0 (−4-7)
Gair Mumkin Johad	3-4	3-4
Gair Mumkin Pahad	318-3	318-3
Gair Mumkin Rada	451-8	451-8
Gair Mumkin Bada	4-10	4-10
Gair Mumkin Bavdi	7-7	7-7
Gair Mumkin Nala	7-14	7-14
Gair Mumkin Chaha	0-11	4-18 (+4-7)
Gair Mumkin Rasta	1-16	1-16
Gair Mumkin Abadi	14-12	14-17 (+0-5)
Gair Mumkin Khalihan	0-9	0-9
Gair Mumkin Mandir	0-1	0-1
Charagah	183-10	98-10 (−85-0)
Total	1,421-15 [1,422-15]	1,421-15 [1,419-0]

Source: Patwari's record (SDO office in Thanagazi).
Notes: Figures inside ( ) indicate gain/loss of land in that particular category.
Figures inside [ ] indicate the actual total.
1 bigha = 20 biswa, 3 bigha + 19 biswa = 1 ha, 32 biswa = 1 acre,
1,365 sq ft = 1 biswa.

categories. The second type of land considered encroached and allotted is the one that is newly formed. The construction of johads, anicuts, bandhs and field bunds has helped in trapping and spreading the silt over these lands, thus making it not only cultivable, but also resulting in the formation of new land. Cultivation of new land, strictly speaking, is not an encroachment. However, this newly-formed land is officially regularised only after allotment. Till that happens, there are no disputes about ownership. Moreover, the lands belong to the people who have land in the water-harvesting structures till allotment. The encroachments have been endorsed twice by the government as already discussed earlier. The allotment rounds have now been stopped.

Changing Livelihood Strategies

One has to place TBS' intervention in the context of the transformation that the village underwent sometime in the last century from a mixed economy—pastoral and agriculture—to a predominantly agricultural economy. Some of the elders in the village talked about a time in the past when each household owned about 100 cattle. They said that the hills were densely forested and there was enough green fodder for the cattle. At that juncture, even the bigger cattle were not stall fed and open grazing was common. During this time, the people grew enough crops for their own use and agricultural produce was usually not sold. Things started to change with the decline of forest cover in the region and increase in population (the elders had no explanation for the decline in forest cover). While this led to a decline in animal husbandry due to a drastic decline in sources of fodder, agriculture continued to be practiced in the same way. It seems that this was the period when the intervention made by TBS in the region helped change the economy into a predominantly agricultural economy. The TBS' intervention to rehabilitate water harvesting structures in tandem with other state policies that encouraged high productivity agriculture since the introduction of green revolution provided a boost to the agricultural economy in this region. This region having semi-arid climate with severe water scarcity as a normal part of the agricultural cycle, it would have made more sense to encourage animal husbandry as a desirable strategy for livelihood diversification; however, no such support has come forth from either government or from TBS. Compared with the considerable support agriculture and water augmentation have received, improving cattle breed and milk production have been almost neglected. The absence of veterinary hospitals and veterinarians in the region consolidates the all round absence of institutional support for animal husbandry.

As a result of both TBS' emphasis on rejuvenation of water harvesting structures and state policies in supporting agriculture, villagers also have come to see animal husbandry as an unviable livelihood option. Usually people rear buffaloes mainly to produce milk and milk products for self consumption. Even when almost every household keeps some cattle for their own milk consumption, milk sales from the village are low mainly because of the lack of access to an organised market. Right now, someone from the village collects milk from various households and then sells it to a nearby village, Kishori, which is the milk collection centre of the local government dairy.

Given the decline in animal husbandry, the importance of forests as a source of fodder has also reduced. Villagers usually bring or buy fodder from non-forest sources. Leaves from the trees and agricultural residue seem to take care of the fodder needs of cattle, while needs of the small ruminants are met by open grazing of wild bushes and grasses. No crops are grown especially for fodder. While the forest protection movement in the region has caught on, it remains, as discussed at length later in this chapter, confined to individual villages. Hill slopes and common lands (other than the protected forest land) have not been treated and there have been no attempts to ban open grazing. Harvesting of forest products (leaves, fallen twigs and dry branches), while regulated in the 'people's forests', is done indiscriminately in forests under the Forest Department.

While TBS' intervention gave a push to agriculture in the village by providing an assured supply of irrigation, the latter seems to be in need of many other forms of institutional support to sustain agriculture as a dependable and remunerative livelihood strategy. Marketing networks in this region appear to be very poorly organised. People in this village sell their marketable surplus of agricultural produce to the local traders who are also more often than not moneylenders. The compulsion to sell the agricultural produce to the local traders-cum-moneylenders however is all encompassing. Due to the appalling conditions of the roads and large distances to nearby towns, people find it more convenient to sell their produce to local traders, albeit at a lower price. For instance, while the government had announced a support price of Rs 1,700 per quintal for mustard (in March 2005), some of the villagers sold their produce to the local traders at Rs 1,550 in the village itself. A few entrepreneurial farmers transport their produce (by hiring a tractor) to the towns of Thanagazi, Alwar and Dausa and get a better deal. Otherwise, for most of the villagers transporting the produce to the nearby marketplace is a distant dream. The government collection network is not a very convenient option for the farmers either (even when it fetches higher price) because of many formalities and the hassle of transporting the produce long distances. Similarly, in the absence of collective storage facilities, the produce has to be sold soon after the harvest which consequently receives a lower price. Yet another reason for agriculturists being pushed into a cycle of exploitation is lack of processing and packaging facilities. People in this village sell off their mustard produce in the market, and then buy mustard oil for cooking from the open market at a much higher price. These problems also discourage villagers from diversifying agricultural options, for instance growing a third crop or cultivating vegetables, especially during the summer. After the rabi harvest in March–April, the lands lie idle till the July rains when fields are prepared for the kharif crops. Some of the land in this village (bandh lands for instance) have enough water in the wells to produce a third summer crop; however, in the absence of institutional support this is often not considered viable. The TBS' forceful intervention in organising the community for augmenting water resources, therefore, addresses only the tip of the iceberg in terms of people's problems.

As a result, migration to nearby towns has increased mainly to meet supplementary cash needs required, for example, to buy various agricultural inputs and other household necessities. Usually, at least one male member from almost all the households migrates for work during the lean period (after sowing of rabi in November and April–June after the wheat harvest). Most of them go for manual labour in nearby towns, but some of them also go for better paid jobs, for example as truck drivers. A number of people work outside the village and regularly send money back home which crucially helps families cope. However, it is important to mention that the incidents of distress migration from the village are relatively low. The TBS' intervention in rejuvenating water harvesting structures has improved water availability and access, consequently reducing large-scale distress migration.

Water, Forests, Agriculture and Community

Traditional Community

The earlier discussion makes it clear that although repaired or newly-constructed water-harvesting structures could be understood as based on traditional knowledge, the relationship between traditional knowledge and community involvement with water management is not as simple and straightforward as often projected in the literature (Mishra, 1993, 1995). It is, in fact, clear that the current form of community in Gopalpura is constituted of social groups whose interrelationship is determined significantly by state policy and the political economy of agrarian change as well as by cultural and historical factors. Most importantly, the state-initiated structural changes induced by redistribution of land in the village were a blow to the fabric of the traditional community. The idea of *parampara* (tradition) entirely or significantly forming communities, therefore, needs to be interrogated closely. We do not deny the importance of parampara in the formation of communities; however, the influence of parampara needs to be understood in the current context of state policy and the relationship between state and community.

Hence, how do we further understand the process and dynamics of community formation around NRM? Two fault lines could help understand the geography of community formation. First, economic notions of loss and gain could be one way to comprehend who participates for what reasons and, further, who gains and who loses. We are, however, very much aware that economic notions of gain and loss may not always be an appropriate concept to understand the motivation behind participation. Neither can gain and loss in a pure economic sense help us understand who benefits in which way. A second way to understand community formation is around conflicts. In addition to comprehending community as an expression of consensus and conflict, we add a third element, reiterating what has been discussed in the previous section: how do community formations around NRM interact with social and agrarian relations and social organisation?

Gain and Loss/Public and Private

To examine the question of gains and losses, we raise two issues based on the information on water-harvesting structures (see Table 4.1). First, 11 water-harvesting structures out of a total of 14 structures constructed in the village in the last two decades were constructed on private lands while only three structures were entirely communal. Moreover, only three bandhs—Mewala, Nava and Ullala—out of these 11 structures constructed on private land have clear irrigation benefits.[13] The other structures have more indirect benefits in terms of reduction in soil erosion and enhancement of soil water and groundwater recharge. Also, the communal structures are mainly meant for providing water to cattle—their contribution to agriculture, the main source of income, is limited. The distinction between private and public, at the same time, may not be without contradiction for the simple reason that all water-harvesting structures enhance groundwater recharge and reduce soil erosion, and therefore even private structures contribute towards communal benefits. For instance, groundwater is always extracted from privately-owned wells and hence even when groundwater is a communal resource the actual access is organised privately. That means that the benefit of the enhanced groundwater table due to several water-harvesting structures in the village depends upon private capacity to access groundwater. The distribution between public and private is thus not neat.

The village has 30 wells, all of which are open wells. Almost all wells are at least two generations old. Roughly six wells have been constructed in the last 30 years. Barring two wells owned by Balai and

Brahmin families respectively, the rest belong to Meenas. Digging a well is a costly affair usually done by two to three families (often brothers) in collaboration. Water is shared among the collaborators corresponding to their share in the cost of construction. Water from the old wells is divided among the contenders based on customary and hereditary agreements that are usually registered in the land ownership and revenue documents. Most of the time, water from the old wells is divided amongst the brothers who have inherited the well. Old wells have been located at favourable spots in the watershed to the extent that the location of new wells is likely to be compromised. For instance, five Balai families built a new well after they were allotted land. The well has been located at a higher position in the watershed for two reasons. First, a favourable lower spot was not available on the Balais' land as it was not the best quality land available in the watershed. Second, if the well was located at the lower and more favourable position in the Balai land, it would not have commanded all 20 bighas of land belonging to five families who contributed to the cost of its construction. As a result, the Balai well usually gives water for four to five hours a day when other wells provide water all day long soon after the monsoon season. The well located in the land irrigated by Mevala and Nava bandhs seems to have substantially enhanced water availability after both the bandhs have been reconstructed. This way, historical and structural inequalities deeply influence the current equation of who gains and who looses.

This is as far as ownership of the wells and their respective location in the watershed is concerned. The issue of public and private takes another turn when the relationship between the construction of water-harvesting structures and enhanced availability of groundwater is unclear. We heard from villagers that storage of water during the monsoon months in Mevala, Nava and Ullala bandhs have generally increased water availability in the surrounding wells. However, we learnt later that three streams flowing down the hills on the eastern side during the monsoon months also have a significant impact on water availability in the wells. It is popularly believed that if these streams flow for at least three months during the monsoon, the wells do not dry up for three years. In fact, good rainfall may result in wells overflowing. The connection between water storage in Mevala and Nava bandhs and water availability in the wells while generally acknowledged is not reverberated with confidence. Some of the villagers think that water availability in wells improves once water is released from the Mevala Bandh. On the other hand, they believe that the Nava Bandh does not have any significant impact on availability of well water. Two of the working johads or johdis similarly are considered to have a general influence on water availability in wells, but not in particular. Several villagers expressed that the repair of the Ullala Bandh would significantly improve water availability in wells located on the downstream of the bandh.[14]

Owning wells with sufficient water is one part of the story, running a well endowed with enough water is another story. Running a well even when enough water is available needs another liquid—diesel—which needs something solid—capital. Wells started to be energised first in the mid-1970s in the village, but between the mid-1980s and mid-1990s, almost all wells were fitted with diesel engines. This was also the time when agrarian relations in the village took a significant turn after Balai tenants became owners of their own land. Tenants' labour used previously for irrigating land was replaced by machines and women's labour around the same time. Running a diesel engine to irrigate wheat entirely with well water would need around 100 litres of diesel per bigha of land, which would cost Rs 3,000–4,000, a very high sum for many people. Mostly those who migrate to cities for a few months a year and those who have a regular income can generate enough money for a diesel engine. Income generated from agriculture alone is not enough to reinvest in irrigated agriculture. As such dry crops are not market-attractive crops. On top of that, the cost of diesel makes it prohibitively costly for many farmers to grow wheat, the most market-friendly crop. As a result, nowadays mustard is grown in the land irrigated by well water. Mustard was previously grown on barani land, that is, rain-fed land.

We would also like to raise another important issue pertaining to gains and losses, namely, equity. The TBS' policy of not treating the entire watershed comprehensively has resulted in upstream-downstream disparity. The organisation's belief that support should only be extended to a village (or a group or person in the village) when the village collectively comes forward to participate[15] means that the benefits of work done in one village, or in one micro-watershed, might be reaped by another village downstream. For instance, in Gopalpura, the villagers believed that they were not getting benefits (in terms of increased water availability in the wells) from the Mewala Bandh. Though this contention has no scientific/hydrological backing as such, villagers' opinion is based on their detailed understanding of their surroundings. In this case, they say that after the construction of Mewala Bandh there was an increase in the level of water in a well in Bhikampura (a neighbouring village), while none of the wells in Gopalpura (except for the one near Mewala Bandh) benefited. The first attempt to comprehensively treat an entire catchment was done in 1995–98 in the PAWDI project when the entire basin of the river Sarsa (till Ajabgarh) was treated. Incidentally, TBS was only entrusted with the 'software' aspect of the programme. Merging of these two types of equity— equity within a community and equity within a watershed—is a fuzzy area, which is not easy to handle.

Furthermore, most of the farmers are indebted to local moneylenders from the nearby town and are forced to sell their produce to the same moneylenders at a lower rate than the market or government rate. Although credit relations are changing in the sense that Banias (the trader/moneylender caste) are no longer as blatant in their manipulation of records (we heard several stories), the rate of interest at which a Bania gives credit is considerably higher than the institutional rate. Farmers still continue to take credit from moneylenders because they provide credit whenever necessary. Second, moneylenders also provide other benefits such as collecting produce to be sold directly from households, saving households the cost of transportation. We found that none of the farmers from Gopalpura sold his/her produce to the Food Corporation of India through the district market organised by the government although it provided higher prices than the local moneylenders.

The owners of the land located in Mevala and Nava bandhs have beyond doubt benefited the most from water-harvesting structures. While previously gram and mustard were grown on most of these lands and wheat was sown only on a part of it, after the reconstruction wheat is cultivated on all 20 bighas of land of Mevala Bandh. There have been several more benefits. As the stream passing through these lands is dammed, water is impounded. As a result, the land receives a fresh layer of silt every year renewing its fertility. While the land outside of the bandh area yields 20 man (1 man = 40 kg) per bigha, the bandh land produces 40 man per bigha and that also with reduced amounts of well water irrigation and fertilisers. This means that the cost of cultivation for the bandh land is substantially lower than the outside land. Furthermore, silt deposits lead to increased cultivable land in the bandh area. In the last two decades, 20 bighas of new cultivable land has been formed behind Mevala Bandh (40 bighas instead of 20 bighas previously). Land in Nava has increased from 12 to 22 bighas. So owners[16] of these bandhs not only have the best quality of land with the best yields, but the amount of such land increases every season.[17]

Similarly, private lands on which anicuts and medhbandis are constructed have also benefited. However, the benefit accrued to them is not as substantial as the benefits to bandh lands. Anicuts, in fact, have reduced soil erosion and made some uncultivable lands cultivable.[18] Medhbandis also help in conserving moisture and preventing soil erosion from agricultural lands.

Thus, it can be said with a reasonable amount of confidence that the owners of the land located in the Mevala and Nava bandhs have benefited and that too considerably. On the other hand, the benefits to the rest of the community as a result of the construction of the water-harvesting structures have been indirect and the actual impact on their possible livelihood basket not unequivocally and significantly visible. Other structural forces, for instance, access to capital, market and location of the land have a large impact on livelihood options.

Whose Forest is It Anyway?

There already were collective efforts to conserve natural resources in the village before TBS arrived, efforts that TBS later supported and nurtured. Around 1984, almost two years before Rajendra Singh arrived in the village, a meeting of all villagers was called in Gopalpura followed by a similar meeting called in the neighbouring village (in the Sarsa river catchment). It was decided in the meeting to impose a ban on cutting green wood from the communal forest. Only leaves, twigs and dry and fallen branches were allowed to be collected. Forest conservation by the community in the subsequent two decades is considered a success story. The common forest owned by the village has been converted from a barren tract to a green one diligently protected. How was it achieved? What made the villagers make and follow rules? We illustrate in this section that the community effort was not simply an 'awareness' raising exercise.

After the decision to protect the forest was taken at a meeting, the forest was first guarded at nights by volunteers from the village. Fines were imposed on defaulters. Within a couple of years, the forest regenerated. During this time, fines were imposed on a few defaulters. However, it is our interpretation that the fear of social ostracism had a greater impact than the imposition of fines on the rules being followed. We concluded this because the fine imposed was only Rs 11—an amount that most people could afford to pay. Ostracism, on the other hand, meant that the men in question were not allowed to sit in the village public space and share a hukka with other men; weddings and deaths in their families were not attended by other villagers and even food and water was not exchanged with them. Economic exchanges continued. Buying and selling, and borrowing and lending continued. Even exchange of labour in the agricultural season was not affected. Social ostracism that generated fear can be interpreted as related to the fear of proletarianisation—being demoted on the social (not economic) ladder of status. Not being allowed to sit in the common public space and share a hukka for a Meena man would mean being demoted to the status of a Balai man as Balai men are not allowed to sit in the common public space and share a hukka with Meena men. Evidence of such social ostracisation exists in other parts of the country as well. Several farmers from Karnataka, Andhra Pradesh and Punjab in recent times have even committed suicide driven by the fear of proletarianisation—fear of losing social dignity and honour by being called names of lower castes (Shah, 2004). Conservation might, therefore, be achieved by reinforcing and entrenching the social perception of hierarchy.

Forest conservation has also been achieved at the cost of adverse gender relations and at the cost of exploiting the forest that does not belong to the village. While meeting villagers in their homes, we found that all villagers bring big logs of wood for house construction and for fuel from forests on the other side of the eastern hills. The ban on

cutting green wood from the village common forest has made the villagers conscious of the wastage of forest produce including wood and
accordingly the consumption of wood has been reduced; however, in
reality the ecological footprint has merely been shifted somewhere
else, in the area under the jurisdiction of the Forest Department. 'How
else would we have saved our forest?' is how one villager quizzed us.
In this forest, which falls under the jurisdiction of the forest department, all forest produce is exploited indiscriminately without any
restriction, even by the villagers from Gopalpura.

The cost of shifting the ecological footprint is paid by women.
Women in this village collect firewood and fuel as elsewhere in India.
Fuel collected from the communal village forest is not enough hence
all women from the village travel up the mountain to collect more.[19]
Climbing a steep hill and returning with a back load of fuelwood in
the hot summer is demanding, almost impossible. A group of women,
therefore, generally travels up the hills in the early hours of the morning
(around 3–5 a.m.) during winters to collect fuelwood for the entire
year. They return with a back load of fuelwood by noon and in time
for their routine household work. If they visit the hill forest every day
for one month, they collect enough fuel for the entire year, which is supplemented by whatever they can collect from the communal forest.

So the community's effort to save the communal forest has happened
at the cost of perpetuating and exploiting social hierarchies, both caste
and gender. Sustainability, which is not even sustainability in the true
sense of the term, therefore, sits at odds with the notion of social equity.
What the experience with forest conservation in Gopalpura shows is
that efforts at sustainable natural resource conservation are deeply
entrenched in the hierarchical social order. This offers some important
lessons for, or concerns to be explicitly addressed in, CBNRM initiatives.

Contours of Community

Ideological Mobilisation

Returning to water, community participation in the construction of
public or private water-harvesting structures had two concerns: the

location and nature of the structure and the participation in the construction process. Location and type of traditional water-harvesting structures for reconstruction, especially in Gopalpura, did not demand much deliberation in the beginning owing to the dramatic initiation of the process by Mangu Baba. A series of traditional structures in disrepair were taken up for reconstruction one after the other. The location and type of the structure became a point of conflict during the implementation of subsequent schemes like PAWDI, a point we discuss later.

In the beginning, when Mevala, Nava and Ullala bandhs were repaired and reconstructed, even those villagers who were not to benefit directly by the bandhs had a stake in the construction. At the time when the Mevala Bandh was proposed to be repaired, the village had experienced severe drought for several consecutive years. Those who laboured for the construction were paid 8 kg of grain by TBS for a day's work which was almost life saving in the drought years. Participation in the repair of the Mevala Bandh, even when the structure was privately owned, was thus communal. Subsequently, TBS consciously adopted a principle that the financial support for construction would be provided only if all the villagers were ready to contribute one-fourth or one-third of the total cost of the construction in the form of labour or cash. The villagers' contribution was called *shramdaan*. It could be argued that those who contributed shramdaan must have had a reasonable idea about the benefits the bandh might bring to them. The entire village was supposed to contribute one-fourth to one-third of the cost only in communal structures. One-third of the cost of private structures, like most in Gopalpura, was shared by the families who were to benefit from the structures (or on whose land the structures were situated). The non-benefiting villagers worked as wage labourers.

We think that enhanced water availability in general has been just one among other motivational factors for the villagers to join collective efforts. Some of the other factors may not even have anything to do with natural resource management. The TBS has successfully employed a strategy, more prominently in other villages than in Gopalpura, to ideologically mobilise villagers by calling the collective construction of water-harvesting structures as *dharam ka kaam*—work inspired by moral sanctity. The TBS usually employs a mobilisation strategy in which a group of sadhus or saints are invited to interact with the villagers. Sometimes, the medium of street theatre is also used in the awareness-raising process. These interactions often project the work for water management as work for God and dharma (divine law). The idea that is projected is that working for water management not only benefits human beings, but also helps wild animals quench their thirst. In an atmosphere of festivity and feeling of social and environmental uplift, it is often hard for non-landholders in the village to refuse to provide shramdaan, although the landless contribute shramdaan only in communal structures. The TBS usually does not make a distinction between landholders and non-landholders while demanding shramdaan. However, this question did not prominently come up in Gopalpura as almost all farmers have at least a tiny piece of land. As already stated, the Mevala Bandh was reconstructed for which all villagers provided labour in exchange for grains as daily wages. At the time of reconstructing the Nava and Ullala bandhs, those whose land benefited from the reconstruction contributed volunteer labour and the rest of the villagers provided their labour in exchange for cash. Similarly, all villagers provided shramdaan and/or paid labour for the construction of three johads and johdis, which are communal structures. In fact, in the drought years, the work provided by TBS for the reconstruction of water structures provided a substantial opportunity to earn extra money. Therefore, even when a bulk of the villagers did not benefit directly from the enhanced water availability due to the construction of water-harvesting structures, they were enthusiastic to participate in the construction activities no matter where and for what purpose the structures were initiated.

This was the case until the mid- to late-1990s. Things started to change, particularly around the issue of shramdaan, in the mid-1990s when the PAWDI scheme was implemented. Since then shramdaan has been a bone of contention.

The PAWDI scheme was implemented by the government in 52 villages located in the catchment of the river Sarsa near Ajabgarh. The TBS was given the task of raising awareness and ensuring people's participation. Serious conflicts emerged between TBS and the line departments during the implementation phase. As per its ideology,

TBS wanted villagers to contribute one-third to one-fourth of the total cost of construction, which the line departments did not think was necessary. In fact, the line departments were seriously opposed to the idea of villagers contributing labour as they felt that that might jeopardise their pecuniary interests. The villagers also did not want to contribute shramdaan. A serious struggle ensued between these three actors. Some villagers complained that as a result of the non-resolution of this tussle, the village did not benefit much from the PAWDI scheme. The resolution of these differences in the gram sabha and the politics around the implementation of the PAWDI scheme are discussed in further detail in the next section. The project later introduced a rule of 10 per cent contribution from the individuals/groups benefiting from any structure.

Currently, one more such conflict around the issue of shramdaan is going on. The 20 bigha of land allotted to the Balais in 1975 is located in the upstream of a watershed area (upstream of the Mevala Bandh). This is not the best quality land in the watershed. The seasonal stream (*nala*) first passes through Balai land which is then embanked in Mevala Bandh in the immediate downstream. Long ago, the Balai farmers proposed that a bandh could be constructed on their land that would improve the quality of their land and improve water availability. The Balai farmers had spoken with TBS leaders about it several times. We did not get a clear picture while talking to Balai farmers about why that request was not taken up. The leaders of TBS highlighted the fact that all Balai lands had already been helped with the construction of medhbandis. But the ecological and economic function of a bandh and a medhbandi are dramatically different. Other villagers said that it had not been taken up because Balai farmers demanded shramdaan from all villagers whereas usually only those farmers whose land is going to be directly benefited are asked to provide shramdaan and the rest provide paid labour. The Balai farmers, on the other hand, argue that when all farmers have provided shramdaan, including those who do not have cattle, for the construction of johads and johdis that largely provide drinking water to cattle, why is it that the farmers cannot provide shramdaan for the construction of a bandh on their land. Obviously, the other farmers consider this proposition frivolously inappropriate. Given that Meena and Balai farmers already have power dynamics that are historically and culturally determined, the issue is very difficult to resolve. Significantly, TBS, in its attempts to forge a community around NRM, has completely ignored this conflict.

Gram Sabha

As already discussed in the previous section, an all villagers meeting was called, before TBS' entry in the village, to take action around forest conservation. This meeting was an informal one. The TBS started to name all villagers' meetings as gram sabha. The gram sabha did not have any formal structure and usually all villagers were invited to participate which de facto meant that only men participated as women rarely enter public space for such matters. We learnt that gram sabha meetings were regularly held in the past. More recently, gram sabha (all village) meetings have been called only in the event of a problem or conflict. For instance, once a meeting was called when it was realised that the rules for forest conservation were openly and widely being violated. A meeting was spontaneously called and all those who had violated forest rules by collecting firewood had fines levied against them. Finally, the confiscated wood was piled up in the middle of the village and divided among all villagers equally.

It is important here to further qualify the term villagers in the context of discussions around participation. Villagers here clearly mean Meena men. As discussed earlier, out of 18 Balai families in the village, five to six have shifted from the main village habitation and settled about 1 km away. The men of these families informed us that they are usually not invited to village meetings and even when they are invited they usually do not attend. In fact, there are social reasons why Balais do not and often cannot attend these meetings. These meetings usually happen in the evenings after the evening meal is over. As it is a social gathering, the men assemble to smoke the hukka and discuss matters of village life. Both Balais and women of the village are not supposed to even sit with Meena men in public, let alone speak openly. Balai farmers themselves never described their non-participation in this way. However, it may not be inappropriate to interpret their non-participation in the village meetings as a mark of assertion of their newly-found independence and a form of resistance to discriminatory social norms. Moreover, it can be interpreted that Balai men refuse to follow such rituals that reinforce their ideological subordination.

Furthermore, the public space in the village has a dubious identity. The distinction between a relatively formal organisation that may ideally deal with issues around NRM and the informal community space on the site of which internal differences among villagers are expressed and resolved is difficult to maintain. The same meeting/public space in the village is used to express discontent about other matters related to everyday social life. For instance, two brothers refused to attend a *shradh* (a ceremony performed on the death anniversary of an older person in the family) feast organised by a fellow villager because of an earlier incident related to their widowed sister-in-law. On losing her husband, the woman in question had decided to send two of her children to Delhi to stay with her brother's family and study there. The brothers-in-law were not happy with the decision and refused to attend the feast demanding that the woman in question should not be invited to the same feast. The matter was discussed in a so-called gram sabha. After hectic and furious negotiations and several back and forth propositions and refutations, the matter was finally resolved and the actual feast took place one day later. This incident is highlighted here to illustrate that community space and decisions are often influenced by social norms. It is this same space that operates in the case of NRM. We reiterate that the good and bad of NRM are thus heavily interspersed with the good and bad of social norms.

The gram sabha, or whatever one chooses to call it, is not able to resolve conflicts around water resources due to social hierarchies. The same village meetings are not able to resolve many other issues around water structures that involve only Meena men. The Ullala Bandh and Chhoti Johdi are examples. Chhoti johdi, located just outside the village, has been encroached upon by a Meena family in the village. Villagers think that the repair of the johdi would improve water quality in the community well of the village—the source of drinking water. The same goes for Ullala Bandh. Ullala Bandh has a special position in the watershed, so much so that villagers think that its repair would improve water availability in several wells located downstream. Five brothers holding land in the bandh area have differing perceptions about the benefits of the bandh. The storage of water for two to three months prevents cultivation of a kharif crop during the monsoon months and hence one of the brothers is not in favour of the construction of the bandh. If the bandh is constructed, he would not be able to grow two crops as he does now. The other brothers think that the impoundment of water due to the bandh will improve the quality of their lands and hence they do not mind if the bandh is repaired. What is important here is that the gram sabha, which has been very successful in imposing forest conservation rules, has not been effective in resolving conflicts around these water-harvesting structures. This is because of the politics of interpersonal relations which are difficult to mediate. The point we are trying to drive home is that a so-called formal public space like the gram sabha cannot exist in isolation of social and human relations. The formal space does not exclude informality of social and human interactions.

The process of decision-making for the PAWDI scheme is an example of this interaction between formal and informal space. We already mentioned that the method of implementation of the PAWDI scheme caused differences between line departments, TBS and villagers. The project had provided for the formation of a village-level committee responsible for decision-making. A school teacher, one of the few literate people from the village, was chosen as the head of the committee comprising seven members from all caste backgrounds, including Balai. During the first three years of the committee's term, four anicuts were constructed on the lands of those who agreed to provide shramdaan as per the insistence of TBS. A compromise was arrived at with the line departments, namely, that the concerned individual on whose lands the structures were being constructed would provide shramdaan for the cemented part of the work and the earthen work would be carried out entirely by the line departments. Kinship relations played an important role in deciding whose lands benefited from these constructions. After the committee's first term, a new committee was selected with a new chairman as was mandatory in the formal requirements of the PAWDI scheme. A huge row ensued with regard to the selection of the new chairman for the scheme. The school teacher could not be the chairman again as per the formal requirements of the scheme and the only new contender for the post did not have much clout with the other villagers. The latter's personal equations with the other villagers did not earn him support. Considerable discussion and persuasion followed in the gram sabha. The illiterate wife of the literate school teacher was made the chairperson in order for the teacher to de facto continue as the chairman of the committee.

Whose Awareness Raising?

It may also be pertinent to understand why the entire village has pledged to abstain from the consumption of alcohol and meat. For almost a decade, the village has declared itself as an alcohol free and vegetarian village. The onus of this action is often put on awareness raising by the external agency alongside other achievements like forest conservation. It is often declared that the village has been a successful case of saving the forest and also imposing a ban on alcohol and meat consumption as a result of the intervention of the TBS. We have already discussed the visible and invisible facets of forest conservation. About the ban on alcohol and meat consumption, one also finds an interesting story. The story goes like this. A decade ago, the village discovered the arrival of a God—in modern language, something like a tribal totemic God or a cult figure—via the body of a man suffering from unusually high fever and a few other ailments. The God was later named as Tejaji Maharaj whose arrival in the village was marked by several miracles and unusual happenings. The legends around the arrival of Tejaji Maharaj were narrated in front of us with gusto, unshakeable belief and reverence by all villagers. On arrival, via the body of a villager, Tejaji Maharaj demanded a complete ban on alcohol and meat consumption, which has been diligently followed ever since. The presence and vigilance of Tejaji Maharaj even prevents villagers from consuming alcohol and meat outside the village limits, away from the watchful eyes of fellow villagers. The influence and commitment to a pledge thus is not entirely social, borne out of a fear inflicted by the presence of fellow villagers. It is ingrained in the spiritual and religious belief and world view that are the basis of identity formation of individuals. These beliefs and world view are culturally acquired and nurtured and are also fundamentally formative of the individual's self-identity in society. We do not mean to suggest that these beliefs and world views do not change. We also do not mean to suggest that external agencies do not or can not have an influence on changing world views and beliefs. We are arguing that the change occurs as a result of interaction, exchange and even clashes between different world views and belief systems.

In fact, we were told an interesting story about how the mutually contradictory world views of TBS workers and villagers came face to face. One of the TBS leaders was once bitten by a snake; he was taken around for treatment and finally was brought to the village in a critical condition to be treated by Tejaji Maharaj after it was clear that nothing else had worked. The same leader in the past had disapproved of the villagers' healing practices prompted by Tejaji Maharaj. The rituals to invite Tejaji Maharaj to heal the TBS leader were performed amidst hysterically frenzied drum beating. Finally, via the body of one of the villagers, Tejaji Maharaj sucked the poison out of the body of the TBS leader and saved his life. The villagers took enormous pride in narrating this story of how their belief system saved the life of someone who did not believe in their belief system in the first place.[20] Paradoxically, when TBS, on the one hand, may claim a ban on alcohol consumption in the village as an achievement of their intervention, the villagers, on the other hand, wanted to claim a 'life saving' triumph of their belief system over that of TBS workers. To sum up and to repeat, rule making and rule following are always integral parts of a culturally-ingrained belief system.

Conclusion

The staggering scale of TBS' work in Rajasthan is focused on the rejuvenation of the parampara of community participation in building, using and managing the *paramparik* (traditional) heritage of water-harvesting structures. The focus of TBS on organising communities, which incidentally started in Gopalpura, thus intends to rejuvenate traditional knowledge and traditional modes of water management. Hence, the nature of community-based management practiced by TBS centres more on reviving this knowledge and these systems of management than on introducing 'new' livelihood enhancing measures. What we have illustrated in this chapter are the complexities involved in reviving traditional knowledge by organising the village 'community' in a highly-segmented social milieu. A number of issues pertaining to dynamics and process of community formation and participation in NRM have been raised in this chapter.

First, we have not only shown how the community space in Gopalpura is fragmented, heterogeneous and hierarchical, but also how the social relations that significantly shape this community space are formed as a result of the deep influence of state policies and historically and culturally specific forces. The idea of traditional communities autonomously in charge of their natural resources in the spirit of brotherhood, goodwill and fraternity seems to be inherently flawed. Further, in pure economic terms, rural communities are incapable of standing on their own. A closer look at the type and amount of investments made from external sources, other than that made by TBS, in Gopalpura reveals a rather substantial figure. Rural communities thus could hardly be imagined to be autonomous and independent of the state and other forces like the market.

Second, we have illustrated that TBS worked through the existing social organisation and cultural and religious beliefs. While spreading the message of the community's role in the management and construction of water-harvesting structures, TBS has not rigorously engaged with the issue of equity. Its explicit focus on building water-harvesting structures to solve the problem of water shortage, although involving the community, appears to be more of a technological fix that has not addressed concerns related to the highly-segmented nature of the rural society. As a result, we see a differential impact of its intervention across the community. The historically dominant Meena farmers, traditionally owners of three bandhs in the village, have benefited beyond doubt due to external interventions. The lands of historically disadvantageous groups like the Balais are located at unfavourable spots in the watershed and hence benefits accrued to them are limited as compared to benefits to other farmers. Even when availability of water is communally organised, the access is privately arranged, influenced by factors like access to capital, which imposes serious hurdles on many farmers in the village to reap benefits.

Furthermore, participation in the gram sabha is marred by the observance of ritual purity between the Meenas and the Balais and also by the imposition of cultural norms that prohibit women's participation in the public space. The consequences, as we have shown, is that the community's efforts at successful forest conservation are achieved at the cost of perpetuating social hierarchy and at the same time shifting ecological footprints beyond the village boundaries. Besides, the gram sabha has a dubious degree of success in terms of imposing rules for conservation and resolving conflicts around NRM. While spontaneous meetings of the gram sabha to address issues of forest conservation achieved enormous success, the same gram sabha has failed to resolve conflicts around other water-harvesting structures amongst Meena farmers and also between Meena and Balai farmers.

Finally, public space such as the gram sabha in the village has an ambiguous identity. The informal–formal configuration of the public space makes it difficult to hang the success or failure of development programmes on simple equations like awareness raising, rule making, institution building and rule following. The so-called formal public space does not exist in isolation of social and human relations. The informality of social and human interactions ultimately makes institution building around natural resources a fluid process.

The limitations of TBS' work, therefore, seem to stem primarily from the fact that its main agenda of reviving the 'traditional' has resulted in a form of community-based development that is unable or unwilling to tackle the complex social dynamics of Gopalpura. Decisions by TBS to not engage with panchayat structures and the 'political' and the social fragmentation within Gopalpura, so as not to fracture the 'community', has limited its vision and its impact. There is a need for TBS to engage more actively with these concerns, like some other NGOs are doing more explicitly, if it is to foster community-based forms of development that are socially and ecologically sustainable.

Notes

1. There are four types of water-harvesting structures: anicuts, medhbandis, bandhs, johads or johdis. An anicut is usually a cemented embankment constructed to check soil erosion by reducing the velocity of the water flow in the monsoon.

The size of an anicut usually does not allow water to be impounded behind the check dam and permits water to pass over the structure. Medhbandis are short earthen embankments of a height of roughly 2–3 ft constructed as boundaries of the fields that again check soil erosion by reducing the force of the monsoon flow and retain moisture in the field. A bandh is an earthen embankment with a height of between 5 and 10 ft or more built on a stream to impound and store rainwater. A johad is a crescent-shaped embankment which is similar to a bandh except that no overflow structure is provided in a johad. The land in the water storage area of a bandh is used for cultivation after the excess water that is stored for a few months is let out through the overflow weir. Water in a johad, on the contrary, is kept until the post-monsoon season to supply drinking water mainly to cattle and even to wild animals. Water impounded in a johad recharges the ground-water. Both johads and bandhs are constructed on seasonally flowing streams. A johdi is a smaller johad constructed not directly on a seasonally flowing stream, but somewhere else in order to harvest and preserve water from a small independent catchment.

2. In order to counter the reductionist notion of community employed by both the state and civil society organisations in development planning, Agrawal distinguishes between community as shared understanding (including action orientations) and community as a social organisation that accounts for difference, hierarchy and conflict (Agrawal and Gibson, 1999). One of the important aims of this chapter is to show how community as social organisation—community made of social relations among different caste, class and gender in a historically and culturally specific context, heavily influenced by state policy—interpenetrates with community formation for sustainable and equitable distribution, use and management of natural resources.

3. We learnt that an independent researcher from the University of California at Berkeley has been stationed at Mandalwas and a team of researchers from the Institute of Development Studies, Jaipur, have been studying Bhaonta-Kolyala.

4. Mangu Baba, by virtue of being the eldest in the direct line of descendents of Lambardars in the village, was considered the leader of the village till very recently.

5. The villagers, however, believe that there are more Banjara households and the reason these are not enumerated in the census is because the Banjaras migrate to other places for a good part of the year.

6. There is one more Meena caste known as chaukidar Meena who are not a land-owning caste (de Graverol, 2003). There are no chaukidar Meenas in this village. It is important to point out here that the Meenas are actually a middle-level dominant agrarian caste in Rajasthan. In fact, the Meenas have been classified as STs based on their lifestyle and social organisation. However, as argued by some scholars, the Meenas enjoyed prestigious positions at the centre of the princely state of Jaipur in the late colonial period (ibid.). They were allotted significant military posts and also jagirs and thus were not only involved in the security of the kingly state, but also benefited as a result of the organisation of land tenure (ibid.). The validity of the classification of a caste as a tribe by the Indian state needs a detailed discussion, which is not possible to handle here. What may be relevant for this chapter is that several tribes, the Meenas included,

claim a Kshatriya status in current times, indicating their affinity and closeness to the Rajput rulers.

Two more caveats pertaining to the distinct nature of agrarian society of Rajasthan are important. First, it is argued that Rajput domination continued for a thousand years (Narain and Mathur, 1990). This dominance was challenged mainly by the Jats, but also by other middle-level agrarian castes especially during the period of princely rule (Sharma, 1998). Unlike other states, where irrigated agriculture was practiced for centuries and where Brahmin domination ushered the process of Sanskritisation, whereby Brahmins provided a model to be emulated by the non-dominant castes, it is argued that Rajasthan has experienced a Kshatriya model of Sanskritisation wherein the lifestyle, values and norms of the Rajputs have been imitated by the non-dominant castes (Narain and Mathur, 1990; Sharma, 1998). The challenge to Rajput domination during the princely period was paradoxically about the emulation of the practices of the dominant castes. Rajasthan's history is replete with several peasant uprisings that challenged Rajput domination by means of resisting certain ritually-important social restrictions imposed on the non-dominant castes (Sharma, 1998). For instance, riding horses and elephants during wedding ceremonies was prohibited for non-dominant castes, but was repeatedly challenged by the Jats and other non-ruling castes, the Meenas included. Several violent conflicts took place between the Jats and Rajputs on the issue of Jats wanting to ride horses and elephants during their weddings (ibid.). Most of these uprisings against Rajput domination in the early twentieth century have been reported by historians as conflicts between the Rajputs and Jats because the Jats have been in the forefront of demanding higher social and ritual space in the social rung. However, as Sharma (ibid.) argues, several middle-level castes other than Jats participated in the challenge to Rajput domination, the Meenas included. Challenge to this domination resulted ironically in following ideal Rajput type of cultural behaviour. The Meenas could also be found following feudal Rajput practices wherever they could acquire such space. We found ample evidence of such feudal practices followed by Meenas in our interaction with the villagers in Gopalpura that we elaborate upon later.

7. 1 bigha = 20 biswa, 3 bigha + 19 biswa = 1 hectare, 32 biswa = 1 acre, 1365 sq.feet = 1 biswa.

8. Savai chak is the land owned by the government, but given to the village for communal purposes such as grazing cattle and growing trees for fuel.

9. The villagers used the term 'army', though the army's intervention is not generally sought in such civil situations. Most probably, a battalion of the Armed Constabulary of Rajasthan Police might have been sent.

10. This bandh is close to a very old bavdi (step well), and hence the name Bavdiwala Bandh.

11. The TBS has contributed from time to time to the construction of nine structures, while eight were built under the PAWDI programme in which the TBS was also involved.

12. All the land in the various categories of 'gair mumkin' (literally mean 'what is not possible') falls under the category of 'land not available for cultivation'. Charagah is the grazing land and there is a significant decline (85 bigha) under

this category owing to the fact that a part of it has been converted to cultivable land and allotted to the landless in the village. Part of this land has also been converted to barani dwayam and part of this land is used for growing protected forestry.

Chahi lands are irrigated/double-cropped lands. Despite the construction of water-harvesting structures by TBS and the government, the amount of land under this category remains the same. Chahi pratham is the land on which a second crop (rabi) of irrigated wheat is grown, which needs to be watered 5–7 times. Chahi dwitiya and swayam are lands on which a second dry crop like mustard or gram are grown that do not require much water. The sub-categorisation of chahi land is also based on the quality of soil. Chahi pratham is the best and chahi swayam is the most inferior soil type of the three.

Barani lands are largely unirrigated land on which one kharif crop is grown. It is further sub-categorised as barani swayam and barani dwayam. Barani swayam is the type of land (also soil type) on which a dry rabi crop like chickpea or barley is grown during the monsoon seasons. The rabi crop is usually rainfed, however, it may also be periodically provided with irrigation. These lands may also receive some water in years when wells receive good amounts of water. Barani dwayam lands are inferior soil type, with a predominance of pebbles and sand and grow only one rainfed kharif crop.

Banjar lands are inferior lands which do not support any vegetation because the clay content in the soil is relatively less. Banjar beed supports some trees, while banjar pratham lands have no vegetation except some bushes of the local variety. From Table 4.2 it becomes evident that there has been a significant reduction in the banjar pratham land (110 bigha = 1 biswa) between 1971 and 1997 perhaps because this land has been converted into barani dwayam (178 bigha-1 biswa).

13. While the three wells in the bandh area have direct benefits in terms of augmentation of water, other wells which lie along the main village stream benefit because the water released from the three bandhs in September flows through the stream and augments the water supply in these wells.

14. We are aware that we have based this discussion on anecdotal and circumstantial evidence. Therefore, we must emphatically clarify that the discussion in the previous paragraphs about water availability in wells is not meant to find out the truth and falsity of water availability in the aftermath of the construction of water-harvesting structures, something that would require a different methodology. The previous discussion is meant to capture perceptions of villagers on likely benefits from the construction of water-harvesting structures and also to understand the motivations of different groups of villagers to participate in communal efforts.

15. Quoting Kanhaiyyalal Gurjar (general secretary, TBS), 'We work for the people and our approach is participatory. So unless the people themselves come forward we cannot get any work done.'

16. All the members of the three main branches of Meena families own some land in the bandh area. The rest of the villagers (including the Balais) do not directly benefit from the bandhs in terms of accrual of new lands.

17. These lands are, of course, encroached upon by the owners of adjacent lands and have been subsequently regularised.
18. It is very evident now that the main aim of the villagers in Gopalpura is to increase their agricultural holdings. Since there is a scarcity of land with the increase in population, there is no selling and buying going on. The only option now for the people in this village is to find ways to extend cultivation to other lands. Water-harvesting structures like anicuts are now being sought to help increase cultivable lands. The demand of the Balais to construct anicuts on a stream running through their land should also be seen in this context.
19. The slopes on the Gopalpura side of the hill are completely devoid of forests. So the women have to gather wood and leaves from the other side of the hill (towards Mandalwas).
20. We were also quizzed at the end of the story whether we believed Tejaji Maharaj could have healed the TBS worker or not. Puzzled, confused and amused at the same time on hearing the story, we smartly went around the issue by giving a 'naro va kunjaro va' answer—it may be possible, it may not be.

5

Community-based Natural Resource Management in Bhutan

THE CASE OF THE LINGMUTEYCHHU WATERSHED

Introduction

In 1997, the Renewable Natural Resource Research Centre (hence-forth RNRRC, Centre) in Bajothang, Bhutan, initiated a programme of integrated renewable NRM in Lingmuteychhu Watershed, Wangduephodrang *Dzongkhag* (district), in an attempt to promote CBNRM. This experiment was supposed to be heavily process driven (in the spirit of CBNRM) and aimed at conservation and sustainable use of natural resources. Over the years, Lingmuteychhu Watershed has acted as a benchmark for other CBNRM experiments in Bhutan in terms of approach.

The importance of the Lingmuteychhu Watershed experiment and the seriousness of the government towards CBNRM can be gauged from the fact that CBNRM has become part of official government policy. In 2002, the Royal Government of Bhutan published a report titled 'Community-Based Natural Resource Management in Bhutan— A Framework' that explicitly recognised the state's limited capacity to 'effectively monitor and manage Bhutan's natural resources' and the need to enable 'dynamic partnership arrangements' that build on 'capacities and self-interest of local communities in combination with the technical and institutional capacities of the state' (DRDS, 2002: 4).

The Lingmuteychhu Watershed is different than other case studies in a number of ways. For one, it is part of the state's wider initiative at operationalising CBNRM within its wider development agenda. Hence, the implementing agency is a government research centre not an NGO. Our decision to look at this case study despite it not being an NGO-driven experiment was not only to ensure greater South Asian coverage, but also because the research centre has 'autonomy' in a way similar to that of the other NGOs studied.[1] Moreover, it is this form of CBNRM that is being 'mainstreamed' in Bhutan. Second, Bhutan's socio-political context is different than most of the other South Asian countries. It is a monarchy that has recently embraced the idea of 'development' and is currently embarking on a gradual route towards parliamentary democracy in the context of a largely pre-capitalist social formation. Third, it is a multi-village experiment in a watershed context and hence highlights the difficulties of inter-village and upstream–downstream dynamics.

As we shall illustrate in this chapter, each of these factors has influenced the nature of and approach taken to CBNRM in Lingmuteychhu and in one sense scripted the possibilities and limits to CBNRM. The manner in which CBNRM has emerged, like many of the other CBNRM experiments, is to a great extent shaped by local social dynamics and the desire to 'make an impact' despite these dynamics. The approach has focused more on household-level interventions than village-level ones and within the village as opposed to the watershed as a whole, partly because it was felt that household-level interventions would create interest amongst households to later participate in watershed activities.

Situating Lingmuteychhu Watershed

Bhutan, a landlocked country in the eastern Himalayas, is a country that comprises high mountains and deep valleys. The southern part of the country, near India, is more low lying with elevations as low as 100 m above sea level whereas in the north, the mountains rise to a height of over 7,550 m. This means that there are significant variations in climate and topography. Forest cover constitutes approximately 64 per cent of the total area (DRDS, 2002: 1).

Bhutan is primarily an agrarian society. Almost 80 per cent of its population depends on mountain agriculture and livestock farming. As only about 8 per cent of the total land area is under cultivation, largely because of the terrain, agriculture is small scale and there is considerable pressure to increase its productivity to make agriculture viable. At present, most households depend to a significant extent on off-farm livelihood sources as well (RNRRC, 2000c). There is considerable diversity across the country with central and western Bhutan being predominantly a paddy-based economy and eastern Bhutan being a maize-based economy. Agriculture depends significantly on natural resource products such as firewood, leaf litter for farmyard manure (FYM) and even pasture land within government-owned and managed forests. Being a mountainous country in many parts, stream water is the backbone of irrigated agriculture.

The Lingmuteychhu Watershed is located in west central Bhutan (see Figure 5.1). The Lingmuteychhu Stream originates at an altitude of about 2,400 m and flows southward until it drains into the Puna Tsang Chu 1 km upstream of Bajo town. The watershed is located on the east bank of the Puna Tsang Chu and covers an area of 34 sq km across three districts, three *geogs* (blocks) and ranges in elevation from 1,200–2,600 m above sea level. The watershed is defined by the neighbouring hills of the Antakarchu and Darchulaa ranges. The upper watershed is well forested with evergreen and deciduous (broad-leafed) forests in the upper reaches and chir pine in the lower reaches. The lower watershed, on the other hand, has much poorer chir pine forest

FIGURE 5.1

Location of the Lingmuteychhu Watershed

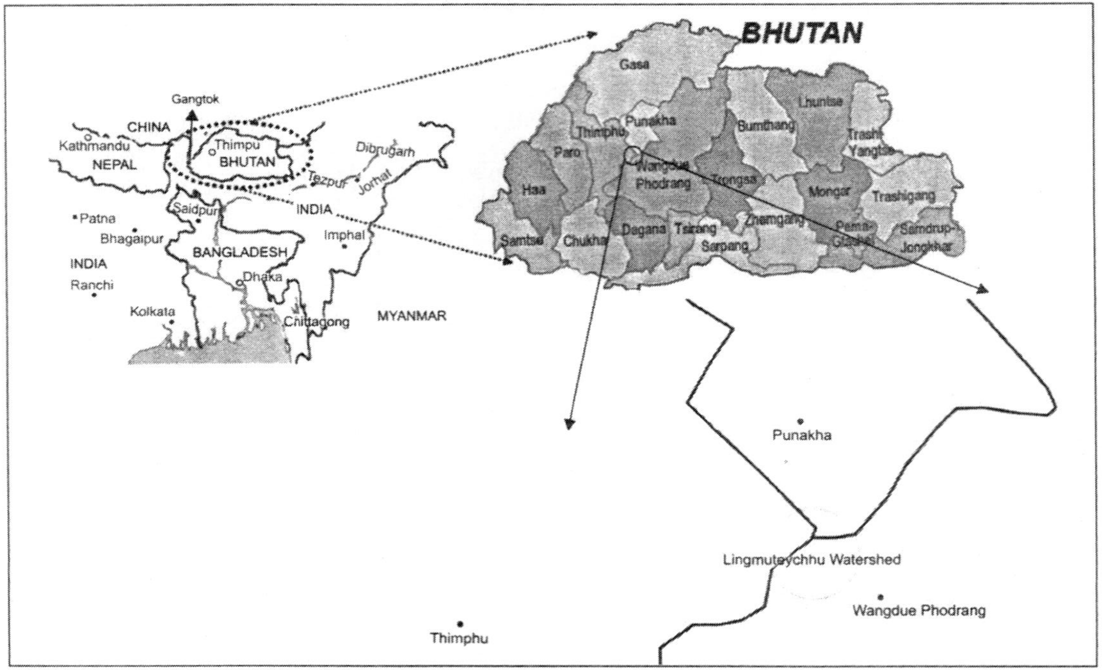

cover. The upper and lower parts of the watershed are also separated by vast tracts of the same type of forests (RNRRC, 2000b: 6).

There are seven villages within the watershed (see Figure 5.2) with a total of 175 households. Six villages—Limbukha, Dompola, Omteykha, Matalungchu, Thangu and Wangjokha—depend mostly on irrigated agriculture and Nabchey on dryland farming. Agriculture in Lingmuteychhu is small scale (though variations range from less than 1 ha [hectare] to more than 5 ha) like elsewhere in Bhutan,[2] but most households do have some amount of irrigated land in wet-land systems.

FIGURE 5.2
Location of Villages and Canals in the Lingmuteychhu Watershed

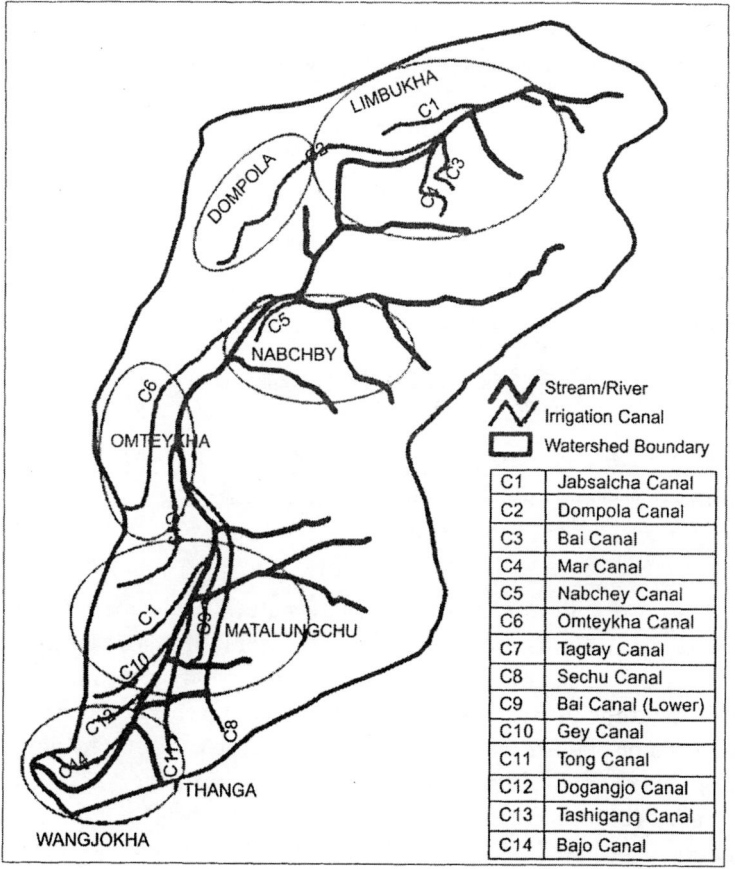

The irrigated villages are rice-based wet-land systems while Nabchey has a maize-based dry-land system (RNRRC, 2000b: 4–15). By and large, people (barring those in Nabchey) own three types of land, namely, wet land, dry land and a small plot of homestead garden or kitchen garden (which is part of the dry land). Most households have less than 1 ha of wet land. In most villages, during summer, paddy is grown on wet lands and vegetables on dry lands. During winter, wheat and mustard are most often grown on wet lands. However, there is considerable diversity in the cropping patterns across villages. For example, potato has become a common winter crop in Limbukha, whereas there is more maize in Dompola. Common vegetable crops include chillies, radish, turnip, beans and spring onions. A few fruit trees are also generally grown in kitchen gardens. Some essential features of the villages are summarised in Table 5.1.

TABLE 5.1
Pertinent Details of Villages in the Lingmuteychhu Watershed

Parameters	Limbukha	Dompola	Nabchey	Omteykha	Matalung-chu	Wang-jokha
Altitude (msl)	2,170	2,100	1,870	1,600	1,500	1,300
Households	35	35	20	28	20	37
Wet land (ha)	34	16	1.5	16	58	40
Dry land (ha)	12	2	6	8	2	6
Main crops	Rice, Potato	Rice, Wheat	Maize	Rice, Wheat	Rice, Wheat	Rice, Wheat, Mustard
Forest resources	Very good	Good	Very good	Moderate	Poor	None

Source: Unpublished data provided by RNRRC, Bajothang.
Note: Data provided under Wangjokha includes data for both Wangjokha and Thangu.

The possibilities and constraints of the natural landscape as well as market opportunities that have emerged for selling produce have influenced the diversity of cropping pattern. While paddy is the mainstay of agriculture, there is considerable variation in terms of the variety sown. Some of the more popular new varieties such as IR-64 are not as productive at high altitudes. Downstream villages, on the other hand, are more concerned with untimely irrigation, as very often transplantation gets delayed because of delayed irrigation water. As we shall detail later, upstream villages have customary rights over water and hence delayed availability of water downstream is common. It is during the rice transplanting season that water is scarce, making matters worse. Crops such as potatoes have become common in Limbukha because of market opportunities, but in Dompola they have been susceptible to pests.

Agriculture depends on a mix of farmyard manure and fertilisers and pesticides. Traditionally, FYM has been the main source of manure. The villages on the upper reaches (Limbukha, Dompola and Nabchey) all have access to *shokshing*, government forest land, which is privately registered. The women clean sweep the shokshing and collect the leaf litter. The leaf litter is spread in the cattle shed and used as bedding for cattle. The cattle's urine and cow dung get mixed with the leaf litter and after some time the litter starts decaying. Then the bedding is removed and stocked; before the ploughing starts it is spread in the fields. The FYM is then worked into the soil through ploughing (RNRRC and Sustainable Soil Fertility and Plant Nutrition Management [SSF & PNM] Project, 2001). Villages in the middle and lower reaches of the watershed like Matalungchu, Thangu and Wangjokha, however, do not have access to shokshing. As forests are less diverse and more degraded, the availability of FYM is scarce there. Hence, while chemical fertilisers such as Suphala and urea are used mostly for 'topping up' in the upper reach villages, they are used more as a substitute for FYM further down the hill. Pesticides, and more so herbicides, are also commonly used throughout the watershed.

Livestock is an important part of the economy—draught power, processing of cheese and butter, FYM and transportation. Poor farmers depend on livestock more as rich farmers can earn some money by selling rice. Farmers get a subsidy to buy livestock. Also, there is an exchange programme in which people can exchange four local cattle for one improved one so as to dispose of unproductive cattle. However, the programme is yet to be implemented. Almost all households have cattle and many of them sell butter and cheese balls in the local market. Livestock provides them with ready cash income and plays an important role in meeting the livelihood requirements of the family. Tshering is one of the better-off farmers in Limbukha. He churns milk twice a week and in one churning he gets 0.5 kg of butter and 13 balls of cheese (1 ball = 250 gms). The current market price is BTN 200 for 1 kg of butter and BTN 10 for one ball of cheese.[3] Thus, he gets about BTN 460 in a week from the processing of milk. Access to fodder is, therefore, of central concern to households.

Access to forests is largely determined by the location of the village. The upstream villages of Limbukha, Dompola and Nabchey are located close to the dense sub-tropical forests. Villagers are entitled to two standing trees each for fuelwood and fodder purposes after paying a royalty of BTN 40 per tree. The upstream villages also have community grazing lands available to them. The collection and sale of ferns and mushrooms from the forest is a supplementary economic activity for about five to six months in a year (after the monsoon begins) for villages in the upper reaches of the watershed (Anonymous, n.d.b). For example, there are households in Nabchey that earn up to BTN 5,000 per year from the sale of ferns.

Shifting cultivation is officially banned from government lands. It is now restricted to people's registered lands. This move of banning shifting cultivation has affected the regeneration cycle. For example, earlier a typical household had access to about 20–25 ha of land and each year the household used to bring about 2 ha under cultivation. Thus, the household operated shifting cultivation in a 12-year cycle and then revisited the first plot. This allowed time for regeneration.

The old social structure, however, remains very much part of the agrarian dynamics in Lingmuteychhu. This is most critical in the context of rights to water. In most villages, farmers are divided into three (sometimes four) categories with different entitlements to water: *thruelps*, *chheps* and *chathos*. While a thruelp can take the full flow of the water in the canal on his/her rotation day, a chhep gets half of what a thruelp gets and a chatho half of what a chhep gets. There is also a category of *hangchu* or 'water beggars' who only get water if the others give it to them. They have no customary rights over water as the other three categories of farmers have. The important point is that rights to water are intricately linked to the social structure 'of the past'. Thruelp in Dzongkha refers to the original tax payer (Van den Brand and Jamtsho, 2002). What is also relatively clear is that the thruelps have more land and hence water rights are linked to extent of land owned. There are also cases of people belonging to a higher category moving to a lower category because of division of property. Wangmo from Dompola was a thruelp earlier, but after she subdivided her land with her sister she became a chhep. An added dimension is that water rights are linked to contribution to village pujas and labour contribution at the time of canal restructuring, repairs and maintenance, again factors that are linked to land and assets more generally.

There is some more complexity to this picture. Historically, water rights have been allocated on a first come first serve basis. Hence, farmers who settled in the area got first rights to the water. The extent of land farmed was mediated by availability of labour. While labour is even more of a constraint today, so too is water. Households that are located further down the canal or at the tail end of the system sometimes complain of shortage. Wangmo has been a victim of this tailender deprivation. Her land is located at the end of the system. Apart from the fact that she gets her turn only after all the other farmers, who are located upstream from her, the actual quantum of water that reaches her plot is very little because of conveyance losses.

The other principle that governs the use of water between villages is that villages upstream have rights to use as much water as they need. Thus, downstream villages, in fact, have no legal entitlements to claim a greater share of water. Within the Lingmuteychhu Watershed, Limbukha at the head reach has first rights to the water. The time of transplanting rice in Limbukha for the summer crop (kharif) has a direct bearing on water availability downstream, more so because the transplanting season is also the period of water shortage—affecting yield as grains do not mature in time.

Rights over the seepage water are generally with the downstream villages and because of this the downstream people very often object to the lining of canals. People from Omteykha went to court once with regard to the issue of canal lining by the upstream village of Dompola. The court decided that Omteykha's approval has to be obtained to line the canal. The sharing of water between villages is all the more complicated in the context of Limbukha and Dompola as the Dompola Canal intake lies in the middle of paddy fields in Limbukha. Furthermore, the water released to different villages is done according to a date as per the Bhutanese calendar. The issue of double month in the Bhutanese calendar creates problem of sharing as sharing gets delayed because of this. A suggestion to fix the date according to the Julian calendar (on a fixed date) has been tabled, but no progress has been made thus far on this issue.

There is no separate irrigation tax or water charge. The tax on irrigation is fixed as part of the land tax. Thus, there is a difference in the tax for wet lands and dry lands, the tax being higher in the case of the former. Each household contributes three to four days of labour for the maintenance of the system prior to transplantation. Generally, meetings are held to decide the details with regard to maintenance and contribution. A *yupen*—something like a waterman—monitors the canal system, sees to the maintenance and also calls for meetings. The irrigators pay for his services. In Dompola, the farmers formed a Water User Association (WUA), but it functioned only for two years. The reason for this was partly due to the fact that there was tension between the old rights and new aspirations over the question of access to water. Moreover, the policy framework was ambiguous. Though the government has come out with a National Irrigation Policy in 1995, primarily with the idea of making irrigation self-sustaining, the modalities of how this is to be done remains unclear. In Bhutan, water falls under the ministry of agriculture and there is no independent ministry for it.

Sharecropping is widespread throughout the watershed, more so in the lower reach villages. Although there is considerable diversity in terms of the sharecropping arrangements, the most common arrangement is that the summer crop produce is divided 50–50 between the sharecropper and landowner whereas the winter crop belongs entirely to the sharecropper. Inputs for the first year are generally provided by the landowner. Contracts, as highlighted earlier, vary significantly in terms of duration and specificities.

Exchange labour also remains an important part of the local economy. Exchange labour is common during much of the agricultural season, mostly during the transplanting and harvesting periods. Scarcity of labour is a major reason for exchange labour. In most households, many of the younger generation have moved to towns and cities for employment purposes. Joining the monastery or the army is very common. Larger farmers tend to engage in more exchange labour because they require more labour themselves during critical agricultural periods.

Agriculture, however, does appear to be at a transitory stage. Prior to the 1950s, land was far more concentrated and bonded tenancy was common. Also, farmers paid a number of taxes in kind to monasteries and to the local rulers to the order of 25 per cent of their total produce. However, the land reform programme implemented in 1952 led to the abolishment of serfdom and made all farmers 'free farmers' despite the fact that land reform was voluntary and hence not too successful. Similarly, the movement to a monetary taxation system has resulted in a uniform and significantly reduced tax burden (Van den Brand and Jamtsho, 2002: 11). Wage labour has also become more common and non-farm employment has become an important dimension of household livelihoods. Nonetheless, pre-capitalist social relations continue to be very central to agrarian relations in the village economy. Later in this chapter we will refer back to these social dynamics within and between villages and illustrate how they have had a major role to play in the evolving form CBNRM has taken in Lingmuteychhu.

It is also important to highlight the nature of gender relations. In west-central Bhutan, where Lingmuteychhu is located, the inheritance of property follows matrilineal norms. After marriage in most cases, the husband comes to stay at the wife's house. Decisions, especially those related to household affairs, are mostly taken by women. Even if there is a fight or misunderstanding between a husband and wife that leads to separation or divorce, it is generally the husband who leaves the house. Similarly, if in case the husband marries again or lives with another woman, he has to leave the house. The wife is normally entitled to the property. Moreover, unlike in India and most of the other South Asian countries, men are the focus of family-planning operations (through vasectomy).

But matriliny does not mean that women have attained an equal role in decision-making either in the private or public sphere. If we compare the different tasks men and women do on a daily basis, it is very clear that women's contribution both in the productive (agricultural) sphere as well as in the household is much higher than that of the men. Women, in fact, joked with us about how easy men's lives were. Moreover, women continue to play a relatively minor role in terms of public administration. For example, most of the political-administrative heads right from the village-level upwards, such as the *tsopa* (village head) and *gup* (block head) are men. Only recently, the first woman gup was elected in Samste district.

Finally, it is important to point out that over the last three decades or so, Bhutan has consciously chosen to 'modernise' and develop. The state has taken a more proactive role, going after foreign investment and extending its own reach to previously isolated areas of the country through its extension officers. This is evident in the Lingmuteychhu Watershed as well. One of the most noticeable changes in Lingmuteychhu due to state intervention has been the advent of electricity approximately two years back. The coming of electricity to this area has had a great impact on people's lives and lifestyles, most visible in terms of electrical gadgets in the kitchen—right from electrical stoves to food processors, massive rice cookers and fridges. Women have found these developments beneficial as it reduces their time in the kitchen. The coming of CBNRM in Lingmuteychhu must be seen in the context of these wider developments.

CBNRM Processes and Intervention in Lingmuteychhu

The idea of CBNRM in the Lingmuteychhu Watershed emerged in the mid-1990s. Unlike in some of the other South Asian countries at this time, CBNRM was a foreign concept in Bhutan. Development interventions were mostly implemented through line departments or through the renewable natural resource (RNR) research centres. Moreover, decentralised bodies such as geogs and dzongkhags had not been entrusted with many responsibilities. The Centre's involvement in Lingmuteychhu was that of a research centre concerned with giving technical inputs that emerged from its own research. The Centre traditionally adopted a commodity-oriented and cropping systems approach (RNRRC, 2000b). The adoption of a community-based approach in Lingmuteychhu, according to the Centre, was based partly on the need for it given a move towards a sub-sectoral approach. However, the larger influence seems to have been the role of the donor community in highlighting the need for a watershed approach and the importance of CBNRM to it. In particular International Development Research Centre (IDRC), which along with International Rice Research Institute (IRRI) had been funding the Centre from its inception in 1982, was a major supporter of CBNRM.

While external agencies promoted CBNRM, the Centre has internalised CBNRM in its own philosophy. The Centre's take is that unless local communities are actively involved in articulating their own felt needs and central to the implementation of policy, development interventions will fall short of their intended goals. In fact, at times the Centre is criticised for overemphasising 'community-based' and not focusing enough on 'development'.

Criticisms notwithstanding, the emphasis in Lingmuteychhu has been very much on 'process'. The Centre's planning document on CBNRM is called a 'process document'. The document states that CBNRM is 'a process by which people themselves are provided the opportunity and/ or responsibility to manage their own resources, define their needs, goals and aspirations, and make decisions affecting their well-being' (RNRRC, 2000b: 6). The CBNRM approach taken in the watershed continues to be very much research driven and thus provides a contrast to most of the other initiatives discussed in this report. The project implementation agency, RNRRC Bajo, has embarked on CBNRM for a number of specific reasons: (*i*) to strengthen farmer participation in the research process, (*ii*) to strengthen the interdisciplinary approach to research, (*iii*) to strengthen linkages between research, extension and farmers in managing natural resources and (*iv*) to extend agricultural research in its scope and clients (ibid.: 6–7). On paper, the CBNRM intervention is also watershed-based and aimed at understanding the relationship between off-farm and on-farm resources.

The main stated purpose of the diagnostic survey, which lasted for three days, was to gather physical and socio-economic information for each of the villages. The team from the Centre gathered at the end of the first day to take stock of activities and then the next day returned to the villages to discuss what they felt were some of the critical issues to be addressed in the villages and how to go about it. More details were gathered from farmers, on-site visits were made and farmers' suggestions were elicited (Bhujel, n.d.). Based on these discussions, a number of tentative plans were made in terms of follow-up interventions around a number of key concerns: (*i*) cropping practices and productivity, (*ii*) diversification of cropping pattern, (*iii*) fodder production, (*iv*) vegetative stabilisation, (*v*) fertilisers (including FYM) and (*vi*) forest produce. For each of these issues, local communities were consulted, research trials planned and intervention strategies charted out.

The 'interactive process' adopted in Lingmuteychhu resulted in a number of specific interventions in the watershed starting in the late 1990s. These interventions differed from village to village according to the stated preferences of the community. As highlighted earlier, the broad range of priorities available to different communities were bounded by the research strengths of the Centre. A major intervention that straddled almost all communities was the provision of new and better varieties of paddy seeds. In the upper reach villages of Limbukha and Dompola, the Centre introduced new varieties (some short duration) such as Yunan-13, Yunan-16, Hexi-24, Kunming-217, Kunming-830, Machapuchray and Chumro. Improved seed varieties were introduced not only for productivity gains, but also in many cases to reduce the necessary transplanting period due to the shortage of irrigation water. In Matalungchu and Omteykha, where water shortage is a major issue, trials were undertaken in 1997 itself to assess the possibility of introducing mostly improved local varieties in the late part of the season that require less water than the longer duration *maap* varieties. Newly-introduced varieties included Domkaap, IR-64, Botoli and Zakha. Trials were conducted in the case of rice, to test for yield, non-stickiness, softness, fodder quality, etc. In Nabchey, new varieties of maize were introduced in order to address concerns of low and decreasing yields, taller plants and barrenness and lodging with traditional varieties (RNRRC, 2000a).

The introduction of new crops has also been a major part of the Centre's intervention. The Centre's major intervention, predominantly in the last three years, has been the introduction of fruit trees, mostly peach, persimmon, apple and pear. The emphasis has been on group production and marketing. Most farmers have kitchen gardens and fruit trees are grown there. Some people were also taught the technique of grafting. Given the centrality of butter and cheese to the culinary habits of the local people, it is little surprise that the livestock economy has also received considerable attention. The Centre, under the aegis of its 'crop–livestock interaction research' has attempted to address both winter and summer fodder needs. Winter fodder needs are more acute both because fodder from the forest and crop stubble are less readily available. Attempts were made by research staff to convince farmers to participate in fodder trials. Only a few farmers agreed to partake in the trials as it involved setting aside some land. Nonetheless, the Centre distributed kits to these farmers containing 10–15 gm of seed of different species depending on the extent of land set aside. These trials continued for two years and the participation of farmers increased incrementally (RNRRC, 2000b: 42).

It is in Nabchey that the Centre's livestock intervention was most significant. As Nabchey is a dry village, income from agriculture is substantially less than in any of the other villages. The Centre decided to promote the community management of a Mithun bull with the financial help of the Bhutan–German Sustainable Renewable Natural Resource Development Project (BG-SRDP). Research and extension staff held a meeting with Nabchey households to explain the function of the bull and also the need to have a committee and bull-keeper to manage the bull. The bull-keeper was to be exempt from other developmental work in the village. Although this idea was proposed during the diagnostic survey in 1997, the bull was provided only in 1998.

In the lower reaches of the watershed, in Matalungchu, Omteykha and Wanjokha, the lack of forest produce has been a major constraint. The main reason for this is that soils are heavily eroded, vegetative cover is poor and deep gullies have been formed. The diagnostic survey illustrated that forest degradation had resulted in a lack of fuelwood and timber, a lack of preferred timber species, poor availability of fodder and, in general, less ready access to forest resources (ibid.). Hence, from the outset of the Centre's intervention in the watershed, community forestry was a major thrust. As the Centre did not have the necessary expertise in forestry, BG-SRDP acted as a partner for forestry activities in the watershed (ibid.: 34–35).

Like with other activities, a series of meetings were organised with villagers from Matalungchu and Omteykha over a six-week period. The first few meetings were joint meetings for villagers from both villages, while later separate meetings were held for each village. Each household was represented by one member—approximately an equal number of men and women attended. The major concern raised by villagers was that of rights, namely, whether or not protecting forests would result in ownership rights being transferred to them. The divisional forest office assured villagers that both ownership and use rights would be vested with the communities.

Community forestry management groups (CFMGs) and CFMG committees were formed in Matalungchu and Omteykha in accordance with forest service division guidelines. In addition to this, a management plan was submitted in both villages. This plan was to contain details of rights and responsibilities to be entrusted upon villagers and the rules and regulations with regard to management. The broad guidelines given to the villagers were that the plans should respect the three principles of protection, production and social equity that guided the state's forestry initiatives.

There were two major thrusts that were pursued in relation to community forestry: the selection of multi-purpose tree species and the establishment of well-functioning user groups. With regard to the former, two major initiatives were taken, the establishment of community forestry nurseries and the setting up of on-farm multi-purpose trees species screening trials of species selected by farmers. The purpose of these initiatives was to closely involve local communities in the selection of particular species and also train them to manage and look after the plantation in the long run. Villagers were taken on a study tour. Twenty-seven members (15 males and 12 females) went on a two-day trip to the Dawakha Community Forest, Punakha Flood Protection Plantation, Gaselo chir pine plantation and the Lobesa and Centre's forestry nurseries (RNRRC, 2000b: 29–34).

Species were to be selected for fuelwood, timber and fodder as well as to help prevent further gully erosion through hedgerow planting and reforestation. Major species selected were *Cypressus*, *Quercus* and *Salix*. Financial assistance came from BG-SRDP and seedlings were provided from the nursery. A total area of 16 ha was initially planted. The visit to Dawakha exposed the villagers to the functioning of user groups. Subsequently, they were able to draw up their own management plans.

These were the initial activities undertaken in the Lingmuteychhu Watershed. As is evident, all of them involved a period of consultation with the villagers, a major research/trial/exposure visit component and then specific interventions based on the 'felt needs' of the villagers. Moreover, these were not one-off interventions. Being a research centre, new varieties are continuously tested and new ideas floated in Lingmuteychhu. In all cases, past and present, an attempt has been made to build local capacity so that local people do not become dependent on the Centre and that initiatives end up being self-sustaining. Thus, for example, villagers have become proficient at grafting fruit trees.

Over the last few years, the Centre has also focused on promoting self-help groups (SHGs) in a few villages. There is a women's and mixed SHG in Limbukha and mixed SHGs in Dompola⁴ and Nabchey. The National Conservation Division of the Royal Society for the Protection of Nature provided the impetus for the formation of an SHG in Limbukha. Their interest was to generate awareness about NRM. In 2004, women from Limbukha underwent a one-week training programme on group formation. Subsequently, eight members of the group were taken on a study tour to Paro and Thimpu to see a post-harvest centre in the former and solar driers in the latter. These visits were aimed at exposing the women to possible alternative forms of employment and income generation.

Livelihood Gains

The Centre's interventions in the Lingmuteychhu Watershed have resulted in significant improvements to many households. The introduction of new and better crop varieties, both local and improved, has resulted not only in productivity gains, but also in a more diverse cropping basket. Potatoes are now a major cash crop in Limbukha and to a lesser extent in Nabchey. Villagers in Nabchey, Matalungchu and Omteykha, especially, have benefited significantly from fruit trees. With local markets in Punakha on Saturdays and Bajo on Sundays, households are able to sell their produce there and earn some cash income.

Improvements in the livestock economy are equally significant. In particular, oat as winter crop has helped alleviate the fodder shortage. Not only does growing oats in the nearby fields reduce travelling time to harvest fodder, but also the denseness of oat allows it to be cut quickly and hence helps farmers save time. The adoption of oat has been most prominent in Limbukha. In Dompola, Matalungchu and Omteykha, quite a few farmers are growing Sudan grass as a summer fodder crop. In the case of Nabchey, the Mithun bull (see Box 5.1) has been a relatively lucrative investment with more than BTN 5,000 being collected for the bull's services. The community forestry initiatives,

---

Box 5.1
Mithun Bull

The Centre and the villagers of Nabchey together have evolved a very elaborate institutional arrangement for the management of the Mithun bull. The need for a bull was expressed during the diagnostic survey. The idea of having a bull of this type is to gradually improve the quality of local breeds. All the households in Nabchey village, who have cattle, have received the service of this bull. So far there have been more than 30 progenies from the bull. Villagers in Nabchey get the services of the bull free. For outsiders, if the progeny is a male then BTN 1,000 is charged; if it is a female then BTN 1,200 is charged. Thus far, Nabchey has earned money from three progenies—two at the rate of BTN 1,000 and one at the rate of BTN 750 as it died. One acre of pastureland has been provided exclusively for the Mithun bull. Thinley, a member of one of the poorer households in the village, takes care of the bull. He is supposed to feed the bull and also see to its grazing needs and maintenance. He is paid BTN 200 per year by each of the households in the village. The earlier arrangement was that every household contributed one day's labour to him so as to compensate him for the time he lost in looking after the bull. For the last two years the cash compensation has become the norm. The shift from labour to cash has been mainly because of labour shortage. Apart from this, each household in the village is supposed to contribute 4 *dreys* (1 drey = 1.5 kg) of maize flour, one bottle of mustard oil and one *pita* (250 grams) of salt in a year for the bull's feed. A three-member committee, all men, looks after the bull fund. They have also started lending money from the bull fund at an interest rate of BTN 5 for BTN 100 per month. The fund has now increased to BTN 5,000. If any other cattle enters the one-acre pasture land reserved for the Mithun bull, then the owner has to pay a fine of 3 dreys of maize per cattle. The money earned through selling this maize is put in the bull fund. Apparently, BTN 1,000 has been collected in this manner. The villagers also feel that the time has come to replace the present bull with a new one, which would cost about BTN 10,000.

while aimed more at ecological stabilisation, have also resulted in grazing benefits.

Finally, SHGs have, in practice, played an important role in promoting new activities that have attempted to explore market opportunities. Solar driers are now being used in Limbukha and the SHG is considering the possibility of marketing these solar driers. The promotion of mushroom cultivation in Dompola is aimed at fulfilling a significant market for mushrooms in Bhutan given the fact that mushrooms form an important part of the local diet. Similarly, fruits and walnuts, though not central to Bhutanese diet, are increasingly in demand. In Nabchey, the SHG looks after the piggery house. The SHG has managed to raise over BTN 8,000 through the lending of money at 5 per cent interest per month.[5]

Although not all interventions and experiments have been successful, the fact that a research centre is involved allows for a trial and error methodology. Hence, many varieties of rice were tried before particular ones were adopted in the watershed. A similar story holds true for new fodder species that were tried and evaluated within the villages. The Centre's village-based approach, moreover, has allowed for different experimentations in different places—for example, the promotion of piggeries in the dry-land village of Nabchey.

It is perhaps not surprising, therefore, that villagers collectively and individually expressed interest in the Centre remaining active within the watershed. Most households interviewed were happy with the Centre's role in providing new varieties of foodgrain as well as new crops such as fruit trees. On some occasions, villagers from one village expressed interest in activities undertaken in other villages. For example, in Omteykha, during our informal village-level meeting, villagers said that they too would like to have an SHG. Villagers in Dompola felt that Nabchey benefited more from the Centre's interventions than other villages, which is perhaps true given the fact that it was specifically targeted as the 'poorest' community.

Yet, despite the Centre's best efforts and the enthusiasm with which it is received by villagers, agriculture and livelihoods remain vulnerable in Lingmuteychhu. Perhaps the strongest indication of this is that many families have to borrow rice from other households for four to five months a year (usually from May–September till the first crop is harvested). There are also households that sell rice immediately after the harvest when the prices are low (distress sale) and then borrow rice later. Though this type of a system helps the needy households tide over their immediate requirements, it results in many families living 'on the edge' and being continuously in debt as for every 20 dreys of rice people borrow they have to pay 3 dreys as interest every year till they repay the original quantity that they had borrowed. In Limbukha, we were told that there are five households who lend rice this way and all the five are thruelpas (in terms of water access). Perhaps one way forward is for the Centre to experiment with the idea of grain banks. But what also emerges is that structural measures aimed at redistribution are necessary if not sufficient to tackle the precarious state of agriculture-based livelihoods.

Contextualising the Making of Interventions: Inter- and Intra-Village Differences

The benefits accrued from the Centre's intervention also need to be disaggregated more carefully in terms of the equity and sustainability dimensions thereof. Critiques abound of the problems, both generic and situational, that exist in implementing CBNRM in practice, specifically with regard to the segmented community (see Agrawal and Gibson, 1999; Agrawal and Ostrom, 2001; Mosse, 2003a, 2003b). The case of Lingmuteychhu provides some insights into these concerns.

The Centre's intervention in Lingmuteychhu has been deliberately cognisant of the social dynamics that might allow for CBNRM both at the inter-village level and intra-village level. A combination of pragmatism, shaped by natural constraints, and a vision of enhanced livelihoods (which, of course, has social consequences) best explains the nature of interventions across the watershed (inter-village). At the village level, potatoes, for example, have been promoted in Limbukha, but not in Dompola because of the existence of a pest in the latter village. New varieties of paddy such as IR-64 are more prominent in the middle and lower reach villages because it is not as productive at higher altitudes. While fruit trees (including walnuts) are found throughout the watershed, a larger percentage of farmers in Nabchey have planted them. The introduction of piggeries and the Mithun bull in Nabchey were a way to enhance livelihoods in the dry village. Similarly, community forestry in Omteykha and Matulumchu were aimed at meeting fuelwood needs and stabilising soil erosion in the middle reaches of the watershed.

Within villages, planned interventions were for the most part aimed at the household level. Households selected for individual activities such as on-farm trials or the introduction of new fruit variety trees were for the most part based on their willingness to participate. Enthusiastic villagers, that is, those who were more willing to take risks whether by adopting new varieties of crops, diversify their cropping pattern or venture into new commitments such as joining SHGs or protecting community forests were selected either at the initial stages of intervention during focus group discussions or in subsequent visits by field staff. Village-level interventions were mostly imagined in the form of SHGs community forestry, activities that were seen to be win–win situations.

While there was a logic to the Centre's approach both at the village and inter-village level, the approach adopted has had certain consequences in terms of distributive concerns that require attention. First, choosing households that are willing to participate could have a size bias. Our fieldwork suggests that interventions have reached most farmers, but that generally those with more land have been more adventurous in adopting new measures at livelihood enhancement. Occasional tension has also emerged. Sangay, one of the farmers with the most land in Limbukha (he has 9 acres of wet land and 1 acre of dry land) mentioned that he was the only farmer in Limbukha who took up walnut grafting on his land. However, when other farmers witnessed the success of it, they deliberately damaged his trees. Also, in the case of SHGs, for example, smaller-scale farmers have on occasion found it difficult to make the monthly payments. Wangmo, in Dompola, who has only about 2 acres of land, and that too poorly irrigated, had to leave the SHG group because she could not participate in a meeting of the SHG and pay her monthly contribution as it was transplanting day. She was fined BTN 60 though she had indicated that she could send the money through another member. So she decided to withdraw.

The flip side of the coin, of course, is that unlike watershed interventions that have an inherent location and size bias, the advantage of the household approach is that it has the potential to target all households. In Lingmuteychhu, however, there has been no targeted approach aimed at smaller farmers specifically. Thruelpas, chheps and chattros have all benefited in some way or the other. There are only a few landless households in Wangjokha that have not benefited at all directly and a few households that belong to Thangu but are located just below Matulumchu who seem to have been effectively excluded from any of the Centre's interventions because they appear to be 'invisible' due to their location. These are problems that the Centre is aware of and is slowly starting to address.

As we have highlighted in some detail, the agrarian economy of Lingmuteychhu to a large extent rotates around access to water that is determined according to customary rights. There are tensions (if not open conflicts) around access to water due to these rights.[6] Yet, the Centre is only just starting to address these concerns through the formation of a watershed-level organisation. Its silence thus far has had consequences not only in terms of its own interventions, but also in terms of the exclusion of some. Water shortage is important in the context of the Centre's efforts to introduce more productive crop varieties, both local and non-local. The shortage of water at particular times and in particular seasons has meant that yields fluctuate quite a bit. In the case of Nabchey, it altogether precluded the introduction of particular crops that may have been more remunerative. The households from Thangu mentioned earlier, that have land in between Matulumchu and Thangu, have no claims to water. The same is true of the water beggars from Limbukha who depend on the goodwill of others. Kinley is a water beggar in Limbukha. He has 1 acre of wet land and 1 acre of dry land. However, half of the wet land is left fallow because the canal does not reach that portion of his land. In other villages, there are no water beggars, yet there are farmers who are situated at the tail end of canals who often do not get adequate water supply, especially during the transplanting period. Very often the people from Bajo Thangu have to either pump in water from the river using their power tillers or buy water from a contractor-cum-farmer who has installed a pump set. The cost of this is BTN 200 per hour.

Out of the 20-odd households in this village, eight households get water from the Bajo Canal. But for the other 12 households, the only source is the Lingmuteychhu Stream, and since they are at the tail end of the system, they hardly get any water.

Not all problems of 'shortage' of water, however, are linked to the social structure. Mahesh Ghimeray, from the Centre, who has been leading the research efforts at improving rice varieties, pointed out that farmers often keep local varieties for up to 90 days in the nursery, which is not necessary given the climate, do not puddle properly and irrigate between the furrows—hence using too much water. Studies undertaken by the Centre also highlighted that farmers do not face lower yields due to water shortage when they plant short-maturing and high-yielding varieties. Local red varieties, which are preferred, require more water and hence shortages occur when farmers have preference for these only. Thus, although the Centre's relative silence on questions of water sharing means that these inequalities remain intact, they do not seem to explain inter-village differential benefits in terms of productivity much.

Privileging or Silencing of Sustainability Concerns?

Farmyard manure is mostly used as a green manure in agriculture. Customary practices allow households to use shokshing. Farmers are allowed to collect leaf litter for manure purposes, but they do not have the right to fell trees. Most households in Limbukha, Dompola and Nabchey have access to shokshing, whereas households in the lower reaches do not have such access. This is so because the forest cover is denser in the upper reaches and there are many more patches of moist semi-deciduous forests—most shokshing are located in these forests. Leaf litter collected from shokshing is used as cattle bed and then converted into FYM. In general, FYM is largely used on lands that are located close to the homestead.

Chemical fertilisers and pesticides, however, are not completely foreign to local agricultural practices, that is, they have been used for many years prior to the Centre's intervention. A farmer in Wangjokha told us that he started using single super phosphate 24 years ago. Fertilisers that are commonly used today are urea and Suphala, which is a combination of nitrogen, phosphate and potassium. Only farmers in Nabchey depend exclusively on FYM. In other villages, chemical fertilisers and FYM are used in tandem. In general, Suphala is used as a basal dose and urea as a top dressing. Some farmers, who do not use chemical fertilisers, have chosen not to because of the costs: Suphala costs BTN 465 for a 50 kg bag and urea BTN 330 for the same quantity.

The use of chemical fertilisers and pesticides seems to be increasing over time. Urea and Sùphala are household names in the Lingmuteychhu Watershed and easy to procure though the costs are high. New short duration varieties such as IR-64, according to scientists at the Centre, require chemical fertilisers to maximise their yield. Hence, the scientists argue that the 8–10 tonnes of FYM used on average per hectare is not enough for high-yielding varieties and is best supplemented with 35–40 kg of urea just before the grains begin to flower. Urea is used more for potatoes and cereals. Suphala is used more as a topping dose. Apparently, the local varieties of paddy are particularly susceptible to blast, a fungal disease. Butachlore, a herbicide, is very commonly used by almost all farmers especially in the paddy fields. It is applied at the rate of about 25–30 kg per ha to suppress aquatic weeds. Other weeds such as *sochum* (*Potamogeton distinctus*) grow as a result and need to be weeded manually. Butachlore is applied to the paddy fields through the irrigation water within one week of transplantation. According to Mahesh Ghimeray, sochum can reduce the yield by 37 per cent. Discussions with the farmers indicate that they do not seem to have given much thought to the possible negative impacts of butachlore.

It is perhaps too early yet to see what the impact of using chemical fertilisers and pesticides is. As we have highlighted, the use of chemical fertilisers especially is largely a function of access to forests for FYM. Hence, in villages such as Limbukha, Dompola and Nabchey, where the shokshing is more easily accessible, the use of chemical fertilisers on average is much less than in the middle reaches and in the downstream villages. Of course, there are variations within villages and even amongst households with land scattered in different locations: in general, more fertiliser is used on lands located further away from the homestead. Moreover, chemical fertilisers are largely supplementary to FYM except in downstream villages where access to forests is much more difficult and the use of FYM almost zero.

In fact, the Centre has recognised the possible harmful effects of chemical fertilisers and is attempting to promote integrated nutrient management with the prime emphasis on FYM (RNRRC, 2000a). A study undertaken by the Centre in 2001 highlighted the fact that there were significant nutrient deficiencies in the soil profile of much of the cultivated land. Seventy-four per cent of the households interviewed in this study from the villages of Dompola, Nabchey and Wangjokha felt they could increase their use of FYM (RNRRC and Sustainable Soil Fertility and Plant Nutrition Management [SSF & PNM] Project, 2001: 28). Soil samples taken by the Centre's research staff highlight in particular the deficiencies of phosphorus, calcium and magnesium. These deficiencies are more prominent for paddy than for other food-grains and non-foodgrains.

Yet, despite the emphasis on FYM, there are a few reasons why the use of fertilisers and pesticides could pose a bigger problem in the future. Despite the general feeling that there is little to worry about with regard to the use of chemical fertilisers and pesticides, there are indications that its use is haphazard and that it is likely to increase over time. Excessive fertiliser use is more likely to become a problem in the downstream villages where FYM is less available and in fields which are further away from the homestead. Another reason that the use of chemical fertilisers might increase over time is the scarcity of shokshing land. The use of shokshing has resulted in the ground cover becoming almost totally bare in some areas as the entire leaf litter is removed. Although the implications of this for tree growth and density are not clear, it is possible that the productivity of these lands will decrease over time. In fact, Yeshey, who is the soils expert at the Centre, feels that though no systematic studies have been done yet, complete removal of leaf litter can cause nutrient deficiency as well as compactness of soil. She also feels that there has been a decrease in the canopy cover of these shokshings and hence there is a need to replant the old shokshings.

Farmers are cognizant of the possible hardening impact of fertilisers on the soil. In a survey conducted by the Centre in three of the villages in Lingmuteychhu in 2001, almost 25 per cent of households interviewed in Dompola and Nabchey said they did not use fertilisers because it hardened the soil (RNRRC and Sustainable Soil Fertility and Plant Nutrition Management [SSF & PNM] Project, 2001: 33). Some farmers mentioned to us that the continuous use of fertilisers and pesticides had a multiplying effect, that is, the more one uses fertilisers and pesticides, the more they are needed the following year to maintain the same productivity. Pema of Omteykha says that she does not want to apply chemical fertiliser because if she starts then every year she has to apply it and thus she would become totally dependent on chemical fertiliser. The Centre's 2001 study highlights the poor nutrient balance in most plots and has tried to respond to this scenario by promoting integrated nutrient management. It is working with the supposition that urea, the most frequently used fertiliser in Lingmuteychhu, has a negative impact on soil properties and hence FYM needs to be promoted.

The main reason for inadequate usage of FYM is scarcity of labour. Given the returns to agriculture, households do not want to invest their limited labour supply in spreading FYM on their fields. The Centre has, therefore, invested considerable time trying to address farmers' concerns by developing 'farmer-relevant options' to increase the supply of FYM. Two initiatives are relevant here: the development of on-farm fodder crops and attempts to improve the supply of leaf litter through raising the productivity of shokshing systems in the watershed. The on-farm fodder can be cut and used to stall feed the cattle and cattle need not be sent out for grazing. This way they can save the cow dung (which would have been lost in the case of open grazing) and use it as FYM. The Centre has focused more on on-farm fodder crops. This is because increased productivity of shokshing does not address the concern of labour scarcity whereas better fodder availability at least partly does. The Centre has also recognised the importance of improving economic returns to agriculture through better marketing of fruits, vegetables and milk products, as a means to encourage farmers' willingness to invest in FYM collection.

Whether or not an FYM-based strategy to improve agricultural productivity has ecological consequences of its own remains to be seen. There are signs that shokshings are completely devoid of any undergrowth which might have long-term consequences in terms of the potential of these forest areas to regenerate. At present, there are no signs that farmers are abandoning old shokshings for new shokshings, but continuous exploitation of the land might have such consequences in the long term if not regulated properly. According to Yeshey, the dependence of the farmers on shokshings is on the increase.

Another major concern in Lingmuteychhu is soil erosion. In the middle and lower reaches of the watersheds around Omteykha and Matulumchu, we observed significant erosion of the hill slopes and even around the main Lingmuteychhu stream. There are a number of reasons for this degradation: excessive exploitation of the forest resources in the area, local cultivation and water management practices, the nature of soil properties and open grazing practices. The middle reaches of the watershed are mostly composed of red clayey soil which is more prone to deep cracking and where gully formation is a frequent occurrence. Forest areas are also heavily degraded, in fact, much more than in the upper reaches of the watershed. Why this is the case is uncertain, but the likely reasons for it are greater exploitation of the forests, more population pressure on these forests (people from outside the watershed also use these forests) and poorer 'natural' vegetative cover. We also noticed during our visit to Omteykha that water is running down the hill slopes in multiple places which is forming deep gullies on sloping lands. Moreover, at the tail end of the canals (distributaries/secondary canals) water is simply let out without proper channelling (for drainage) and this has aggravated the problem of soil erosion and gully formation. This problem is very serious in Omteykha and Matalungchu.

The Centre has taken some initiatives to address these concerns. The major initiative is community forestry in Omteykha and Matulumchu. Community forestry initiatives have a stabilising effect on possible soil erosion. Also soil stabilising initiatives have been undertaken with the planting of Napier grass in areas near canals where gully formation has taken place.

Only time will tell if these measures are able to stabilise the watershed in the long run. What can be observed at present is that the focus on meeting the community's needs often conflicts with the 'bigger' concern of stabilising the watershed and preventing further gulley erosion. The fact that there are large areas directly below the community forests that suffer from severe gully erosion suggests that community forests on their own will not prevent this process of degradation. Initiatives will also have to address concerns related to agricultural practices and water use. We visited the Gangchukha Canal, which is a tributary of the Dompola Canal. Despite efforts by the Centre to grow Napier grass there as a soil stabiliser, there is severe erosion around the canal. The main reason for this is that the meshing at the opening of the pipes has been torn and hence water is leaking from the pipes. Even on the lower reaches of the eastern side of the watershed, there are signs that gully erosion is increasing over time. All these developments require concerted effort on a larger scale—beyond the household level and often at the inter-village level or even watershed level. Moreover, the Centre needs to put in place monitoring mechanisms to see the extent of processes such as soil erosion or forest regeneration, otherwise it is difficult to assess the extent of environmental degradation or the success of interventions aimed at regeneration.

A number of initiatives have also been taken at the household level, especially in Nabchey but with mixed success. Nabchey's agricultural land is mostly on very steep slopes, sometimes even at a gradient of more than 50 per cent, making cultivation difficult. This is due to the fact that most of Nabchey's residents were resettled in Lingmuteychhu from the east of the country and other less steep lands were not available. The Centre has intervened in three main ways: (i) construction of trash lines, (ii) promotion of contour bunding and (iii) terracing. The construction of trash lines using maize stalks to stabilise the lower bunds of the fields has been very successful and has been welcomed by farmers. Farmers in Nabchey have traditionally burned maize straw and stover in the dry winter season. The resultant ash generally blows away and hardly anything is left on the ground. Some farmers, instead of burning the straw and stover just leave it on the fields.

The contour bunding and terracing initiatives have been less successful. Potato, a major crop in Nabchey, is mostly cultivated vertically (along the slope) as opposed to along the contours. Although staff from the Centre have encouraged farmers to cultivate potato along the contours, farmers are worried that their potato crop will be washed away if contour ploughing is undertaken (as the water would accumulate along the contours). But cultivation along the slopes means that the likelihood of soil erosion is high. In Nabchey there is, therefore, a need to look for alternative methods of balancing the production requirements with the environmental factors. The current practices, which are a result of resource constraints and pressing livelihood requirements, are taking a toll on the soil conditions.

Water is also another major sustainability concern, but not only because of local practices. Little effort as yet has been made at improving water use efficiency apart from several trials aimed at reduction of on-farm water use (Van den Brand and Jamtsho, 2002: 29). Discussions with the Centre staff indicate that not much effort has been placed on experimenting with some of the interesting alternative rice cultivation methods (like the system of rice intensification or SRI as it is popularly known) that use much less water as compared to the conventional method of rice cultivation. The water conflicts highlighted earlier suggest the need for a much stronger intervention in terms of community management of water and the resolution of water conflicts. Although a watershed committee has recently been formed, the challenge is how to resolve water conflicts in ways that address concerns of distribution and sustainable water use. The Centre has made efforts by involving farmers in role playing games that are aimed at developing better coordination between farmers and developing negotiating mechanisms for the sharing of water. Such efforts might have to be supported by line departments interested in enforcing more equal water sharing arrangements between upstream and downstream villages. The problem is that at present the irrigation department largely plays a maintenance role.

In the downstream villages, questions of sustainability take another form. The villages of Thangu and Wanjokha receive water from outside the watershed. During the time of transplanting, they receive water mostly from the Bajo Canal, which brings water from Bay Chu that lies to the east of Wangdue. This water is first used by the town of Bajo and then by the Royal Bhutanese Army camp for drinking water. The Centre's water also comes from the Bajo Canal. Hence, Wangjokha and Thangu are at the tail end of this water supply system. To compound the problems further, the canal is damaged in parts because of use of bulldozers for the construction of a tank nearby. The people would like to get some resources to repair the canal.

The impact of unmanaged water usage on sustainable agriculture in Lingmuteychhu plays itself out at present in the form of water conflicts and less than optimum yields. Given the fact that access to water is determined by customary claims makes it all the more important to manage the existing water supply properly. In the long run, unless action is taken to use water more efficiently, it is possible that such water conflicts will escalate. Moreover, agricultural land is at a premium given the expansive nature of forest cover and the state's commitment to maintain it. Therefore, it is unlikely that water scarcity will result in the conversion of forest land into agricultural land as a strategy to increase production given the current political disposition in Bhutan. This will in all likelihood result in continuing declines in yield or in increased usage of chemical fertilisers in the hope of increasing productivity. Finally, the outcome of these pressures on agriculture depends partly on alternative sources of employment, both on-farm and off-farm. New agricultural opportunities in the form of horticulture have helped most households earn additional income from their kitchen gardens. We also mentioned that in most households, many members have left the village for other employment opportunities. Whether these employment opportunities provide sustainable forms of non-farm employment remains to be seen. Questions of sustainability are very much tied up to how future developments in the wider economy work themselves out.

To a large extent, the problems Lingmuteychhu faces are because interventions thus far have been primarily aimed at livelihood improvement. The Centre's contention is that household-based livelihood interventions were necessary to induce interest. Moreover, watershed activities require more planning and more 'meddling' with existing social practices, something that is not easy. However, some of the important sustainability dimensions and issues related to resource flows can be captured only in the context of a unit of organisation. This is not an issue that can be tackled in an isolated manner; it needs a policy orientation at a macro level. Unfortunately, in Bhutan there is no watershed development policy or any guidelines as yet; nor is there any resource allocation to take up watershed programmes at a larger scale.

Finally, while access to land (the means of production) is at one level critical to the agrarian landscape in Lingmuteychhu, non-agricultural employment has become central to livelihoods as well. Almost every household has one son who is a monk (or training to become one). Many families also have sons in the army. On average, almost half of the registered Census family is no longer in the village with urban employment appearing to be much more attractive than remaining in the village. In lower-reach villages such as Thangu, non-agricultural employment is more common as it is situated closer to towns such as Bajothang. Almost all households have people working outside the village—in the army, forest department, saw mills (making wooden boxes), etc. Moreover, except for two households, all other households go in search of work outside in the lean season (September–May). At one level, this illustrates the limited possibilities of not only agriculture in Lingmuteychhu, but also intervention strategies in the context of small-scale mountain agriculture in Bhutan.

CBNRM, Devolution and Democratisation

As suggested at the outset of this chapter, the Lingmuteychhu experiment was the impetus behind the formulation of CBNRM policy in Bhutan. It is, therefore, important to assess the community-based 'participatory' dimensions of the Lingmuteychhu experiment in more detail, especially in the wider context of decentralisation. The Centre's involvement in Lingmuteychhu, as detailed earlier, started with an interdisciplinary diagnostic appraisal in each of the villages. The Centre acknowledges that this process 'was more rapid and less participatory than desired' (RNRRC, 2000b). The diagnostic survey, which was undertaken by a team that included two resource persons from IRRI, was cut short to four days instead of the planned 10 days. The range of choices open to villagers was also bounded by seven predetermined themes central to the Centre: agro-forestry and community forestry, crop establishment and management, crop–livestock interaction, integrated plant nutrition systems, water management, institutional and social analysis and resource mapping. To a large extent, these themes mapped the institutional expertise of the Centre. The focus, therefore, has been largely aimed at providing 'technical' solutions to livelihood constraints with a little emphasis on society and resource use. This itself has limited the possibilities in which interventions can be made. The project leader at the Centre mentioned to us that they have no social scientists and 'all of us are trying to become social scientists'. Although that is a good thing, it could be argued that it had a limiting impact on understanding the social nature of CBNRM in Lingmuteychhu at least in the short run and that this explains some of the shortcomings of the intervention. In the immediate future, therefore, the Centre could benefit from efforts to strengthen its capabilities in understanding and analysing social concerns and its capacities in the area of community mobilisation. In the long run, the proponents of the larger CBNRM programme would have to think of whether and how such capacity can be made available when implementing CBNRM elsewhere.

The other important issue is with regard to the sustainability of the initiative. The Centre's presence in Lingmuteychhu is ongoing. Its sectoral staff visit the watershed intermittently to take stock of existing interventions, to conduct on-farm trials for possible future interventions and to promote new interventions. At one level, the Centre's presence, and particularly the presence of staff, who regularly visit the watershed, is held in high esteem by villagers. Villagers organise meetings and consultations when requested by the Centre staff even at a very short notice and invariably these meetings are well attended. The normal practice is that the Centre staff contact the village tsopa and he/she mobilises the other villagers. Tsopas are chosen by the village collectively and on a rotational basis. During our stay in the watershed, we observed that most of the focus groups discussions had a high level of participation by most. For the most part, however, villagers looked to the Centre for initiatives—almost everyone expressed concern that the Centre should not withdraw from the watershed. Even those who expressed dissatisfaction with the Centre did so in terms of articulating their concern that the Centre's staff focused more on some villages than others or that Centre's staff, as one villager put it, 'were not to be seen'. The question then is 'what happens if the Centre withdraws?' To what extent is the process that the Centre has started sustainable in a social sense?

Here the wider process of democratic decentralisation is important. First, can the Centre's activities increasingly become part of a decentralised NRM strategy carried out by government departments with the active participation and lead from village communities as opposed to being a 'top–down' approach? The question of local body capacity is central to this question, even more so than in other South Asian contexts. At present, most of the interventions in Lingmuteychhu have been Centre-driven and departmental extension work plays second fiddle. It is not clear to what extent the geogs and dzongkhags have the necessary expertise to act on their own. The Centre is officially in charge of activities within the watershed, but in the long run the plan will be for geog departments to carry out interventions on their own with research input mostly from the Centre. The other concern apart from expertise is that of the extension staff paying adequate attention to interacting with villagers, something which is not the case at present. But there are signs that this is improving. Relationships between the Centre's staff and some extension staff are good and hence it is possible that the extension staff will become more sensitive to the issues being addressed by the Centre. Extension staff are participating in many of the Centre's interventions now and are also promoting their own activities.

Second, and equally important, can CBNRM graduate from a programme to a system of governance? For this to be facilitated, efforts will have to be made to reconcile CBNRM ideas with NRM law. Bhutan, as we have highlighted earlier, has historically had a centralised political structure where lower-level institutions have played little role in the actual decision-making process. Nonetheless, the Bhutanese government appears to be grappling with these issues and responding to ideas put forward mainly by the RNR centres. For example, its CBNRM policy document is quite visionary in its outlook. Recent moves to give more power to the geogs and dzongkhags suggest that efforts are being made to devolve powers (Anonymous, 1999: 79). The new Geog Yargye Tshochung (GYT) and Dzongkhag Yargye Tshochung (DYT) *chathrims* (rules and regulations) at the block and district levels have a number of provisions that grant authority to the DYT and GYT to undertake a number of activities related to NRM. For example, the geogs will have administrative powers to care for community lands and community forests, help conserve and protect water sources and promote cooperatives and community-initiated and managed activities. Having said this, the line departments continue to have significant powers. A case in point is forest administration. While the GYT has regulatory power over 'edible forest products', this is not the case for all non-timber forest products. In other words, it remains to be seen how a policy like CBNRM influences the making or changing of law in Bhutan.

What the Lingmuteychhu Watershed experiment highlights are the complexities of issues around decentralisation at a number of levels. The first challenge is galvanising communities to participate in the first place. The second challenge is to create suitable decision-making institutions at different levels (individual village and multiple villages) that are sufficiently representative. The third challenge is to make them last beyond the Centre's involvement, by linking them to the geogs or such official structures. Related to that is the challenge of building the geogs' capacity and ensuring them adequate powers. Given the pre-capitalist nature of society, these questions become even more pertinent in the context of Lingmuteychhu. In other words, will decentralisation in the given social context address adequately socio-economic differences in local society and is it more or less likely to do so than centralised initiatives?

Finally, it is necessary to remember that CBNRM in Bhutan is located in a political context that has historically been centralised. The question is less as to whether CBNRM and a constitutional monarchy can go hand in hand, but whether or not CBNRM itself can help the process of democratising the state. While this might not be the intent of CBNRM, ultimately its ability to do so will influence its own destiny. The indication thus far is that initiatives such as Lingmuteychhu have helped highlight the potential of not only community-based management (with all of its limitations), but also the need for wider processes of democratisation at supra-local levels.

Conclusion

A process of CBNRM has started in the Lingmuteychhu Watershed, the first of its kind in Bhutan. The involvement of a research centre with professional expertise is also important as it shows the potential of research-driven interventions to support community-based processes that enhance livelihoods. Also, while CBNRM is being promoted by a research centre, it appears to be part of a wider initiative at devolution. In fact, it would be fair to argue that the Lingmuteychhu initiative has played a significant role in influencing the direction of decentralisation within the state—the policy document on CBNRM being testimony to this. The unanswered question is whether decentralised bodies such as the geogs will continue to engage with village communities in the way the Centre has if the Centre scales back its activities.

The Lingmuteychhu experiment has its limitations as well—many of which are linked to wider critiques of community-based development. First, while the Centre has emphasised the importance of process, interventions have been largely determined by its own expertise. The privileging of agriculture and agricultural extension is the hegemonic discourse being articulated. It has come, in the short run at least, at the expense of addressing wider structural impediments to change within the agrarian economy. Second, interventions have tended to be win–win ones and although households in all villages and with different incomes and assets have benefited, inequalities between households, and the reasons for them, remain unaddressed to a large extent. The case of customary water rights is a case in point. Recent watershed-level initiatives are welcome, but it is too soon to comment on them. Finally, it is important to note that international organisations play an important role in initiatives such as the Lingmuteychhu Watershed. Will these initiatives be sustainable if and when such organisations make an exit?

There appears to be a willingness within the Centre and amongst extension workers to learn. This bodes well not only for developments in the Lingmuteychhu Watershed, but for CBNRM and decentralised development in Bhutan. Ultimately, the future of community-based development and devolution will depend on the space provided in the wider political arena. Will CBNRM sit easy within a monarchy? Will it have the space to grow into a wider system of decentralised governance?

Notes

1. Proponents of NGO-driven CBNRM have invariably referred to the greater autonomy of NGO-driven CBNRM as opposed to state-driven CBNRM.

2. In 1999, the Centre conducted a wealth-ranking exercise in four of the villages, namely, Limbukha, Omteykha, Nabchey and Wangjokha. It emerged from the survey that the largest landholdings are found in Wangjokha and the smallest in Limbukha.

3. 1 BTN (Bhutan Ngultrum) = 1.04604 Indian Rupee; 1 BTN = 0.02251 US Dollar (as on 31 July 2006).

4. The SHG in Dompola is called the Dompola Village Development and Small Farmers' Savings Group. Except for three households, all the others are members of this SHG. Interestingly, two households from Nabchey, who are originally from Nabchey and have not come from the east as in the case of others in Nabchey, are members of the Dompola SHG, though there is an SHG in Nabchey itself. It gives an 'ethnic' tilt to the SHG membership. Or probably the people who have come from the east and settled in Nabchey are still considered 'different' from the rest. The one-time membership fee is BTN 200 and the recurring monthly charge is BTN 60. The safe that contains the SHG money and the records is kept in the temple. The safe has four keys. The four keys are kept with four different people (three office-bearers and the messenger) and only if all the four use their respective keys one after the other can the safe can be opened. The SHG meets once a month in the village temple. After they transact the formal business, they all eat together at someone's house. Thus, SHG meetings have also become a social event for them. Earlier, the fine for missing a meeting was BTN 60, but now it has increased to BTN 100. Currently, the total money saved is around BTN 56,000. If a person defaults on repayment then that person is not given a second loan. The interest rate is about 13 per cent. The upper limit of the loan amount is BTN 300 and only in exceptional cases are larger amounts sanctioned—amounts up to BTN 5,000. By and large the members take loans to meet educational expenses, electrification, to buy agricultural inputs, etc. This SHG has taken up mushroom production. There are four sub-groups—one each in each of the four hamlets. Members were trained and given basic material free from the Centre. Voluntary labour on a rotational basis is common. Thirteen kg of mushroom has been sold from the first harvest, an earning of BTN 1,300. The group is not planning to distribute the money they generate from mushroom sales amongst the members, but instead plan to plough it back into the SHG funds.

5. This is a pretty high rate of interest as the annual interest rate comes to 60 per cent. People are also aware that it is a pretty high rate, but they say that they have kept a high rate of interest for faster 'accumulation' of capital. They are confident that later they can reduce the interest rate. They are thinking of giving about BTN 10,000 without interest for a couple of months to a household if there is a death in the family. The SHG also want to initiate a group piggery in the village.

6. Earlier we referred to the conflict between Dompola and Limbukha. A similar problem exists between Omteykha and Matalumchu. Villagers in Omteykha believe that they have the right to take the full flow of the stream into their own canal. Other examples exist. In the case of Nabchey, the issue is slightly different where the priority remains getting a secure source of water to irrigate their lands. There is a canal that runs through parts of Nabchey, but only two 'original' households of the village are entitled to water. The 'new' settlers in Nabchey, who came from eastern Bhutan, have to grudgingly accept the fact that they have to manage without irrigation water.

6

Community-based Natural
Resource Management in the
Central Himalayas

THE WORK OF DOODHA TOLI
LOK VIKAS SANSTHAN

Introduction

The Central Himalayas of Garhwal and Kumaon in India are associated
with the world-famous Chipko movement. As a movement in which
environmental conservation and decentralised resource control were
central concerns, Chipko was in many ways the philosophical pre-
cursor of the concept of CBNRM. Although the movement died down
in the 1980s, several offshoots emerged and took root in different parts
of Uttarakhand, attempting to translate the vision of Chipko into

practice in different ways, inspiring local communities to take up what we now call CBNRM-type initiatives. The Doodha Toli Lok Vikas Sansthan (DTLVS) is one such offshoot. Their work exemplifies a particular approach to implementing CBNRM, namely, the idea of grassroots voluntarism and patient, low-key work with communities over a long period using limited funds, focusing on evolving ideas suited to local needs. This case study throws light on the strengths and weaknesses of such an approach and also serves to highlight the opportunities and challenges to CBNRM-type initiatives in a socio-ecological context that is historically considered favourable to community management, but is also changing rapidly.

Social Geography of the Paudi Garhwal Region

The Doodha Toli region lies at the eastern edge of Paudi Garhwal district, adjoining Almora district, in Uttarakhand state of India (see Figure 6.1).[1] This region falls in the middle Himalayas, with altitudes ranging from 1,500–3,000 m above mean sea level (msl). The Doodha Toli peak is 3,045 m high. The terrain is rugged and mountainous, the climate temperate with mild summers and cold winters (including snowfall) and rainfall around 1,500 mm.[2] The geology is a mixture of sandstone, limestone and metamorphic rocks, with flint and sand conglomerates (Khan and Tripathy, 1976: 7). The natural vegetation of this region is classified as Himalayan moist-temperate forest dominated by *banj* oak (*Quercus leucotrichophora*), chir pine (*Pinus roxburghii*), and at higher altitudes deodar (*Cedrus deodara*), *burans* (*Rhodendron* sp.) and other oak species (for example, *Quercus semecarpifolia*). The region is rather distinct in that the slopes of the Doodha Toli Mountain are the source of two–three major non-snowfed rivers (such as the Nayar and Poorvi Nayar), in contrast to most of the rivers in Uttarakhand that originate from the upper Himalayan glaciers.

The Doodha Toli region, as is the case in most of Uttarakhand, is inhabited mainly by the so-called 'Hindu' communities who speak the Garhwali or Kumaoni dialects of Hindi.[3] People of the two language

FIGURE 6.1
Location of the Study Sites in the Doodha Toli Area

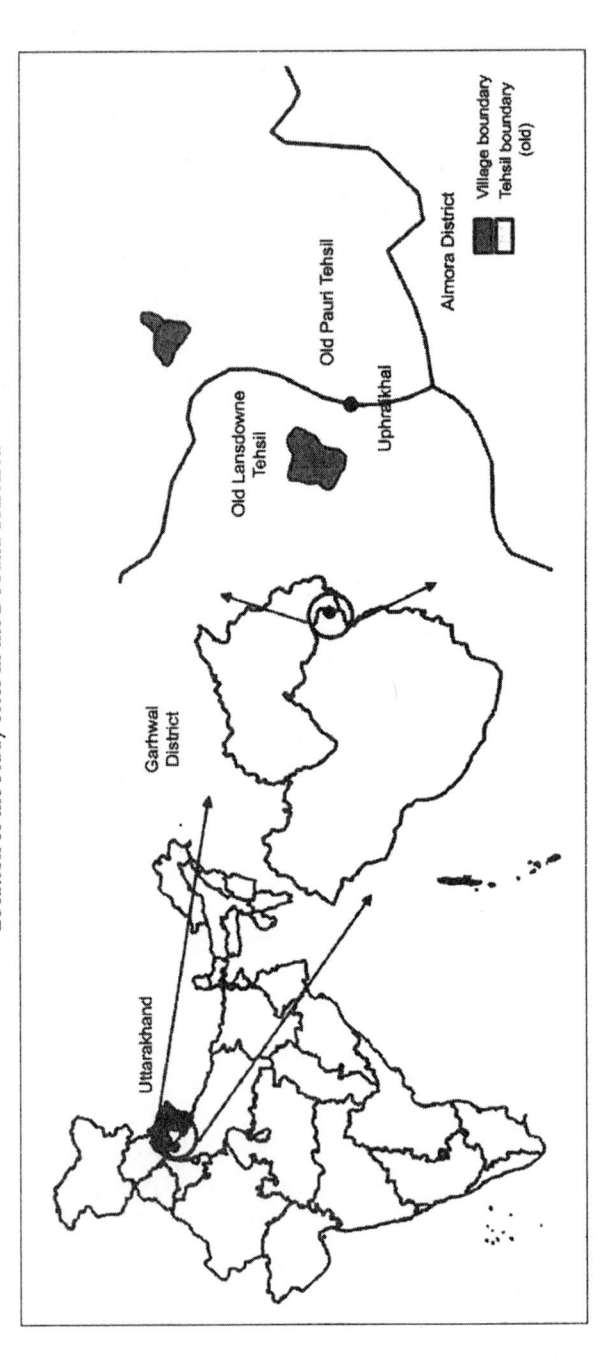

groups are together often termed as *Pahadis*, that is, the hill people.[4] Terraced agriculture, mostly rain-fed and snow-fed, forms the primary occupation. The availability of moisture in winter through snowfall and the presence of a multitude of springs and streams on the hillside mean that farmers are able to produce two crops even though the quantum of rainfall is not high. Broadly speaking, paddy, *mandua* (*Eleusine coracana*) and *jhingora* (a millet: *Oplismenus frumentaceus*) are the main kharif crops whereas wheat, *jau* (barley) and mustard are the main winter/rabi crops; small amounts of *choulayee* (amaranth), vegetables and fibre crops are also cultivated during the kharif (monsoon) season.[5] Animal husbandry, involving cattle, buffaloes, goats and some sheep is an important component of the livelihood system, providing draught power and dung for agriculture and milk and wool for self-consumption and for sale.

Forests and other uncultivated lands play a key role in the livelihood system, as they provide firewood (which is the main fuel in the region), grazing and cut fodder for livestock, timber for buildings and agricultural implements and a variety of medicinal plants and other non-timber forest produce for local consumption and sale. Pine needles or other leafy matter used as bedding in the cattle shed is an integral part of the manure that is applied to the fields. Generally, cattle and goats are grazed during the warmer months and stall fed during the winter months, while buffaloes are stall fed year round. The main sources of fodder during the winter are the lopped leaves of oak trees as well as dry grass, supplemented by agricultural crop residues. Till a few decades ago, local migration to the upper reaches of the Doodha Toli Mountain and other forested areas for grazing livestock was quite common: individuals from each household went off and stayed at specific locations, called *kharak*s, where there were stone huts traditionally occupied by specific clans.[6] This practice continues today in some of the villages.

Ethnically, Brahmins and Rajputs constitute the overwhelming majority of the population in the Doodha Toli region (and most of Paudi Garhwal as well). Scheduled Castes (SCs) comprise about 15 per cent of the population, consisting of *luhar*s (blacksmith), barbers, drummers and other castes (although occupational mobility is quite high). Unlike in the plains, there are no 'middle' castes, that is, those located in between the upper-caste Brahmins and Rajputs and the lower castes or SCs. And the population of pastoral nomads and other communities coming under the category of Scheduled Tribes (STs) in Paudi Garhwal is virtually zero.[7] Moreover, the villages being small, they may have almost no lower-caste households and several villages may be entirely single-caste villages.

Demographically, the female to male sex ratio is tilted significantly towards females (54:46 in Thali Sain tehsil as a whole as per the 2001 Census), an indication of the high outmigration of males.[8] Indeed, male outmigration to the plains has been a key feature of most of Uttarakhand for the past several decades, the phenomenon intensifying after the rapid growth of population in the post-Independence period leading to declining per capita availability of agricultural land. Consequently, women bear a higher burden of work in the villages, as they have to now manage virtually all agricultural, livestock-related as well as domestic activities.[9] The status of women in Pahadi society is somewhat paradoxical. On the one hand, in comparison to the situation in the plains, women have historically enjoyed greater freedom—there is no seclusion (*purdah*), no dowry, divorce is by mutual consent, widow and divorcee remarriage is allowed and overall there is freer interaction between men and women (Berreman, 1963: 261). Moreover, the lower number of men in villages with high male out-migration might suggest that women would have a greater voice in village decision-making. But this does not appear to be generally the case (Manjari Mehta, 1996). While the burden of work has increased enormously, limited voice in decision-making in the household and outside, and poor access to reproductive healthcare still seems to be the norm.

Another feature of the culture of the Pahadis (in common with the cultures of many other mountainous regions) that has been noted by many scholars and is also visible to the common observer is the presence of a strong sense of community in these villages. Part of the reason for this is the heavy dependence on the commons and the need for collective action in many contexts. Women need to go in groups when they go to the forest, because of the dangers from wild animals. Households need to exchange labour at the time of sowing and harvesting. Public irrigation channels or *kuhls* are in need of constant repair because of the fragile nature of the soils and the hill slopes. The kharak system of local transhumance with livestock requires movement in groups. The sense of community is strengthened by the fact that there is relatively less socio-economic differentiation. Caste does create social divisions and lower castes are clearly under the control of upper castes and consequently economically disadvantaged. But there is greater inter-caste interaction and many more similarities of custom across castes than in the plains. On the whole, the Pahadis pride themselves on having a strong sense of village community that overrides or limits the divisions along kin and caste lines.[10]

Historically, life in this region, though very hard, was largely self-sufficient in terms of food. Till two generations ago, villagers wove blankets and clothing from their sheep's wool. External imports were mainly salt and jaggery, which were obtained by bartering amaranth and potatoes. Consequently, the Pahadis were less vulnerable to famines, floods, or market fluctuations that plague rural communities in the plains (Berreman, 1977, 1978). One reason was the reliability of rains and snow and hence availability of soil moisture for two crops. The second was the relatively low level of economic differentiation: very low incidence of landlessness and the continuation of some traditional systems of supporting the lower castes through a share in the crop harvest. Social relations were in a pre-capitalist stage: hiring in or out of wage labour was almost non-existent and sharecropping also quite limited.[11] The third reason has been the availability of fairly assured supplies of drinking water and of fuelwood and grazing from the uncultivated lands (especially in less densely populated regions and inaccessible regions like Doodha Toli).

Given the nature of the physical terrain, the forces of modernisation and 'development' have been generally slower in penetrating Uttarakhand than the plains to its south. This is, of course, now changing. A major impetus came in the 1960s with the building of all-weather roads for military purposes. This coincided with rapid population growth. Educational facilities have spread dramatically, resulting in high literacy rates. However, the Doodha Toli region remains one of the more inaccessible regions (in spite of it not being in the greater Himalayas), showing up as a blank area even in recent tourist road maps. Many of the roads have been recently built and are in a fairly precarious condition. Electricity is yet to reach many of the villages in the region. But interestingly, television is penetrating rapidly, with people using electricity from batteries for operating them if necessary. Wireless telephones have also reached the ridge-top shops at several points.

One consequence of these recent developmental processes in the Doodha Toli area has been reduced self-sufficiency and increased integration into the larger economy. Modern education has further stimulated outmigration. Thus, there is a much greater dependence on food imports into this region, with little corresponding expansion of commercial crops locally.[12]

History of NRM in the Region

It is generally believed that, till the advent of British colonial rule in the early 1800s, local communities exercised direct control over cultivated and uncultivated lands with little interference from the rulers (Sarin et al., 2003a: 98). Between 1823 and 1971, the British introduced a series of changes in the control over land and over forest resources that had dramatic consequences for the management of these natural resources. They introduced the permanent land settlement and then the reservation and heavy exploitation of forest lands by the state for timber, softwood and resin. This led to a series of intense and even violent protests (Guha, 1985). Eventually, common lands got split into three categories: Class II reserves (commercially valuable forest lands) that were directly controlled by the forest department, Class I reserves (commercially less valuable forest lands) that were open to all residents of Kumaon and civil *soyam* lands controlled by the revenue administration.

In 1931, as a direct consequence of the protests of the preceding decades, the British passed a law (Kumaon Panchayat Forest Rules) enabling local communities in Kumaon to apply for and create *van panchayats* (VPs; literally, forest councils) to autonomously manage specific patches of lands from the Class II reserves and civil soyam lands. These VPs regulated grazing and harvesting of fuelwood, fodder and small timber, had the authority to grant limited amount of timber for domestic use and to fine those who violated its rules. The VPs were slowly set up across Kumaon division, which included Paudi district. By the time of Independence, about 400-odd VPs had been set up.[13] Although not without lacunae, the institution of VPs is considered to be one of the rare examples of state-supported yet substantially autonomous community-level forest management institutions in South Asia till recent times.

The VP arrangement ameliorated some of the adverse effects of earlier British forest policy in pockets, but it did not modify the overall approach of 'modern' forest policy—a policy that continued after Independence. This approach was not only geared towards the generation of revenues for the state (through commercial extraction), as against the meeting of local needs, but also the manner of its pursuit deprived local communities from a significant share in the employment or profits generated from such commercial use. The emergence of the Chipko movement in the 1970s was correspondingly driven by a multiplicity of concerns—over environmental problems such as floods being triggered by unbridled forest exploitation and devastation due to mining, over inadequate access to and availability of forest products such as fuelwood and fodder for subsistence use and over the exclusion of local communities from sharing in the benefits of the commercial use of forests (Guha, 1989; Krishna, 1996; Mawdsley, 1998). These were also linked to a broader concern about the lack of voice for the Pahadis in the administration and development of the hill region, which was seen as having a distinct ecology and cultural identity from the plains of Uttar Pradesh. This broader concern translated into a movement for a separate state and several of the younger supporters of and activists in Chipko went on to play an important role in the agitation for Uttarakhand. Other concerns of importance to some constituents of Chipko and of the Uttarakhand movement included concern over proliferation of alcoholism and untouchability.

The Chipko movement had significant, though somewhat paradoxical, impacts on forest and environmental policies. The forest contractor system was abolished and a Uttar Pradesh State Forest Development Corporation was created to take on all extraction activities. A complete (but temporary) ban on all green tree felling was issued in 1979 and a 15-year ban on all commercial tree-felling above 1,000 m was issued in 1981. Using the environmental awareness created nationwide by (among other things) the wide publicity given to Chipko, the Government of India was able to get the Forest Conservation Act 1980 passed, which removed state governments' autonomy in converting forest lands to non-forest uses, whether for roads, school buildings or reservoirs. These changes, however, were lopsided in that they focused on the 'ecological' demands of Chipko without addressing the questions of forest-dependent livelihoods and incomes, let alone decentralised decision-making in development as a whole. Indeed, the Forest Development Corporation became an umbrella under which the earlier forest contractors simply resurfaced as sub-contractors of the Corporation and the same old exploitative practices continued. The initial ban on all green felling, imposed in response to protest fasts undertaken by a Chipko leader, Sunderlal Bahuguna, prevented even the exercise of traditional rights of local communities to timber for house construction or repair—even in well-organised VPs. Another Chipko leader, Chandi Prasad Bhatt, then intervened and got the ban modified to ensure that some local needs could still be met (Krishna, 1996: 160). But these timber rights are based on settlements made during 1910–17 and are quite inadequate to meet today's needs (Sarin et al., 2003a: 100). Local communities began to resent this adverse fallout of Chipko and there was even a *Ped Kato Andolan* (agitation to cut trees) by the Uttarakhand Kranti Dal in 1989 because the Forest Conservation Act was apparently holding up several developmental projects. In parallel, major changes were made to the Kumaon Forest Panchayat Rules. They included a reduction in the forest area from which the VP could be carved out, a reduction in the autonomy of the VP vis-à-vis the state bureaucracy, an increase in the state share in VP incomes and a reduction of the democracy within the VPs by increasing the powers of the VP sarpanch. In certain parts of the hilly region, the establishment of national parks and wildlife sanctuaries further restricted local access to forests and pastures.

Since the mid-1990s, a period which has seen the formation of the state of Uttarakhand, the environmental and development policies in the region have continued and expanded the trend of co-opting the

the mainstreaming of 'decentralised' forest management has led to the adoption by the state of Uttar Pradesh of joint forest management. However, in the hill region, it took the perverse form of village forest joint management, an attempt to bring the VPs (hitherto regulated by the revenue department, but autonomous in their day-to-day management) under the control of the forest department. Funds provided by the donor-supported Uttar Pradesh State Forestry Project were used to get VPs to agree to these changes. Simultaneously, the revenue department went on a drive to create more VPs using only the civil soyam lands within the village boundaries, to 'rationalise' existing ones by splitting big (multi-village) VPs into single-village VPs and to provide funds to them ostensibly to improve their management. Second, watershed development, drought-prone area programmes and soil and water conservation activities have been taken up vigorously through the gram panchayats,[14] all ostensibly with 'community participation', but all with equally slipshod conceptualisation and implementation. The formation of Uttarakhand and its status as a 'backward' state have meant a disproportionately large inflow of donor funds into the state for all these activities. Over the past few years, water scarcity in small towns of Uttarakhand has drawn much attention and the Uttarakhand state government has talked of devoting 40 per cent of developmental funds to work on water resources. Third, the state is also pursuing vigorously a 'developmental' agenda that includes attempts to spread modern education and health care, rapid expansion of the road network, commercialisation of agriculture and intensive efforts to promote 'eco-tourism' as the major revenue earner for the region. Thus, the context in which the intervention being studied began in the 1980s has changed quite significantly over the past 25 years.[15]

Overview of the Work of DTLVS

The founder of DTLVS is Sachchidanand Bharati. In the mid-1970s, Bharati was a student in Gopeshwar College in Chamoli district. This is close to where the Dasholi Gram Swarajya Mandal started its agitation that eventually snowballed into the Chipko movement. Bharati was drawn into the movement and worked closely for several years with Chandi Prasad Bhatt, one of the key figures in the Chipko movement. Bharati was also involved in the Uttarakhand Yuva Sangharsh Vahini, a student movement for a separate state. Bharati travelled widely across most of Uttarakhand, spreading the message of Chipko and initiating environmental activities such as tree planting.

Bharati then got a job as a teacher in the government 'inter-college' (standards 11 and 12) that had been started in Uphraikhal, a hamlet at the ridgetop of his natal village, Gadhkharak Malla, located in the Doodha Toli region on the border of erstwhile Lansdowne and Paudi *tehsils* (the lowest revenue unit). Bharati accepted the job and saw it as an opportunity to return to his native region and channel the inspiration provided by Chipko into constructive work for the environmental rejuvenation of the region. The initial trigger for community mobilisation was the opposition generated by the forest department's plans to fell several hundred precious silver fir and deodar trees on the slopes of the Doodha Toli Mountain. A successful agitation against this move led to the founding of DTLVS in 1980 as a registered society dedicated to the people-oriented development of the Doodha Toli region, with members drawn from surrounding villages. Since then, Bharati has been the leader of DTLVS, directing a network of volunteers based in the villages where the organisation carries out its work.

The philosophy of DTLVS is linked to that of the Chipko movement, particularly the approach of Chandi Prasad Bhatt, and has been one of acting as a catalyst to bring about a transition to sustainable management of the natural resources, namely, forests, water and land, by the local communities themselves. It is conscious of the need for increasing women's voice and for the uplift of the SCs, while retaining the community spirit that it feels is the essence of the hill region. Although it has agitated from time to time on specific issues such as timber smuggling or liquor, the main focus of its work has been on what it calls *nirmaanaatmak* (constructive) work of regenerating natural resources and improving living conditions and livelihoods in the villages through community mobilisation and minimal technical and financial support.

The DTLVS has worked with and experimented with various kinds of technical interventions. These include:

(a) Facilitating access to state-subsidised solar photovoltaic (PV) cells for lighting.
(b) Afforestation and forest regeneration.
(c) Setting up tree nurseries.
(d) Planting of fruit orchards for generating income.
(e) Enhancing water conservation and recharge through different measures.
(f) Constructing/repairing traditional structures (*naulas*) to protect drinking water springs.
(g) Building walls to protect agriculture from wild animals.
(h) Installing improved/smokeless *chulhas* (woodstoves).
(i) Improving sanitation through the construction of toilets.[16]

The DTLVS has tried to implement these activities by mobilising the community in various ways. First, in the initial stages, Bharati and a few colleagues undertook extensive *pad yatras* (walking tours) of the villages to establish contact with the villagers and familiarise themselves with the issues. Second, DTLVS has been conducting regular environmental awareness camps in different villages in the region (at least two camps a year), which are attended by participants from other neighbouring villages as well as from villages outside the region. These camps involve some tree planting or water conservation activities, some exposure to outsiders to the work done in the host village and some exposure to the villagers to the ideas and debates brought in by the outsiders. Third, right from the mid-1980s, it has organised *mahila mangal dals* (MMDs: women's self-help/welfare groups) in all the hamlets or villages with which it has had some contact. Apparently 133 such groups have been organised so far. The women members of these MMDs are supposed to meet every month and contribute a small amount towards a fund. This fund is meant for providing support to needy members in times of difficulty or need and also for use in various community development activities. The MMDs are thus not micro-credit groups, but more like welfare groups. Fourth, DTLVS has established a network of volunteers (one or two in each village) who liaison between Bharati and the villagers, including the MMDs.

The DTLVS strongly believes in encouraging voluntarism and has consciously chosen to stay at a low level of finances. It has no paid staff at all (including Bharati himself) and it expects villagers to contribute towards the activity through some *shramdaan* (voluntary labour), although it does supplement this with some wages for the rest of the work. It also believes strongly in not getting 'NGO-ised', that is, becoming a 'professional' organisation with paid staff and infrastructure, which it believes will inevitably lead to more attention being paid to the survival of the organisation than on addressing the most important local issues. Bharati has, therefore, rejected several offers of large funding from various sources in order to avoid getting entangled in 'project implementation'. The DTLVS has survived financially through small grants and donations. The initial work of setting up nurseries was done with funds from the National Wasteland Development Board and the sale of excess seedlings generated additional funds for a few years (Pahadi, 2004). For the past decade or so, the Uttarakhand Seva Nidhi, which channels funds from the central ministry of human resource development under various programmes including environmental education, had been giving an annual grant of Rs 50,000 to DTLVS. Some additional funds have come from agencies for whom DTLVS has organised special awareness or training camps.

Interventions and Impacts: Two Village Case Studies

The work of DTLVS covers 100-odd villages spanning both sides of the ridge that Uphraikhal is located on, although intensive interventions are limited to about 20–30 villages. The ridge is not only the old administrative boundary between erstwhile Paudi tehsil and Lansdowne tehsil (see Figure 6.2), but also represents a significant socio-cultural divide. The villages on the east by north east side, locally known as *Chauthan patti*,[17] are mostly Rajput-dominated, historically

FIGURE 6.2
**Location and Approximate Boundaries of Sample Villages and
their Van Panchayats**

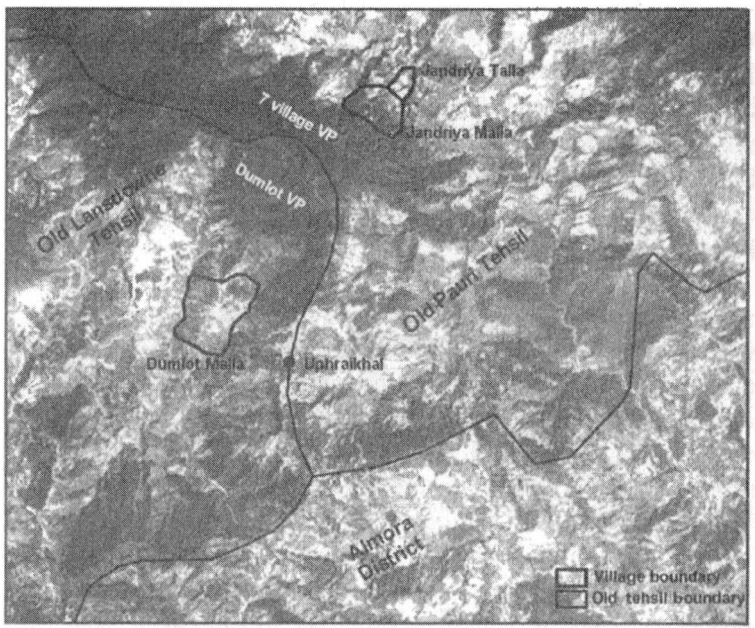

more inaccessible and perhaps thereby with lower levels of education
and correspondingly lower rates of outmigration. The villages on the
west by south west side, historically known as *Dhondiyalso patti*,[18] have
a much higher fraction of Brahmins, have been better connected by
roads, have higher levels of education and (probably therefore) higher
rates of outmigration. The average annual population growth rate
during 1971–2001 for a sample of five neighbouring villages was more
than 1 per cent in the Chauthan patti whereas it was zero or slightly
negative for the Dhondiyalso patti.

Having observed this variation in social and demographic conditions during our initial field visits to different villages, we decided to
study the nature and impact of DTLVS' interventions in one village
on each side of the Uphraikhal Ridge, rather than a single village as
has been the approach in most of the other case studies.[19] In selecting
the village in each cluster, our main criterion was that the village should

have several kinds of interventions by DTLVS and that it should perceive these interventions as having been reasonably 'successful'. We eventually selected Dumlot Malla village in the Dhondiyalso patti and two neighbouring (almost 'sister') villages—Jandriya Malla and Jandriya Talla—in the Chauthan patti. The locations of these villages are indicated in Figure 6.2 and their salient features are given in Table 6.1. Since the households in the two Jandriyas share strong kinship bonds and a history of joint action in many spheres, we will refer to the two villages jointly as Jandriya except when there are differences between them.

TABLE 6.1
Main Features of the Study Villages in the Doodha Toli Area

Vilalge	Cluster	2001 House- holds	Caste Composi- tion	Altitude and Aspect	Road Access and Electricity	Technical Interventions Carried Out by DTLVS
Dumlot Malla	Dhondi-yalso patti	53	Brahmins-14, Rajput-29, SC-10	~1900 m, east facing	Good road link to Baijro, connected to power grid	A, C, D, E, F, H
Jandriya Malla	Chauthan patti	23	All Rajput	~1,900 m, north facing	Road link at Talla is recent, power grid connec- tion being set up	C, D, E
Jandriya Talla		9				B, D, E

Notes: A = solar PVs, B = afforestation, C = tree nurseries, D = orchards, E = water recharge structures, F = naulas, G = walls for crop protection, H = improved chulhas, I = toilets (see previous section for details).

After several days of reconnaissance, we spent three to four days in each of the two villages interviewing individuals, holding group discussions with the MMDs, visiting the sites where DTLVS interventions had been carried out and collecting as much secondary data as possible. The initial interviews were conducted along with the DTLVS volunteer in the village, whereas subsequent interviews were conducted without any such mediation. While this imposed some constraints (because we were not able to fully understand the local dialect and Hindi was not spoken by many of the villagers) it provided us with an opportunity to better understand the villagers' perceptions of DTLVS' work. Of course, it is possible that, with activities spread over 20 years and with public memory tending to be short, villagers today tend to downplay the significance of many interventions. We have tried to keep these methodological limitations in mind while seeking to understand the impacts and processes.

The Village Contexts

As indicated in Table 6.1, Dumlot is a bigger village than the two Jandriyas and also a mixed-caste village. Jandriya Talla and Malla are entirely Rajput villages, although there are a few SC families located in a neighbouring village, which have traditional ties with the Jandriya villagers and who are given some share in the harvested crop by the latter. Dumlot village has an SC population of 15 per cent, including the blacksmith (luhar) and the drummer (*das*) castes. All the SC households in Dumlot seem to own some agricultural lands, although their landholding is significantly lower than that of the upper castes (in the 0.15–0.30 hectare [ha] range, as against the upper-caste average of 0.50–1.00 ha). Our observation was that the socio-economic distance between the SCs and upper castes in these villages varies depending upon the context. The economic distance may be reducing as some members of the SC community have got government jobs.

As can be seen from Table 6.2, the rate of population growth has been higher in the two Jandriyas than in Dumlot. The female to male ratio indicates that there is significant outmigration in both villages. However, the nature of migration is somewhat different. Several households in Dumlot have left entirely for the plains. Moreover, the jobs that the Dumlot migrants have obtained are generally higher paying or more white-collar ones than those obtained by migrants from Jandriya, who have either entered the army or are engaged in casual labour in places such as restaurants. This seems to be related to the level of education in the villages, which in turn might be related to the caste composition.[20]

The agricultural systems in the two villages are on the lines described in the introduction. The main kharif crops are paddy, jhingora and

TABLE 6.2
Demographic Details of the Study Villages in the Doodha Toli Area

Village	Population in 1981	Population in 2001	Annual Compounded Growth Rate (%)	Female to Male Ratio	Literacy in (2001) (%)
Dumlot Malla	212	287	2.2	156:131	57
Jandriya Malla	96	157	2.8	81:66	48
Jandriya Talla	29	50	1.5	24:26	32

Source: Census 2001c.

mandua, whereas the main rabi crops are wheat and mustard. Water or soil moisture availability is higher in the Jandriya Malla fields, enabling more paddy cultivation, although all villages have some small areas under 'proper' irrigation using kuhls (irrigation channels) that bring water to the fields from nearby streams. There is a significant area of fallow land in Dumlot, but less so in Jandriya Malla and Jandriya Talla, reflecting the relative endowment of land versus labour in the villages (see Table 6.3).

TABLE 6.3
Land Use and Cropping Patterns in the Study Villages in the Doodha Toli Area

Village	Total Geographical Area (ha)	Cultivated Area (ha)		Uncultivated Area (ha) within the Village Boundary (Civil Soyam)	
		Irrigated	Unirrigated	Forest	Barren
Dumlot Malla	134	0.7	67.2	30.5 (including 5 ha of pine plantation)	36
Jandriya Malla	~60 (est.)	~3.3	43.0	8.5 (oak forest)	~8
Jandriya Talla	~22 (est.)	~1	16.4	2 (afforested oak)	~3

Sources: Village accountant, Census 1981 and villagers.
Note: In addition to the forest area within the village boundary, each village is part of a VP that gives access to forests outside the village boundary. See Figure 6.2 and text in this section for details.

Animal husbandry is a significant subsidiary occupation. Exact figures on the numbers of livestock in the villages could not be obtained, but it appears that almost every household has at least two bullocks (for ploughing), a few cows and one to two buffaloes. The number of sheep and goats varies significantly across households—most of the households in Jandriya have five to 10 goats, but some keep as many as 30–40. The tendency to go to the kharaks has declined for two reasons. First, some of the kharaks have been settled permanently. Second, with male outmigration, there is not enough labour in the household to send cattle to the kharak for several months at a stretch. Villagers in Dumlot have virtually abandoned the kharak system, whereas it continues to some extent in Jandriya.

It is important to note that all three villages are fairly well endowed in terms of forest resources. They are all part of some VP. These VPs were formed several decades ago out of reserve forests that are technically outside the village boundaries (see Figure 6.2) and hence not included in the total geographical area of the village in Table 6.3. Dumlot is part of a 'Dumlot-Dandkhil' VP that was formed in 1959[21] and covers 800 acres (320 ha) of Class I reserves located above the two villages (and extending all the way to Uphraikhal[22]). The VP earlier used to mobilise household labour for forest protection, but at least for the past 20 years it has been paying two men to act as watchmen. The payment used to be in kind, with each household being required to contribute a certain amount of grain, but is now in cash. Although the watchmen are unable to adequately protect some of the more remote portions of this 800-acre patch, nevertheless most of the VP forest has fairly dense oak or oak–pine vegetation and provides ample supplies of fodder and firewood to the villagers. The sarpanch of this VP is typically from Dumlot and five of the remaining eight panchs are also from Dumlot. In addition, Dumlot has a 5 ha pine plantation created several decades ago in its civil soyam lands, which is managed by the gram panchayat and provides significant quantities of pine needles for bedding material for the cattle sheds, pine bark pieces for lighting fires and some building material (the trees are yet to reach maturity in terms of timber).

Similarly, Jandriya Malla and Jandriya Talla are part of a seven-village VP, whose forest area lies in Class I reserves outside (above) the village boundary. The exact extent of this VP could not be ascertained, but it seems as large as the Dumlot VP, although it probably has a larger population of users in its general body. The forests in this VP are also quite dense and consist of a mixture of oak, pine and rhododendron vegetation. In addition, Jandriya Malla has (in the civil land bordering the reserve forest) significant area (~8.5 ha) under oak forest, which it has begun to protect and manage over the past four decades. This has been recently converted into another VP (a mahila VP). Although there are no pine plantations in the Jandriya villages, there are pine plantations in the non-VP reserve forest close by and the villagers say that they do manage to get some timber from these forests upon payment to the forest department.

Activities, Processes and Impacts

The DTLVS has been working in Dumlot and Jandriya since the early 1980s. It must be noted at the outset, however, that their style of working is different from that of typical, programme- or project-based NGOs. They have not focused specifically on one village, but across a large number of villages. They have not developed any 'integrated' village-level resource management plans, but rather have experimented with various interventions at various points in time over the past 20 years. Therefore, rather than discuss each village separately, we will describe the nature and impact of individual interventions—technological and social.

Solar PV-based Street and Home Lighting

One of the early interventions by DTLVS was in facilitating the provision of solar PV cell-based lighting kits for the villages. In the 1980s, this region had no electricity supply. Households depended entirely on kerosene-based lamps for lighting. A scheme promoted in 1984 by the Non-Conventional Energy Development Agency of Uttar Pradesh involved providing solar PV-based street lighting for villages for free. The DTLVS liaised with the government to ensure that some of the villages got this facility. In fact, villages as remote as Daera got street lighting in 1986. Subsequently, starting in 1988, the government offered similar kits (which lit two fluorescent bulbs each) to individual households at a subsidised price. A few kits were made available per village at a time. Many households in Dumlot and a few in Jandriya got these kits by paying Rs 500 between 1989 and 1993, which was the contribution required by the government since the kits were for private use.[23] Over the years, more households have adopted these kits. The contribution required for the kits has also risen over the years (it now stands at Rs 2,500—not an amount that all households can afford). The technology appears to have been fairly robust, with batteries generally lasting more than six years. Of course, the household has to pay for the replacement of the battery. Most of the original PV sets are in use and more are being installed every year. With electrification from the grid progressing at a fairly slow pace in this region and the connections being quite prone to breakdown, stand-alone PV sets are still the dominant technology for domestic lighting in this area.

The DTLVS' role here was to liaise with the government and (in the early days) to convince the villagers of the technical viability of these kits.[24] It also trained some of its volunteers in the maintenance and repair of these kits, although over the years, the need for maintenance appears to have declined. In today's context, this intervention may look very insignificant, but in the 1980s, it was a significant contribution to improving the quality of life in this region. The scheme was poetically described as having created 'solar fireflies' (Anonymous, 1998a: 101). It should be noted that collective action was required for this activity only in its early format, when it was meant for village street lighting. Subsequently, there was no need for collective action and in that sense it is not a community-based strategy. But it did generate enormous goodwill amongst the villagers towards the organisation, which facilitated future activities. One can say that DTLVS, at least in its early days, was more or less a 'service NGO' providing an important link and conduit between the state and its programmes and the remote villages in this region where the presence of the state was almost negligible.

The limitations of the technology are that it is expensive and for a subsistence-level household in the 1980s, even Rs 500 was a major investment. Consequently, the benefits accrued in a lopsided manner, namely, to those who could afford this technology. The penetration was slower in Jandriya—only three households came forward to take these kits in 1989—probably indicating the lower level of disposable income in those villages. Much later, another eight households received these kits through the block development officer. Some of the SC households in Dumlot do not have these lamps yet. The PV-based street lamps, on the other hand, although fully subsidised and providing wider benefits, did not sustain as the villagers were not willing to collectively bear the cost of replacing the batteries for the street lamps once the original battery wore out. One also needs to keep in mind the fact that the 'public' benefit of a street lamp is bound to be limited in a village where the houses are not tightly clustered together.

Afforestation

Although both Dumlot and Jandriya have access to large VP areas, in both cases the distances to the areas from which winter fodder (oak leaves) and firewood can be harvested is quite significant. The walk is several kilometres and involves a climb of at least 200 m. In the case of Jandriya Talla (which is located below Jandriya Malla), it is a longer climb. Moreover, Jandriya Malla has significant civil soyam forest area within its own boundary (just above the settlement), whereas Jandriya Talla has no such area. After the setting up of MMDs in Jandriya, the women of Jandriya Talla came up with the suggestion that an oak forest should be created within their village boundary to supplement what they get from the VP forest. They identified a small patch of 2 ha of open civil land within their village. The DTLVS provided monetary support in the form of paying wages for the labour, and the women of Jandriya Talla (just nine households) did all the work of collecting oak seeds and raising oak seedlings, building a protective stone wall around the patch and protecting and replanting as required. The work was carried out around 1989 and today the patch is a dense banj oak stand. The harvesting of leaves and twigs from this patch is not yet as intense as that from a full grown forest, but the thinnings and trimmings make a significant contribution to the household requirements. Equally important, the women of Jandriya Talla clearly have a strong sense of ownership and pride in the forest that they have raised in what was a barren piece of land. They jointly protect the patch and jointly decide when to harvest and how much.[25] Their effort prompted the villagers of neighbouring Teolia village to emulate this idea: they also identified a patch of degraded civil forest (in fact, larger than the one in Jandriya Talla), built a stone wall around it and regenerated it to the point that it is now fairly dense and becoming a significant source of oak leaves and twigs.

These efforts are small and their contribution to alleviating women's burden of collecting fuelwood and fodder cannot be enormous. Nevertheless, they constitute significant community-initiated efforts at meeting a felt need of a weaker section (women).[26] Banj oak vegetation is also seen as having significant spillover benefits in ecological terms. Banj is considered a 'source' of or at least conserver of soil moisture and a species much preferred by various animals and birds. There is clearly careful regulation of harvest in these patches, so the sustainability question arises only in the context of the protection arrangement. The women of Jandriya Talla indicated that the protective wall has broken down in a few places and that they would like to approach DTLVS for financial help in rebuilding the wall. While this might indicate a continued dependence on external help, it is quite likely that the women can carry out the repairs on their own if necessary. More important, perhaps, is the ripple effect as seen in the case of Teolia's emulation of Jandriya Talla's efforts. On the other hand, one does not observe any engagement by DTLVS with the question of how the VP (the much bigger source of forest biomass) itself is managed, either in Dumlot or in Jandriya. This has implications that will be discussed next.

It is also interesting to note that while there has been a willingness to augment fuelwood and fodder sources in this manner, there has not been an acceptance of smokeless fuel-efficient chulhas or wood-stoves for several of the same reasons that have dogged the chulha programme elsewhere—smoke (from the traditional chulha) being seen as having the side-benefit of warming the house, the new chulhas requiring fuelwood to be cut into small pieces, requiring too much tending and modifications in the cooking practices to really result in fuelwood savings, they being too slow to meet the requirements of a large household, their chimney causing leaks in the roof and so on.[27]

Orchards and Nurseries

In line with its efforts to promote afforestation, DTLVS started nurseries in several villages. The main purpose was to produce seedlings for their own afforestation programmes, but as the nurseries grew, they also sold seedlings to government agencies, including the forest department, and thereby generated funds for supporting other activities. In the case of Dumlot and Jandriya, the nursery activity was combined with the idea of trying out horticulture on abandoned agricultural plots. So DTLVS entered into an agreement with villagers to 'take over' a certain patch of abandoned/fallow agricultural land, built a protective stone wall around it and planted various tree species, including walnut, apple, pear, deodar, silver fir, etc. To ensure seedling survival during the summer, trenches or pits were dug next to or above the pits in which the seedlings were planted. Simultaneously, a nursery was created in one corner of this enclosed patch and seedlings were grown for sale and use in other afforestation efforts. The local communities, especially the MMDs, were involved in building the stone wall and digging the tree pits and water conservation trenches (*jal talais*) in both Dumlot and Jandriya (both Malla and Talla), for which they were paid wages. Although there was no explicit shramdaan component, the wages paid appeared to be modest. The size of the orchard in Jandriya is about 6.5 ha and about 6 ha in Dumlot. The sites are somewhat different in that the orchard in Dumlot is on a steeply sloping patch, whereas the one in Jandriya Malla is on a gentle slope. In both cases, however, the patches are located rather far away from the settlements, at the edge of the village boundary.

The nursery activity in Jandriya lasted for several years. The villagers benefited directly in the form of wage employment for two to three persons who were employed by DTLVS as gardeners-cum-watchmen. In later years, as the nursery activity tapered off, the number employed came down to one. But this employment went to males, not to members of the MMDs. In Dumlot, DTLVS has employed one watchman, but the women also take turns in watching over the orchard. The villagers, however, benefited indirectly because the protected patch produced large amounts of grass that the women were able to harvest as fodder. In Dumlot, the women are able to harvest one headload each for a period of seven to nine days during each winter and this saves them a lot of effort as they would otherwise have to go to the VP forest for fodder collection. In the case of Jandriya, as some of the trees grew, the grass production reduced, but it is still substantial—women reported collecting one headload each for about seven to eight days during the year. In addition to the problem of tree canopy closure, this orchard faces a problem of declining protection that is leading to grazing by cows and thereby a decline in fodder productivity.

The idea behind introducing horticulture was to augment incomes, especially from lands that are falling into disuse because of shortage of labour as outmigration increases. Several of the other species, such as apples and pears, did not survive at all, but many walnut trees have survived and grown. The DTLVS had particularly high hopes on walnut, as walnut cultivation had earlier succeeded in Gadhkharak village. It was felt that this tree could bring about the kind of revolution that apple cultivation had brought about in Himachal Pradesh.[28] Unfortunately, this experiment has not succeeded in these two villages for different reasons. In Jandriya, the few hundred surviving walnut trees are, in fact, yielding some fruit today. But the quantity is small and in the absence of protection there is no control over who harvests them and when. Recently, a dispute between Jandriya Malla and the neighbouring village of Sasou led to Sasou villagers allegedly setting fire to the grass in the orchard, which led to the death of many fruit trees. In Dumlot, the survival has been fairly good (around 75 per cent), but the growth of all species, including walnut, has been very slow, possibly due to the much poorer soil quality on the steeper slope there.[29] After more than eight years of effort by the NGO and contributions by the villagers, grass remains the only sustained benefit received from the orchard-cum-nurseries. Even here, the benefits have not always been fairly distributed. Although women from Jandriya Talla and Teolia contributed to the initial wall-building, digging and planting in the orchard in Jandriya Malla, they do not get a share in the grass that is now harvested.[30]

The reasons behind the very mixed results of the orchard-cum-nursery activity are not difficult to understand. First, there are clearly technical limitations to the idea of horticulture in abandoned fields. The fields that are abandoned are logically those that are furthest away and have the poorest soils, where the cost–benefit ratio of agriculture becomes quickly unfavourable when labour is scarce. But villagers pointed out that even fruit trees need good soils and some care (manuring, etc.) if they are to be productive.[31] Second, there is confusion about ownership. In both cases, but particularly in Jandriya, there is a feeling that the land (and therefore its commercial produce) now belongs to DTLVS, because a written agreement was reached between the farmers who owned the fallow fields in that patch and DTLVS allowing the organisation to take over the land for the nursery operations. The fact that the profits from the nursery activity went to DTLVS (no doubt to support other activities) and the hiring of male (not female) wage labourers to take care of protection and cultivation in the nursery phase strengthened this feeling. The benefits from grass harvest are significant, but are seen perhaps as temporary and as a largesse from DTLVS rather than as a 'right' from a common resource regenerated communally. Third, even if DTLVS were to make it clear that the nursery operations are over and hence the villagers can now take over the orchard, there is a lot of confusion amongst the villagers as to whether the economic returns from the fruit trees (if any) would be distributed in proportion to the land contributed by the original landholders of the fallow fields or equally amongst all villagers. Many villagers seemed to think it would be the former. Clearly, there had been no discussion on this subject prior to the creation of the orchard.

Water Conservation and Drinking Water Naulas

In the early 1990s, many parts of Uttarakhand began to experience serious shortage of water for domestic use. Although largely a problem of the towns, this problem was also felt to some extent in rural areas during the driest months of the year. As a response, DTLVS began to experiment with water conservation strategies. They came up with the idea of jal talais, that is, small water-harvesting pits or trenches to reduce and store run-off from the slopes. These talais are generally 2 m long, 1 m wide and 1 m deep (less wide if the slope is steep). Hundreds of such talais are dug in a particular patch, usually in combination with tree planting. The soil excavated is used to line the edge of the talais with mounds. The idea was to increase recharge, while also ensuring some local moisture that would facilitate the growth of the saplings.[32] In Dumlot, several hundred talais were dug in two locations— in the orchard described earlier and more recently on the west-facing slope where some tree planting work has also begun. In the orchard, there is also a larger size (15 m long, 10 m wide and 5 m deep) percolation tank. In Jandriya Malla, several hundred talais were dug inside the orchard. More recently, some were dug on the slopes adjoining and above the orchard. In each case, the villagers were paid a modest amount for the labour involved.

Clearly, digging pits across the slope leads to some water harvesting and its recharge and will also benefit soil moisture in the plantation. But there are technical as well as social issues about which there is not enough clarity. Technically, it is not clear what the quantum of recharge can be from pits of this size, which might easily overflow in case of a big downpour. It is also not entirely clear whether inadequate recharge can be a serious factor in a region where the geology is sedimentary and therefore the hydraulic conductivity of the soils may be naturally high. Indeed, the cause of most landslips in this region is the saturation of the slope, after which the soil mass slips along some plane where the bonds are weak (such as a soil–stone interface). With the major portion of the catchments being forested, the infiltration rates in the catchments are likely to be fairly high to begin with and the incremental benefits from the talais may be fairly small anyway. Equally important is the question of who benefits from the recharge and how. In Jandriya, the increased recharge would augment the flow of a stream that is hardly used by Jandriya villagers and the fraction of the catchment of this stream that has been treated in this manner is probably less than 20 per cent. Not surprisingly, when villagers were asked pointedly how the talais benefit them, those in Jandriya categorically said that they do not benefit in any way except for the wage income from the digging of the talais. In Dumlot, although the stream is used for irrigation, the fraction of the stream catchment thus treated is very small, perhaps less than 10 per cent. The DTLVS has so far not focused on the question of water use for livelihood enhancement, which might be the more critical issue. Irrigation, as mentioned earlier, is provided through small channels that are built along the contour, diverting water from the stream into fields on the slopes. Traditional channels were temporary ones, requiring extensive rebuilding every year. The government's irrigation department has attempted to build permanent cement channels in several villages, but we noticed that they were all in disrepair. Addressing this issue of declining or maintaining irrigation systems might be more critical to enhancing local livelihoods by making agriculture more productive.

In recent years, DTLVS has focused more on the question of augmenting or protecting drinking water supplies. It has attempted to renovate or build afresh traditional-style structures called naulas, which are essentially small (~2 sq m) dug ponds (with a roof) that store water seeping out of the hillside. In 2005, one such naula had been built in Dumlot and a few other naulas were built in Bharadidhar village that is downhill from Dumlot.[33] The DTLVS provided Rs 10,000 for the construction of these naulas and the work was carried out under the supervision of a volunteer from this village, with the villagers being paid wages for their labour. In Bharadidhar, the construction of one naula was linked to the digging of talais on the slope above it and the villagers felt that this had led to an increase in the flow of the spring and the revival of a spring in an older naula just below the new one that had been abandoned 35 years ago. It should be noted, however, that both of these villages by now have piped water supply systems constructed by the state water agency which draw water from the major stream a few kilometres away. The villagers find this system more convenient than walking to the naula. But the piped water scheme is also notoriously unreliable: frequent blockages as well as breaks due to landslips are followed by days or weeks of waiting for the agency's repairman to come and fix the problem. The villagers, therefore, see these rejuvenated naulas as a good backup source, where they do not have to be at the mercy of the vagaries of government departments.

Reviving or rehabilitating traditional drinking water sources, however, can also mean reviving or retaining traditional (but not necessarily healthy) social customs around those sources. In the case of Dumlot and of Bharadidhar, this has meant retaining the custom of not allowing SC people to take water from the same naula that is used by upper caste Brahmins and Rajputs. In Bharadidhar, the (upper-caste) villagers cleverly 'resolved' the issue by building another naula on the spring that was used by the few SC families in their village. In Dumlot, the question is still hanging fire. When asked whether they would allow the SC families to use the naula, the upper-caste women in the MMD dithered and said that perhaps they cannot stop them (especially since the mason who had actually done the masonry work on the naula was himself from that community!) but that they (the SCs) had separate drinking water sources that they should continue to use. These sources are, however, not springs, but streams (making the water more likely to be polluted), are not all perennial and are often located further than the naula. If they go to the naula, they have to wait for some upper-caste person to come there who can draw water from the naula and fill their pots for them. Thus, the construction of the naula has exposed and possibly widened the caste divisions within the so-called village community.[34] Whether the friction generated in this context spills over into activities where currently there is no discrimination, such as grass or oak harvest, remains to be seen.

Formation of MMDs

The only village-level institutions formed by DTLVS are the MMDs. In the two sample villages, these were formed around 1989. Although initially the Jandriya MMD included Jandriya Malla, Jandriya Talla and also the neighbouring village of Teolia, they split into three separate MMDs within the first few years. The MMD is supposed to meet every month and each woman member is supposed to contribute a fixed amount to the *kosh* (fund). Initially, this amount was Re 1. It was later increased to Rs 2 and now stands at Rs 5 in both villages. This common fund was used to give interest-free loans to needy members and to spend on community activities. Apart from the financial benefit of this activity, the MMDs were meant to provide a forum for women to meet, discuss their problems and take some collective decisions in terms of planned activities. Over the past 15-odd years, they have collected a fairly large fund—in Jandriya Malla, it is about Rs 15,000. The Dumlot MMD has about Rs 5,000, having spent about Rs 15,000 on renovating the village temple and on a few other community activities. But, as mentioned earlier (see note 34) sometimes these community activities can end up providing benefits in a lopsided manner.

The details of the process of MMD formation in these villages could not properly be elicited, since they were formed so long ago. However, it seems clear that DTLVS was involved (mainly through its village-level volunteers) in various handholding activities. First, they helped the women maintain accounts of the transactions—something that is now being handled to an extent by the women themselves.[35] Second, they tried hard to get these MMDs registered with the government. The idea here was that this would provide them some additional financial support from the government. This did not work out because the government officials who were supposed to register the MMDs were demanding large bribes, which the women were not willing or in a position to pay. It appears, however, that at some point the Jandriya Malla MMD did get registered and co-opted into the government's *Mahila Samakhya* programme and has been receiving some minor benefits from that programme without DTLVS playing any role. This raises two issues that we will be highlighting later in the discussion, namely, the lack of systematic follow-up from DTLVS on certain issues (in this case strengthening of the MMDs) and the expanding scope and reach of government programmes that are co-opting locally-initiated activities and institutions.

In recent years, the functioning of the MMDs is going through some ups and downs. In both Jandriyas, the MMDs reported that they were not meeting regularly and had stopped contributing to the fund. This seems to be partly because the erstwhile president of the Jandriya Malla MMD, Ruma Devi, had emigrated and partly because in both MMDs the members had not come across any new initiative that could generate concrete benefits for the members. In Dumlot, the members indicated that they would meet after the harvesting season was over and make up all the backlog of contributions (because they had not met for a few months). But the women from the SC community in Dumlot have indicated that they have stopped going to the MMD meetings and contributing to the fund as their views are not taken into consideration (and also perhaps due to the controversy over the naula; see note 34). What this suggests is that the condition of the MMDs is still precarious in terms of the women's capacity to organise and run them.

More importantly, the role of the MMDs vis-à-vis the DTLVS programmes seems rather limited: they are involved in subsequent operational issues, but not in planning. For instance, the decision regarding when to open the respective orchard patches for cutting of grass is taken by the MMD in Dumlot and Jandriya Malla. On the other hand, the decision to deepen the percolation tank in the orchard in Dumlot or the decision to construct the naula was not taken by the MMD, nor was the MMD the organisation through which the work was implemented. Similarly, the recent jal talai construction or more recent medicinal plant and turmeric planting activities in Jandriya Malla were also not decided upon or implemented through the MMD.

At the same time, the MMDs are being drawn into a different kind of linkage with state agencies that they may or may not have the capacity to negotiate on their own terms. We have mentioned earlier the relatively innocuous linkage with the state Mahila Samakhya programme. A more interesting development is the formation of a mahila VP in Jandriya Malla in 1997 under the aegis and encouragement of the VP department. The general body of this mahila VP is the MMD membership and the then president of the MMD in 1987 was made the first president of the mahila VP. In fact, according to her, the funds of the MMD were merged with those of the mahila VP. This VP was the formalisation of what had till then been an informal arrangement within the village (starting back in the late 1960s) of protecting and managing the (largely oak) civil soyam forest that lies within the revenue boundary of Jandriya Malla. Locally, such informal protection of civil lands is called *ghar panchayats*. The formation of the mahila VP enabled the villagers to get access to government funds: in 2001, they were sanctioned a grant of Rs 120,000. Of this, about Rs 30,000 seems to have been made available to them for setting up a nursery, building a stone wall and building two small tanks for storing water for irrigating the nursery (the rest of the amount seems to have disappeared into various pockets along the way). Eventually, they were given only Rs 15,000 because they found that the labour involved in getting the stones was too much and so stopped the work. Nevertheless, the nursery was prepared and several thousand seedlings of different species (including chir pine) were grown. For two years these were sold to an agency implementing the government's watershed and drought-prone area protection programmes. A profit of about Rs 15,000 was made in the process (they did not pay wages to the women members who worked to prepare seedlings; only some wages were paid to those involved in watching over the nursery). This profit is now going to be added to the common village fund and used (again!) towards common activities such as festivals.

While this conversion of the MMD into a mahila VP could be interpreted as a sign of the growing strength of the MMD, it is doubtful whether such an interpretation would be appropriate. The MMD members did not seem to know the details about the formation of the VP, nor of the obligations that might emerge as a result of signing up for government recognition. They do recognise that there is enormous corruption in the whole process of obtaining the government grant, but seem to take the position that getting some grant is better than getting nothing. There also has not been any significant discussion about whether the MMD members should get wages in return for the work they put in for the nursery. And the mahila VP is now actually being run by the president's husband, although there probably are some other younger literate women in the village.

Environmental Awareness Camps

The DTLVS has conducted several environmental awareness camps in both villages, although the last ones conducted in Jandriya were several years ago. Similarly, some villagers from each village have been attending camps held elsewhere. As described earlier, the camps include some physical work, discussions and cultural activities (apart from collective cooking and consumption of food!). Most of the villagers we interviewed felt that, during the early years, the camps served an important purpose of building awareness about issues and increasing contact across villages, with DTLVS and outsiders. But they are now beginning to feel a certain amount of fatigue because the camps are becoming repetitive and they do not seem to be addressing any of the issues that are of immediate concern to the villagers, but rather (in some villagers' view) seem to be geared towards showcasing the achievements of the organisation to outsiders.

Understanding the Interventions

The interventions by DTLVS in the case study villages span 20 years, a period during which it has tried to address several issues: improving the quality of life through lighting, augmenting forest resources availability, increasing incomes through fruit orchards, improving water recharge, protecting drinking water springs and building SHGs for women. The quality of life for local communities has improved in various ways, including availability of lighting in houses, easier availability of fodder and in some cases firewood and the setting up of forums for women to come together. There also have been some indirect ecological benefits to the wider public in the form of increased recharge of water or a shift from a focus on chir pine planting to the planting and regeneration of oak species. It is also clear that at least some of these gains are ecologically and socially sustainable, as the local users have seen the benefits of solar PVs and fodder/fuel regeneration and are taking necessary steps—such as protection or maintenance of afforested or fodder patches—to ensure that these gains continue to accrue. Even more important perhaps is the attempt to (re)create and maintain a level of awareness and concern about environmental issues, both those that are directly related to local livelihoods (fodder or fuelwood) and those that may have wider implications (water). Finally, it should be kept in mind that these impacts have come at a very low level of public cost, as DTLVS operates in a very low-cost mode.

On the other hand, there have been some clear failures. In particular, the attempt at income generation through orchards has largely failed and there is no serious programme for income generation operating right now.[36] The approach to water conservation seems to generate rather limited benefits to an unclear set of beneficiaries downstream and is thus not generating deep and sustained interest from local communities. Instead of having a sense of ownership, villagers seem to participate in such activities because they get some wages. The rejuvenation of drinking water sources has exacerbated the social distance between the upper and lower castes. The MMDs, while possibly benefiting the women initially, have only marginally increased women's voice in village-level decision-making. They have also been an arena of converting individual labour contributions towards so-called community activities that often only benefit particular groups or sections of the community.

Perhaps of greater concern are the areas of inattention or non-engagement. First, given the well-known problem of inadequate income-generating activities within the hills, the experimentation with walnut orchards seems both inadequate in scale (a few hectares in each village) and also in depth (without much attention being given to the question of soil quality or inputs, or to questions of marketing). Second, as mentioned earlier, the experimentation with jal talais has neither been rigorously analysed nor been integrated with questions of irrigation. Third, institution building has been limited to the formation of MMDs and there has been little engagement with existing village-level institutions that have a major say in NRM, namely, the VPs, the gram panchayats and the whole host of user groups or committees being set up under various programmes. After 20 years of dedicated work by DTLVS to spread the message of *jal-jangal-jameen bachao* (save water, forest and land resources), villagers are asking whether their environmentalism has come at too high a cost in terms of actual losses (from crop raiding by wildlife) or lost opportunities for enhancing livelihoods. And they seem to be increasingly enmeshed in the state-sponsored processes of democracy (notwithstanding their limits) and institutions of NRM.

To understand these outcomes as seen today and to anticipate the challenges that may arise tomorrow, one has to look at the choice of strategies and the underlying philosophy as well as the influence of the changing local context and the role of the state on the outcomes.

Strategies Chosen: CBNRM Lite

The key issue in any CBNRM effort, as we have said in Chapter 1, is the manner in which the processes of community mobilisation, action and decision-making are initiated and institutionalised. In DTLVS' work, a key (if not innovative) mobilisation strategy has been the extensive visits by Bharati to all the villages in the initial period and the use of environmental awareness camps to further mobilise the villagers. The mood of Chipko was very much in the air at that time and DTLVS continued the campaign mode of functioning to great advantage. Subsequently, however, the processes through which the issues have been selected, solutions identified and activities implemented have varied. The issues chosen to be addressed appear to have emerged in different ways, rather than through any systematic process of identification of needs and priorities. Some were concerns highlighted by local communities (such as the need for an oak patch in Jandriya Talla), some were opportunities presented by government development programmes (such as the solar PVs) and some were ideas conceived by Bharati himself (such as the jal talais).

The decision-making process has the involvement of the villagers, but there seems to be neither any standard democratic forum in which these decisions are taken (such as a meeting of the entire adult population) nor is there always a conscious effort to involve the women as much as the men in decision-making. As mentioned earlier, the MMDs could have been such a forum, but have not been so in every issue. Indeed, looking at the current functioning of the MMDs and their current role in the DTLVS activities, it seems that they have ended up becoming not so much an instrument for gender empowerment, but rather (because of women being responsible for the bulk of natural resource related activities in the village) as a necessary and efficient means for mobilising women's labour for various interventions.

Similarly, the implementation of the activities has followed different routes. Some were implemented at the individual level (solar PVs), some were implemented by the women's groups directly involved (forest regeneration). Some had the entire village involved (hosting of environmental awareness camps) whereas some had only some villagers involved as wage labourers (nursery and orchard protection) and in some all villagers seemed to be involved but more as wage labourers than as long-term investors (orchard planting, digging of talais). Diversity in actual implementation is of course bound to occur because not all work can be done as collective voluntary labour at the level of the entire village. But the inconsistent participation in implementation has compounded the inconsistent participation in decision-making process, as highlighted earlier.

In the absence of a clear democratic forum through which issues can be raised and decisions arrived at in the village, the link between the external intervening organisation and the local community acquires great importance. Here, DTLVS has preferred to function through volunteers. In some cases, such as Dumlot or Bharadidhar, this is a person who resides in the village; in Jandriya and several other villages in the Chauthan patti, the link seems to be through an ayurvedic doctor who is based in Uphraikhal but travels regularly to those villages (he is also the secretary of the DTLVS society).[37] In all cases, the volunteers are male, and where they are from the village they are generally from the better off section of the village. Perhaps it is necessary to have male volunteers because the work requires travelling regularly to and from Uphraikhal to the village and to various other locations for liaison with government offices—something that women would find difficult to do, especially since the women in these villages are completely tied up in domestic and agricultural activities. But it is not clear why, for instance, some women staff could not be recruited on a full-time basis, given the enormously important role played by women in NRM. In any case, choosing persons from the village has both strengths and limitations. While it ensures greater rapport with the villagers, there is also the risk that the persons may be too embedded in the traditions of the village to be able to make a difference. While many NGOs recruit community workers from the villages they work in, they often adopt specific strategies for overcoming the limitations (for example, conscious selection of personnel from the weaker sections, training and exposure visits and shifting staff to villages other than their own). Only some of these strategies (mainly exposure visits) have been adopted by DTLVS.

Similarly, it has been the experience of many NGOs that delegating financial management to one individual rather than to a democratic institution within the village can reduce transparency, can increase the risk of financial mismanagement and certainly reduces the sense of empowerment for the village-level institution.[38] Thus, not surprisingly, in the case of recent interventions such as the construction of naulas, the villagers indicated that they are not aware of what funds DTLVS actually provided or how they got utilised, they were only aware of what wages they got in the activity. This may not necessarily mean financial mismanagement by the volunteer, but certainly shows inadequate attention to processes that ensure transparency.

Intertwined with this strategy of choosing volunteers rather than paid professionals, not engaging them full-time or training them intensively and so on, is the choice to remain small and in some sense underfunded. Again, this choice is a double-edged one. On the one hand, it helps avoid the problem of the organisation becoming a bureaucracy and becoming focused on its own survival.[39] On the other hand, it means that the activities that can be taken up are limited and the attention and follow-up that can be provided also limited. As people's notions of development expand and the state's own programmes of development also expand (as will be discussed next), a small voluntary effort runs a greater risk of getting swamped or sidelined.

Finally, democratisation applies not just to the village-level institutions, but also to the intervening organisation itself, especially when it is constituted as a grassroots organisation and not entirely an 'outside' agency. The DTLVS was registered as a non-profit society with a broad-based membership drawn from many villages in the region. In its initial years, it also functioned in this manner. Over the past decade, however, there appears to be a kind of unconscious drifting away from these processes and norms, even the minimum ones for non-profit societies, namely, holding of annual general body meetings of all the members (which in DTLVS' case are drawn from the villages in which it works), election of office-bearers, presentation of audited accounts and debating about the future direction of the organisation. The DTLVS has stopped conducting these activities for the past decade or so. Again, this is not to say that there is no on-going process of consultation with its members or that there is financial mismanagement. One reason for this drift could very well be that the level of funding and activities has somewhat shrunk in the past few years. But the absence of process leaves room for villagers to misunderstand and make allegations.

In other words, while the Chipko-inspired focus on awareness building has been crucial in the initial stages, the 'lite' mode of functioning, which involves thinly spread investments and inputs, and working in a non-project mode with volunteers may be inadequate in the long run because community-based resource management requires sustained building of local-level institutions and resources.

Organisational Philosophy: A Gandhian Environmentalism

As mentioned earlier, Sachchidanand Bharati entered the field of environmental activism through Chipko and student agitations for a separate Uttarakhand. The leaders of Chipko came from a Gandhian and Sarvodaya background. Their original focus was on decentralising forest control, meeting local needs sustainably, enhancing local livelihoods through small-scale industry and also addressing other social issues such as alcoholism and untouchability. However, as the Chipko movement spread and got 'iconised', the 'ecological' aspects got prominence and the other dimensions were played down (Krishna, 1996; Mawdsley, 1998). In this philosophy, conserving forests, water and soil becomes an end in itself. Bharati was associated with Chandi Prasad Bhatt, who is seen by many as championing the cause of eco-development within Chipko, whereas the ecological dimension was promoted by Sunderlal Bahuguna. However, Bharati's (and DTLVS') philosophy seems to be a combination of the ecological and the eco-development philosophies. Thus, at some level, DTLVS sees resource conservation as an end in itself, whereas at another level the focus is clearly on meeting local needs through sustainable management of local natural resources. There is an emphasis on Gandhian self-reliance, reinforced by the idea that hill communities were till recently quite self-reliant and can, therefore, easily continue to be so. But perhaps because Gandhi always argued for limiting human needs, the emphasis on meeting the 'developmental' aspirations of the poor is not strong. Similarly, although issues of caste-based discrimination and gender discrimination are clearly recognised,[40] there is a strong emphasis on change from within the individual rather than structural change driven from the top. This approach to bringing about the change is non-conflictual and non-violent. And when introducing new technologies, the implicit assumption is that society is not that highly differentiated and that there are enough community institutions to ensure a fair process of integration of the technology. Thus, within the four dimensions of livelihood enhancement, sustainability, equity and democratic decentralisation that were outlined in Chapter 1, DTLVS gives greater priority to the sustainability of natural resource use and on the subsistence rather than income-generation aspects of livelihoods. And while significant attention has been paid to introducing ideas or technologies that fit the micro-level social and ecological conditions, there has been less attention to questions of devolving the governance of the natural resources. Perhaps it is assumed that these village communities were always largely self-governed, which is not actually the case in many spheres of life now.

Social Context: Favourable but Changing

The assumptions about self-reliance as a desirable and feasible goal and the existence of self-governing, relatively homogeneous, village communities may be truer in Uttarakhand (and say Nepal and Bhutan) than in most other parts of South Asia, and may have been true in the past. But they were never entirely tenable to begin with and are becoming more untenable today as even the remote Doodha Toli region is witnessing the effects of modernisation and state policy in various ways.

Everywhere we went, the rising expectations about the material standards of living and a hunger for a 'modern' lifestyle were very clear. While these expectations may certainly not be entirely realistic (given the resource constraints), the fact is that these expectations exist and have to be sifted through by any organisation attempting to bring about 'improvements' in livelihoods. Focusing on meeting subsistence uses alone seemed to work in the initial stages of DTLVS' existence, but its adequacy seems to be increasingly questioned by villagers who find that the combination of population growth and rising 'needs' means that they are not able to meet more than 50–75 per cent of these needs from their own productive assets. The Dasholi Gram Vikas Mandal had started out even in the 1960s on the path of small-scale industries to generate extra incomes for villagers. Today, that organisation seems to be experimenting extensively with promoting fruit cultivation and improving their marketing. This adaptation may be essential in the Doodha Toli region as well.

Similarly, neither were traditional forms of governance entirely democratic (as they discriminated against the lower castes, landless persons and women) nor are they really in existence in their earlier form as the social context (in spite of the remoteness of this region) has changed quite a bit in the last three decades. Male outmigration means women are in the majority in many villages and are doing an even greater portion of the productive and reproductive tasks. Thus, all-male VPs that may have been tolerably functional in the 1930s are likely to be much more problematic today. The formation of gram panchayats under the panchayati raj system introduced in 1992 has created another set of institutions at the local level that have some jurisdiction over natural resources and play a major role in implementing developmental programmes. Hence, an engagement with the issues of local governance now seems imperative.

The case of the VPs provides a good illustration of this point. The institution of VPs is certainly one of the rare examples of state-recognised and relatively autonomous forest management institutions. However, changes are taking place. These include, on the one hand, a slow erosion of the autonomy of these institutions by changes in VP rules and policies, increasing attempts by the forest department to take over these VPs under their village forest joint management (VFJM) programme and many cases of inadequate bureaucratic cooperation with existing VPs, and, on the other hand, inadequate internal transparency and democracy in the functioning of the VPs (Sarin, 2001a; Sarin et al., 2003a). A classic example of the latter is the fact that women continue to be excluded from the entire election and decision-making process of the old VPs (the concept of mahila VPs was mooted in 1997, but seems to be applicable only to the newly formed ones).

Van panchayats are quite common in the Doodha Toli region and they govern the main source of forest produce that villagers have reliable access to.[41] Consequently, their proper management is crucial to meeting subsistence needs and ensuring forest conservation. Today, the large Dumlot–Dandkhil VP faces significant difficulties in protecting its forests from illegal use by neighbouring villages, including Gadhkharak (which has been applauded for regenerating its own civil forest with inspiration from DTLVS) and Bharno. This VP has also faced the threat of closure from the forest department on the ground that the VP area is actually a Class II reserve. Furthermore, if changes are made in VP rules, villagers could better meet subsistence needs and also generate some income through (say) extraction and sale of resin and medicinal plants themselves (currently it is done by contractors). Only through continuous engagement with such issues at the local and the policy level will DTLVS be able to make a broader impact.

Parallel Processes by the State: Synergy or Sidelining?

In 1980, when DTLVS began its work, the presence of the state in the region was relatively weak and issues of sustainable management of natural resources for meeting local needs almost entirely missing from the developmental state's agenda (except perhaps in thinking about modernising hill agriculture). The Chipko movement and the wider national and international environmental movement that it contributed to changed this quite dramatically. Today, state-led 'participatory NRM' programmes have become the norm, whether as JFM in forestry or as Hariyali in watershed development or in drought-prone areas, soil-and-water conservation and minor irrigation programmes, or rural water supply and sanitation. These programmes and projects present a serious challenge to anyone working on CBNRM in the field because they have all adopted the rhetoric of CBNRM while in practice often only tinkering with the rules of the game and, in the case of VPs, actually trying to undermine institutions of CBNRM that were actually reasonably autonomous and well functioning. Moreover, these programmes often seek to involve NGOs in the name of facilitating people's participation, thereby posing a dilemma to the organisations working in the field.

Given that one cannot wish the state away and that in fact the entire politics of the formation of Uttarakhand or even of Chipko was not about rejecting the state but reshaping it to get more space for local concerns, the issue is not how to keep the state out of village-level natural resource management, but one of shaping state policies and programmes so as to better support (rather than undermine) existing village-level institutions such as VPs or newly-created ones such as gram panchayats and to figure out how to push these institutions towards a more socially and ecologically progressive politics.

The challenge is further magnified because these programmes are being funded in a big way by bilateral and multilateral donors. (The flow of funds and programmes is even higher for Uttarakhand

because it is seen as a new and relatively 'backward' state.) Consequently, the villagers now see a plethora of so-called 'participatory'
programmes that bring in a much higher quantum of funds than the
minuscule support provided by DTLVS for implementing its activities.
Moreover, the village elite see these programmes as wonderful opportunities for rent-seeking and convince the general villagers that the
simplest and only possible role they can have is as wage labourers who
can get high wages for doing little and slipshod work. This scenario
seems to be widely prevalent. A *pradhan* (headman) of one of the gram
panchayats in the region openly boasted to us about the 'clean and
simple' system that he follows in terms of the cuts provided at various
points in the flow of funds to various functionaries and how he ensures
that all his elected colleagues in the gram panchayat get at least one
programme in which to get their cut and so on.

The DTLVS' approach to this issue seems to be largely one of non-
engagement. Such non-engagement might be a reflection of the
confidence it has in the villagers being able to navigate through this
maze and flood of programmes without compromising their legitimate
interests and concerns. Certainly, one should not underestimate the villagers' capacity to negotiate with the apparently all-powerful state. It
is also true that a small organisation cannot fight battles at all levels, a
point that Bharati made time and again to us. But given the limited
capacity-building of village-level institutions that has occurred and
the enormous (mostly negative) influence that these heavily-funded
and poorly-designed target-oriented programmes can have on grass-
roots processes, this battle may ultimately turn out to be more crucial .
than the one of getting communities to build more naulas or dig more
jal talais in the Doodha Toli mountains.

Conclusion

How do these observations relate to the larger questions being explored
through this study and the underlying critiques of CBNRM? Clearly,
the 'critique of community' is least apparent in this case—the relatively
homogeneous communities have had a long tradition of collective .

action in general and collective management of natural resources in particular. Though, as we have illustrated in this chapter, there are some forms of exclusion in some of the villages, the community is still able to mobilise for collective action in NRM. Indeed, while outmigration has reduced the commonality of interest in some cases, the political mobilisation of the SCs has resulted in their at least asking some awkward questions that were hitherto unasked. But some voices are less audible than others and certainly the men continue to function with different priorities than women.

Equally, it seems that the critique of being limited by wider discourse and practices of development seems least applicable in the case of DTLVS. The organisation emerged out of a radical social and environmental movement and has consciously strived to maintain its distinct ideology, even at the cost of giving up large grants. It has sought to keep the flame of local and global sustainability concerns alive while striving for improvements in the quality of life and livelihoods for people in this region. The vision has been decidedly different from, if not against, the mainstream. Also, by refusing to get 'NGO-ised', it has evaded the pitfalls of 'projectisation' and consequent 'co-optation' as well.

What this case shows is that even under conditions favourable to environmental concern and collective action, converting such a 'radical' vision into reality can be enormously challenging. First, in spite of DTLVS' efforts to foster sustainability concerns, promote appropriate technology and set personal examples of a Gandhian lifestyle, the villagers' notions of what constitutes a desirable lifestyle and their notions of development are clearly evolving in a different direction. Their increasing contacts with the plains, combined with messages conveyed by the government and the media, are enticing them towards greater material consumption. Even if much of these expectations are unrealistic or impractical, they have to be contended with, which is easier said than done. A sustained focus on improving livelihoods means experimenting with agricultural or agro-forestry technologies, marketing and so on, which requires greater funds and external inputs that in turn might undermine the voluntary nature of the organisation.

Second, the state in this case has played a more adversarial than facilitating role. This may be related to the fact that the context is one in which so-called environmental concerns are juxtaposed awkwardly with so-called developmental concerns. Whereas the environmental concerns are used by the forest bureaucracy to retain control of the forests and in fact to gain control over hitherto relatively autonomous community institutions, the developmental vision of the state, whether in the old undivided Uttar Pradesh or in the newly-formed Uttarakhand, continues to be conventional and perhaps increasingly in tune with the larger mainstream developmental discourse. When the state co-opts the notion of decentralised resource management to deploy a regressive brand of politics, the grass roots civil society organisations can only counter this through networking and advocacy at a higher level and also perhaps by engaging with the political institutions at the village level. How organisations such as DTLVS, which are committed to a different vision of development, rise up to this challenge remains to be seen.

Notes

1. The state of Uttarakhand was formed only in November 2000, so most of the available literature refers to these districts as being part of the larger state of Uttar Pradesh from which Uttaranchal, the hilly portion, was carved out. In January 2007, due to popular demand, Uttaranchal was renamed Uttarakhand. Paudi Garhwal is one of original eight districts of Uttar Pradesh that were said to comprise the hill region. By 2005, Uttaranchal had divided the original eight districts into a total of 13 districts. Similarly, Paudi Garhwal comprised two tehsils (sub-districts) till recently; now it consists of six tehsils. Our study area, which originally straddled the boundary between Lansdowne and Paudi tehsils, is now entirely located within Thali Sain tehsil.

2. We were not able to obtain rainfall estimates specifically for the Doodha Toli region. About 1,500 mm is reported as average for Paudi tehsil (Khan and Tripathy, 1976). A similar figure is reported by Negi and Joshi (2004) for measurements taken in Paudi tehsil. It is not clear whether this includes snowfall.

3. 'Indigenous' communities such as the Bhotiyas and the Bokshas, present in different pockets of Uttarakhand, are entirely absent from Paudi district.

4. The Doodha Toli region borders with Almora district, which is part of what is known as the Kumaon portion of Uttarakhand. Consequently, there is a mixture of Garhwali-speaking and Kumaoni-speaking groups in this region. Administratively also, Paudi Garhwal was part of the British-administered Kumaon Commissionerate during the colonial period.

5. In individual plots, a sophisticated crop rotation system is followed to maintain soil productivity.
6. This is akin to the *chans* of western Garhwal reported by Berreman (1978) and others.
7. For Thali Sain tehsil, the SC fraction is just under 15 per cent and the ST population is zero as per the 2001 Census.
8. The sex ratio for children below six years is 48.5: 51.5, confirming that outmigration is mostly that of working men.
9. The relative scarcity of male labour is one of the reasons for the decline of the kharak-based system of local transhumance.
10. Berreman (1963: 262), in his detailed anthropological study of a Garhwali village, dwells at length on this question. He observes that although divisions along caste, and to a lesser extent kinship, lines do exist, that is

> only part of the story. The nature of caste in this area creates economic and religious interdependence in the village. Every local caste is essential to every other, and ... a strong cohesive bond is formed.... There is certain lore about the village, its locale and people, which is shared almost exclusively among villagers with regard to castes and cliques.... Participation in common enterprises, ownership of common property, and preoccupation with common problems and common antagonisms further bind the community together despite caste, sib, and clique alignments. Community members participate in annual ceremonies, ritual observances, and informal drinking, dancing, and singing parties. Cooperative work on village-owned trails, on the water source, and in certain phases of house building and agriculture also contribute to community identification.... Indicative of a degree of village unity and interdependence is the ownership by the village in common of large cooking vessels and a few large tools available to all community members as needed....

We also observed this last-mentioned practice of sharing community vessels by all castes.
11. Some villagers claimed that they were willing to give land out for cultivation at no charge, but there were no takers.
12. In contrast, in other regions such as Naini Tal and Almora districts, there has been an expansion of cash crops such as apples and pears and vegetables.
13. This section draws heavily on Sarin (2001b) and Sarin et al. (2003a).
14. Gram panchayats have been introduced across the country following the constitutional amendments made in 1992.
15. This has implications for the effectiveness or relevance of the interventions, as we shall see.
16. It has also attempted some work on micro-hydel generation near the Binsar Temple on the Doodha Toli Mountain, but this unit did not work for long.
17. Now called Bungidhar Nyaya Panchayat.
18. Dhondiyal is a common Brahmin surname amongst Garhwalis. This village cluster is now called Meldhar Nyaya Panchayat.

19. Another reason for selecting two rather than one village was that the depth of interventions in all the villages was less intensive as compared to (say) the activities that had occurred in Hivre Bazar. In other words, the interventions seemed more extensive than intensive and hence taking a somewhat wider sample seemed appropriate.

20. For instance, the Census 2001 data show literacy levels at about 75 per cent in villages that have entirely Brahmin populations, whereas they are around 45–50 per cent in villages that have predominantly Rajput populations.

21. Other neighbouring villages can also use this VP forest provided they pay certain charges to the VP, whereas residents of Dumlot-Dandkhil get access for free.

22. Strictly speaking, most of the shops and even Bharatis' house in Uphraikhal are located on this VP land.

23. Note that the total cost of each kit at that time (when PV technology was also at an early stage) was estimated to be Rs 18,000, including the battery (Anonymous, 1998a: 102).

24. The Non-Conventional Energy Development Agency approached DTLVS because they felt that this technology may not be easily accepted by the 'backward' villagers (Anonymous, 1998a).

25. Protection is relatively easy because the patch is small and close to the settlement and so no patrolling is required.

26. Although it appears that the men were largely passive spectators in this exercise.

27. On the other hand, pressure cookers have been quite easily adopted, although again the adoption appears to be higher in the better-off and better-connected villages.

28. Walnut has the advantage of being longer lasting than fresh fruit, being somewhat more protected from pest and bird attack due to its shell and commands a good price in the market. On the other hand, the need to open and dry the kernel immediately after harvest imposes limitations, since harvest often coincides with the late monsoon rains.

29. Only *uttis* (alder) trees have naturally regenerated along the drainage line. The villagers argue that uttis has water retention capacity and helps regenerate streams, although this might be a case of correlation rather than causation, because it is an early coloniser when an area starts getting protected. This species also has some use in the manufacture of household furniture, but is not a preferred species.

30. When the original work on the orchard was carried out, the women of all three villages (Jandriya Malla, Jandriya Talla and Teolia) had formed a combined MMD. Subsequently, there was some disagreement about how to manage the MMD fund and the women decided to form separate MMDs. As a result, the Jandriya Malla women denied access to the grass from the orchard that is located on Jandriya Malla land. It is not surprising that such a dispute should have broken out, but it is a bit surprising that DTLVS did not make efforts to resolve it fairly.

31. In Gadkharak, where walnut cultivation was more successful, it appears that the trees were planted on fields with good soils close to the homes and maintained and protected by each household individually.

32. It should be noted that DTLVS has always advocated the use of tree species that are believed to be less water consuming, such as banj and uttis, as against

chir pine, which is fast growing and quick to proliferate, but supposed to be much higher in its water consumption and more conducive to forest fires.

33. The information we obtained from a shorter visit to Bharadidhar usefully complements the information about naulas in Dumlot and hence has been included here.

34. A few other incidents have also contributed to this ill-feeling among the SCs in this village. One incident is the continued refusal of the upper castes to allow the SCs to enter the newly renovated village temple, even though the SCs had contributed labour towards the renovation. They also point out that half the money for buying the brass bell for the temple was contributed by the MMD's fund, to which the SC women have also been contributing regularly. The SCs also feel cheated because another programme (*Gram Devata*, that is, village deity), encouraged by DTLVS, which was meant to clean up the village and repair the village roads, never covered their side of the settlement. They say that they have also not been invited to participate in any of the camps organised by the DTLVS in other villages.

35. For more than 10 years, the Jandriya Malla MMD was run by Ruma Devi, one of the few educated women in the village. About a year-and-a-half ago, she emigrated to Delhi. Now, the president of the MMD is Shravani Devi, but it is her husband who looks after the finances of the MMD and of the newly-formed mahila VP.

36. Some experimentation with cultivation of turmeric and medicinal plants has begun recently, but it is at a very small scale. Also, like in previous experiments, there is no planning in terms of share in the harvest of these plants.

37. But in these villages too, DTLVS has some local contact persons.

38. Seva Mandir faced this problem in several villages in the course of its work in Rajasthan (see Joshi, 1988; A.S. Mehta, 1996).

39. Which is why Bharati repeatedly makes a distinction between 'truly voluntary organisations' and 'non-profit service organisations'.

40. It is worth noting that the surname 'Bharati' is traditionally a lower-caste surname, which was consciously adopted by Sachchidanand (who actually hails from a Dhondiyal Brahmin family) to indicate his support for the Gandhian goal of removal of caste discrimination.

41. Indeed, many villages, which do not officially have a VP, have some privileges in neighbouring VPs. For instance, many villages adjoining Dumlot and Dandkhil have privileges in the Dumlot–Dandkhil VP, upon payment of some charges to the VP.

7

Sustainable Livelihoods in Riverine Charlands

THE CASE OF GONO CHETONA

Introduction

Community-based natural resource management experiments in Bangladesh have been going on in small pockets largely through the efforts of NGOs. These efforts have gained prominence because of the success of some of the well-known cases of NGO intervention, especially the micro-credit programmes. Non-governmental organisations are also increasingly emerging in the context of state programmes as well as donor-aided programmes. Most of these development programmes that involve NGOs suffer, however, from an absence of a policy framework that guide them, especially in the context of NRM. The Sustainable Environment Management Plan (SEMP) is the first attempt towards the establishment of a conducive policy and legislative and institutional framework in support of improving community participation in, and sustainable management of natural resources aimed at enhancing capacities in government and civil society bodies for mainstreaming environmental concerns. The programme, started in 1998, is a follow-up implementation of the National Environment Management Action Plan (NEMAP) of Bangladesh which was developed between 1991 and 1995 (see Anonymous, 1995a for details on NEMAP).

Sustainable Livelihoods in Riverine Charlands[1] (SLRC) is a programme being implemented in the charlands of the northern reaches of the rivers Jamuna and Brahmaputra in the districts of Gaibanda and Jamalpur. It is being implemented by a local NGO, Gono Chetona (GC). One of the 26 sub-components of SEMP, SLRC is an attempt to demonstrate ways and means to set up sustainable livelihoods in one of the most ecologically vulnerable regions of the subcontinent. The region suffers alternately from the problems of erosion and floods brought on by the rivers Jamuna and Brahmaputra and drought during the dry months. The region is one of the poorest in Bangladesh with a very low land–man ratio (Rahman et al., 2001). The programme aims to specifically support community capacities for sustainable management of environmental resources and enhance access to them.

It is a unique programme in various ways. First of all, the government, donor agencies and civil society organisations have come together to implement the programme. Second, it is the first time in Bangladesh that implementation of participatory environmental management practices on the ground is being attempted simultaneously with streamlining of policies and legislative and institutional mechanisms. This is significant considering the chequered history of Bangladesh with regard to democracy. Third, this programme in the charlands does not exclusively aim at disaster management (unlike other programmes do), but rather focuses on strengthening the coping mechanisms and self-reliance of the poorest section of char society. Fourth, the intervention intends to target the poorest among the population by focussing its attention on forming women's groups from amongst them. The intervention targets households rather than a larger unit because of the socio-economic and ecological context of the chars (described later). Fifth, this is a pilot programme meant to demonstrate 'ideal pathways' to development that can be replicated over a larger scale. To understand the workings of the programme, it is necessary to first describe the biophysical and socio-economic situation of the Bangladesh charlands that form the context of the intervention.

Background: The Bangladesh Charlands

Rivers of Life and Rivers of Sorrow

Bangladesh is located between 20°42′ and 26°38′ north latitude and 88°01′ and 92°42′ east longitude with an area of 147,570 sq km. It has a population of about 128 million, with a very low per capita gross nation product (GNP) of US$ 370. With more than 830 persons per sq km, Bangladesh is one of the most densely populated countries in the world. It is estimated that more than 40 per cent of the population regularly consumes less than the absolute critical minimum of 1,800 kilo calories per day. Poor people are also extremely vulnerable to disaster and disease. Other human development indicators are also low: the adult literacy rate is 37 per cent, life expectancy is 52.2 years, infant mortality rate is 109, maternal mortality rate is 650 and the morbidity rate is 18 per cent for females and 15 per cent for males. Bangladesh has a comparatively poor natural resource base, a high growth rate of population and almost half of the population is below 15 years of age. Most of the poor depend on the natural resource base for their livelihood. Apart from the effects of anthropogenic stress, the low land–man ratio in the country is further threatened by natural hazards (Rahman et al., 2001).

Over 60 per cent of the total land area in Bangladesh is cultivated. Though the country has achieved self-sufficiency in foodgrain production, whether this level of production is sustainable is a big question. Agricultural resources are under severe pressure and environmental strain. Mechanisation of agriculture, emphasis on high yielding varieties (HYV) or high-yielding variety seeds, mono-cropping and increased use of chemical fertilisers and pesticides have caused serious deterioration of soil characteristics and a decline in productivity. There are also competing demands on land for non-agricultural uses (Rahman et al., 2001).

Over 92 per cent of the annual run-off generated in the Ganga–Brahmaputra–Meghna catchment area flows through Bangladesh, although Bangladesh comprises only about 7 per cent of the total catchment. The country is subject to inundation due to the overflow of rivers caused by drainage congestion and rainfall run-off. About 30–35 per cent of the land area is flooded during the monsoon. And this figure goes up to about 65 per cent in years of abnormal floods (like those in 1988 and 1998). Studies on riverbank erosion have revealed that overall erosion is higher than sedimentation along the riverbanks in the Brahmaputra–Ganga river system; a total of 73,552 ha of land eroded between 1973 and 1996 alone, while only 10,628 ha of land was formed. For the poor, loss of crucial land resources affects them economically, socially and psychologically. In particular, loss of land leaves rural people declassed and alienated from the mainstream of society and culture that is based on land as a measure of wealth. Extensive erosion of riverbanks renders thousands of people jobless every year and compels them to leave the affected areas in search of new settlements. Each year over one million people are affected by river-bank erosion. Many of them migrate to cities and are absorbed in the urban informal sector (ibid.).

In Bangladesh, about a million people shift residence regularly (every few years) either because their lands are washed away or to take advantage of newly accreted land (Wiest, 1988, quoted in Indra and Buchignani, 1997). While a person (households are largely headed by males) might have lived in a number of chars, he still has his old address in mind and will inform people about where his original village was. We did not meet any adult who had never moved, except in the more stable chars. Until the char reappears again, which may take 10–30 years (or sometimes even more), a landowner lives the life of a landless person.

There does not seem to be any great ethnic/caste divide between the people living in the charlands. Unlike in India (and Nepal to some extent), class does not tend to correlate with and express itself through caste and ethnicity. But there is a very powerful elite in Bangladesh society which corners all the benefits and which has ownership over most of the natural resources. The rural elite is partly related to those families who were earlier feudal or semi-feudal landlords (Zaman, 1991). As will be apparent in what follows, a substantial portion of their property is illegal acquisitions and maintaining and expanding them involves creating patronage/loyalty exchanges right down to the village level. Kinship, clan and loyalty play a much greater part here than ethnicity and caste (Baqee, 1998; Zaman, 1991).

The rural elites are so powerful that they illegally encroach upon common lands that otherwise are under the control of the government. For instance, while existing laws stipulate redistribution of newly-raised chars, which are categorised as *khas* lands among the landless (see below), large areas of these newly-formed lands may be encroached upon by the landed elites. There are enormous possibilities of leasing out these lands to the poor/landless. Though the poor are aware of the government policy of rehabilitation of the landless in the charlands, they cannot do much in the violence-ridden and *matbar-* (literally, the powerful/dominant—the local Bangla term for the elite) dominated chars. Bureaucratic red-tapism, insensitivity and complex legal procedures further discourage the poor and the illiterate to look for this option (Baqee, 1998: 54–69).

The Shikosti-Payasti Law for the Charlands

According to the Shikosti-Payasti Law (or *nadi-sikasti* law) there are three kinds of land—the land that rises from the river is called *payasti* land, that eroded by the river (and which gets submerged) is known as *shikosti* land and shikosti-payasti land is that which was eroded by the river and has again come up at some later period. The Shikosti-Payasti Law has seen many changes since British rule, but after the latest amendment made in 1994, the following are its salient features:

(*a*)   If privately-owned land becomes shikosti and within 30 years turns into payasti land, then the earlier owner becomes the holder of that land again.

(*b*)   But the person's landholding should not be more than 60 bighas.

(*c*)   The government will become the owner of land that becomes payasti after more than 30 years.

(*d*) If new land rises up from the river or sea bed, the government becomes the owner of that land (Montu, 2003: 15–17).

Thus, the government has under its control, at least theoretically, a lot of such charland. But in practice, these charlands are encroached upon and grabbed by the rich and powerful, who either cultivate it themselves or lease it out to others. The land ceiling (60 bighas[2]) law is also followed more in breach by the rich. A number of local people and NGO activists informed us about how the rich and powerful have cornered these lands. Other studies (Baqee, 1998; Zaman, 1991) have shown how the matbars in the char villages on the right bank of the Jamuna maintain *lathiyals* (private army of clubmen) to gain control over newly-formed chars. After gaining control over these chars, they maintain an exploitative patron–client relationship with the landless by leasing out lands for homesteads or cultivation. Legitimacy over these lands is easily bought by bribing the officials of the local administration and revenue department. At times the poor cannot even stake a claim to their own land that reappears after some time. Such is the hold of the matbars in these remote areas.

People without a Fixed Address

Where do the million-strong environmentally-forced migrants go? The choice of a new settlement is influenced by a number of factors, the most important being kinship ties or old acquaintances who extend help during periods of emergency. Landowners prefer to settle somewhere within the same *thana* (police station) in the hope that they can go back to their own land whenever it reappears. Even the already landless people tend to resettle near their *shamaj*[3] (close to their past employers and patrons) in the hope of social and economic benefits accruing from this relationship.

The process of settling a newly-formed char has been well documented by some scholars. Baqee[4] (1998) has shown how the matbars take control over a new char with the tacit help of the local administration and revenue officials and then settle their own followers in these lands. The first people to be settled are those the matbar considers close and loyal, mostly members of his own *gusthi* (patrilineage[5]), who support him during the violent struggle between different matbars to gain control over the chars. Later, once the news of a new char being settled spreads, people from far away also start coming in, but it is the local matbar who screens the settlement process and gains from leasing out the land. The revenue officials and the local administration turn a blind eye to the illegal settlement process; the settlement of the chars happens a long time after the actual settlement has been made by the matbars. Once the matbar gains full control over the char, he starts getting his share of the profits from the land. The people settled in the char are all at the mercy of the matbar and face various forms of exploitation, including sexual exploitation of women. In chars where more than one matbar holds sway, violent clashes are a frequent occurrence and poor people are sucked into the cycle of violence (Baqee, 1998).

It is common practice that people give out a portion of their land to relatives for homestead purposes. The support system helps the people in the region get some foothold locally in a place so that their families have somewhere to stay while the men migrate to urban and agriculturally-buoyant regions in search of employment. The local term for poor, environmentally-forced migrants living in someone else's land is *uthuli*. Among the locals there is sensitivity towards people who have been rendered homeless by the action of the river. People extend help to others because they believe that they can also become homeless in the future. There is a very complex network-like support system that comes in handy during emergencies. The moral tie signifying closeness and mutual obligation is in most cases because of kinship relations. The kinship relations of the wife[6] also come in handy during times of emergency (Indra and Buchignani, 1997). But uthuli status is also 'associated with marginality, impermanence, shame, disempowerment, wildness, impropriety and lack of voice…but it is widely recognised that often there are no better alternatives' (ibid.: 38–40).

In a number of cases, land is leased out (if available, which in most cases is very improbable) to those who can afford a rent of Taka 5,000–10,000 per bigha. However, there are many people who have no support system or who have exhausted their patriarchal entitlements[7] because of continuous social disruption caused by incessant erosion.

Moreover, their own kin are often displaced and hence have nowhere to go but squat on government/public land like roadsides or the top of embankments.

Community in the Chars[8]

There are two kinds of opposing forces in char villages—one that binds them together because everybody shares vulnerability and the other that is disuniting (or which prevents a more long lasting community to come up). The uniting factor is the casteless Muslim society and a 'shared understanding' of the temporariness of life and property in the region.

The disuniting factor exists because in most chars, especially the newly-formed ones, the villages are usually inhabited by members of 'genealogically unrelated gusthis' many of them coming from different local villages where they lost their homesteads. This is especially true for chars where the settlement was done by more than one matbar and the char is inhabited by members of different gusthis. In these chars, clashes between the different gusthis further lead to weakening of community feeling in the village. At any given point of time, a village might comprise people with no 'fixed addresses' and thus the village has a weak corporate character. The gusthi, otherwise strongly corporate and cohesive, has a very limited role in terms of mutual support due to fragmentation caused by displacement and inter-village migration. But there are also some very old villages in more stable chars and some new ones where a single matbar dominates and hence where these disuniting factors are largely absent. So in these villages the boundaries of a community are much narrower than in more stable and settled villages. This places limits on what an NGO or any other agency can do with respect to a community-based approach. Char villages are also widely dispersed in comparison with the nucleated settlement pattern in the mainland.

Another factor that makes community formation a difficult proposition in these villages is that in most of the households (invariably amongst the poor and landless), the household heads live outside the village for long periods of time. Because of the remoteness of the area and the vulnerability of these chars to erosion and floods, employment opportunities are very few. Most of the people we talked to migrate to far-off places for wage employment for three to nine months a year.[9]

There is one more disuniting factor in these chars—class differentiation. As has been mentioned earlier, landownership in Bangladeshi rural society is highly skewed with about 10 per cent of people owning the bulk of the land. These powerful elites also capture and establish their domain over most of the new depositional land (that should actually come under government khas land). These lands could otherwise be redistributed among the landless. Illegal land-grabbing and violence for control of new charlands is common. The grip of some of the erstwhile local zamindars, *taluqdars* and *jotedars*[10] over char villages is still maintained with the help of a network of political agents and their dependent lathiyals. Though the total landholdings of these elites have decreased considerably due to subdivision of land, a number of them have maintained their prominent position in rural Bangladesh by wielding muscle power and political influence. These landed elites are a law unto themselves as they dominate politics, both at the village and state levels. The fruits of the state rural development schemes and flood relief go to the matbars and their henchmen (Baqee, 1998; Zaman, 1991). The economic differentiation within a particular gusthi, for instance between the matbar (and his henchmen) and the others, is very wide. One should note that these landed/political elites do not belong to any specific caste or community group, unlike in India.

Governing the Chars—Political Wilderness and Misgovernance

These factors, namely, weak community links, endemic poverty and sharp economic differentiation in the chars prevent democratic institutions from finding deep roots in the region. The presence of the state in rural society, especially in the charlands, is minimal and government officials are perceived to be hand in glove with the elites. Community formation, therefore, around NRM is a difficult and time-consuming task in the area. Before we go on to discuss GC's intervention and its methods and means of community formation, we briefly discuss the issues of political decentralisation in Bangladesh and the virtual absence of the state in the chars.

Bangladesh became a separate nation in 1971. After Independence it has witnessed long periods of military rule. Democratic institutions of governance have not found roots in rural areas, especially in the remote charlands. Various experiments have been tried in terms of decentralised governance, but failed because of a lack of political desire and an inadequate amount of power and resources given to local institutions. Bureaucrats manage all the affairs and are all powerful at the district, sub-district and union level. People's representatives at the union level have been given very few powers and even fewer resources (Ahmed, 1997; Karim, 1999; Rahman, 2000; Saqui, 2001; Saqui and Ahmed, 2000).

Most of the union *parishad*[11] (UP) members that we met appeared clueless and powerless at the state of affairs in the area. They too look up to NGO staff for any help. They can only submit a proposal to the UP, but the authority to ratify it is in the hands of the bureaucrats. Even within the UP, all the powers are in the hands of the chairman of the parishad. In short, local people's representatives do not command the respect (from the government officers) or resources to do anything worthwhile for their area (see Box 7.1). With donor agencies also preferring to work through NGOs and user groups, local democratic institutions are being further pushed away from developmental work. Local governance in the chars is worse because of their remoteness and inaccessibility and high levels of poverty.

Box 7.1
Powerless Union Members

Ajibur, a union member from the char village of Algar, said that he is powerless; he cannot do anything for the people he represents. He has several times given proposals (most of the proposals are for roads, culverts, schools), but none of them are passed. The union chairman and bureaucrats decide which project to take on. He does not even have a telephone and his salary is a mere Taka 300. He has to spend his own money even to travel to the union office.

Similarly, Robina Begum is a nominated union member from Kolkihara Bhati village. She also believes that the union parishad as a body, and the members as individuals, are powerless to do anything for the people. She believes that the government does not have the resources to invest in public infrastructure. She herself wants to be helped by the NGOs as she does not expect any help from the state agencies.

Moreover, even the UPs are dominated by the local rural elites. Though in theory development programmes in Bangladesh villages are executed by the UPs, this process favours the interest groups rather than the people at large. There is hardly any people's participation in developmental programmes. The local administration too works hand in glove with the local elites, thus completely ignoring the claims of the poor (Baqee, 1998).

Given the fact that the UPs are likely to be dominated by the matbars and given the high level of corruption mentioned in other studies as well (ibid.), it is likely that development programmes too largely by-pass the poor. Civil society organisations (mostly NGOs) have tried to empower local communities and UPs through social mobilisation,[12] but in the absence of meaningful political decentralisation these small experiments cannot make a big difference. One needs to factor in these developments into one's analysis of beneficiary participation in any rural development programme and how participation in a project can help decentralised governance of natural resources.

The Prominence of NGOs

Another very important factor (highlighted at the outset of the chapter) in Bangladesh, related to the virtual absence of the developmental state in the rural areas, is the prominence of NGOs in developmental work, especially in micro-credit, NRM and disaster management and distribution of relief. Non-governmental organisations first emerged in the country after the violent war of liberation in 1971 to provide relief and rehabilitation to victims of violence, floods, cyclones and famines. During the 1970s, the NGOs were providing services like education, health and sanitation, family planning, etc. There were also several local movements for self-reliance in agricultural produc-tion, culminating in the formation of Swanirwar Bangladesh (SB). Swanirwar Bangladesh played an effective role in mobilising youth in rural areas and small towns to reduce pilferage in the delivery of state relief and development services. During this period, many inter-national NGOs—Terre des Hommes, ActionAid, Canadian University Students' Organisation—were also active. Pilot experiments in the provision of micro-credit to small groups were implemented towards the end of the 1970s. The success of the Grameen Bank in the 1980s resulted in the emergence of a number of institutions, often with indigenous effort, in the late 1980s and early 1990s, even as many of the first generation NGOs continued to engage in the delivery of social services (Zohir, 2004: 4109–10).

There are broadly two kinds of NGOs operating in Bangladesh—those committed to resource delivery and those who promote consciousness raising and literacy training as rallying points for social change. Organisations like Gono Shahajjo Shangstha (GSS) were founded with the purpose of organising conscienticisation groups as catalysts in assisting landless and resource-poor people to fight for their rights. It had very explicit political goals, including efforts to create a democratic space for debate and a means by which to challenge the status quo. This was a radical departure from initiatives that merely sought to enhance the efficient distribution of development resources. The GSS was also against the provision of credit to its members as they argued that credit programmes merely replaced public sector resource distribution with a new form of economic dependence. Unfortunately, GSS has moved away from confrontation with the local power structure after attacks on its offices and members (Feldman, 1997). So while donor assistance has turned most private voluntary organisations into semi-professional service delivery development agencies, the local social and political situation has also discouraged NGOs from taking up confrontational politics. The result is an overemphasis on project outcomes rather than on the process even while the larger issues of socio-economic inequities, political devolution and corruption remain unanswered (ibid.).

The service delivery NGOs are strong and active in Bangladesh; there are some older NGOs that have branches in almost all the districts and thanas[13] of the country. Apart from implementing government programmes, these NGOs implement donor-funded programmes too.[14] Some of the NGOs (for instance, Bangladesh Rural Advancement Committee [BRAC] and Proshika), which have become household names because of the micro-credit movement they launched, have a lot of programmes that are self-sustained. Moreover, the general feeling is that NGOs have ensured that government or donor money reaches the poor and is not siphoned off by elites and bureaucrats. However, criticisms have also been levelled. For example, with regard to micro-credit, some argue that it creates new relations of dependence in the form of privatised credit (Feldman, 1997: 62).[15] This argument is taken even further, namely, that donor support to NGO initiatives is nothing but 'privatization of resource distribution and forms of production and away from locally initiated and locally controlled development activities' (ibid.: 59). What is clear is that in Bangladesh the nature of NGO intervention has largely become 'depoliticised', especially since the 1980s, an argument that we have laid out in detail in Chapter 1. This historical context forms the backdrop to GC's intervention in the charlands.

The Study Area: Socio-economic Conditions in Erendabari and Merur Char Unions[16]

The SLRC programme is being implemented in 19 villages of Merur Char Union (Jamalpur district) and Erendabari Union (Gaibanda district) in northern Bangladesh (see Figure 7.1). Life in the charlands is dominated by the rivers flowing through the area. The area is intersected by several rivers, the most prominent being the Jamuna (the main stem of the Brahmaputra) and Old Brahmaputra. Various other rivers flow in between these two rivers. Ginjiram, which has its source in the hills of Meghalaya (India), and Dashani are the other two rivers flowing in the project area. Erendabari Union and Merur Char Union are tucked in between the Jamuna and the Old Brahmaputra and the landforms constantly change shape and size depending on the mood swings of these rivers. Both the unions are exclusively composed of chars formed by the erosion and sedimentation of the rivers. Merur Char Union is composed of chars of the Old Brahmaputra river and Ginjiram river. The Old Brahmaputra river flows along the west and Ginjiram river flows along the eastern part of the union. All these rivers connect with each other during the monsoon through several drainage channels. But there are some discernible variations in the two unions—Erendabari Union is much more vulnerable to the swift action of the Jamuna and the villages in this union are highly erosion

FIGURE 7.1
Location of the Jamuna–Brahmaputra Charlands

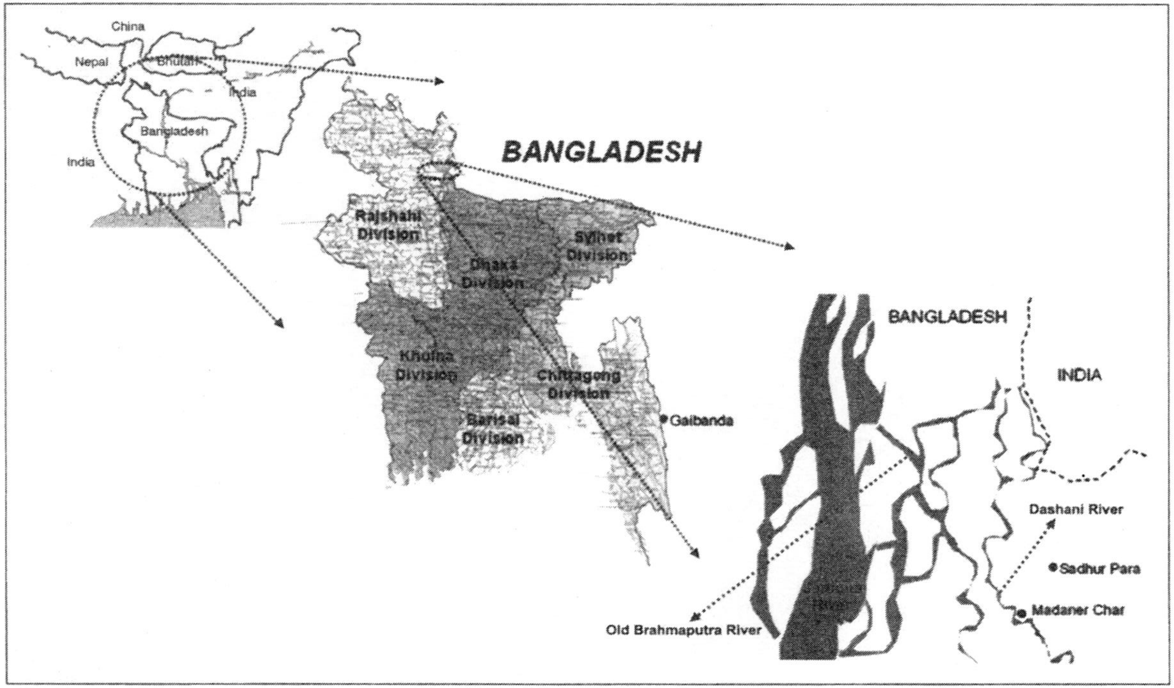

prone. Villages in Merur Char are relatively stable and the soil too is better suited to agriculture.

Erendabari Union is a typical riverine charland of the Jamuna River Basin—situated on the eastern side of the Jamuna. As the Jamuna forms the western boundary of the union, it is separated from the rest of the administrative district of Gaibanda to which it belongs and is better connected with Jamalpur district in the east. The union comprises 13 villages. The land in the union is formed of recent alluvial soil, mostly sand and sandy loam. The chars on the main Jamuna are sandy and open. The northern boundary of the union, particularly Harichandi village, is relatively more stable. However, even in Harichandi, the sandy land mass on its western edge is not suitable for the cultivation of paddy. The land mass is quite undulating in the villages of Char Chowmohan and Anandabari and completely composed of sands that look like sand dune formations. These sandy chars usually remain barren even during the dry months. However, groundnut can be cultivated and in some areas *kanshban* (sun) grass grows. In these chars (Char Chowmohan and Anandabari) tree species are very few. Erendabari and Harichandi village are stable and are well wooded. In some places, there are dead rivers while others retain water ˙ throughout the year.

Merur Char Union comprises the charlands of the Old Brahmaputra river course and is situated in the south-eastern part of Bokshiganj Upazila in Jamalpur district. The union is about 5 km west of the Bokshiganj Upazila headquarters and has 19 villages in its jurisdiction. The terrain of the union is floodplain like. However, the areas influenced by rivers look different; they are typical riverine chars. Some villages (Airmari and Chiner Char) are now stable, but the presence of dead rivers, baor-like water bodies and drainage channels by the sides of old villages indicate that the landscape has originated from the Old Brahmaputra. The Ginjiram has a distributary in the union named Dashani. During the dry months, the Old Brahmaputra does not have any link with the Ginjiram. But during the monsoon, the Gingiram gets connected with the Old Brahmaputra at many points and during abnormal floods the entire union is criss-crossed by water bodies. Ginjiram dries out during the dry months and the Old Brahmaputra narrows down considerably. Except for the river-influenced areas, the soil is loamy. In river-influenced areas prone to erosion, the soil is sandy and composed of fine sand. Paddy and sugarcane can be grown even during the monsoon except in sandy river influenced areas. The stable villages in this union are well wooded, but in villages under the threat of erosion by the Brahmaputra, tree cover and other vegetation is sparse.

In the charlands, mainly two cropping seasons are important—*boro* and *aman*. The boro season starts in January/February and the crop is harvested in April/May. The aman season starts in July/August and crops are harvested in November/December. Boro paddy and a transplanted variety of aman paddy are the most common in the charlands. In Erendabari Union villages, the latter is most prominent. Merur Char Union has both boro and transplanted aman paddy, the former introduced recently. Sugarcane, jute, mustard and lentils (*masoor*) are also grown in Merur Char. In Erendabari Union villages, which are prone to erosion and where the soil is sandy or sandy loam, the other important crops are foxtail millet, *china* (another coarse millet), groundnut, *til* (sesame), *tissi* (oil seed), onion, chillies, jute and transplanted aman paddy. In Erendabari Union, the vegetation is also less than in Merur Char.

During our field visit to the project area we visited 12 villages. We decided to visit as many chars as possible rather than focusing on a single char or a few villages because of various reasons. First, there was diversity in the nature of the chars and we thought it necessary to understand the complexity of the charland context (which was completely different from what we had encountered anywhere in India). This could be done only by visiting different chars. We were also under the impression in the beginning that the NGO had a diverse approach and had made some unique intervention in almost each village which needed to be brought out.[17] Second, language was a big problem that we encountered while doing our fieldwork despite having an interpreter (at times interpreting the interpreter's language and making him understand our questions was a big struggle). Under the circumstances, doing detailed individual interviews and getting nuanced responses was not possible. Hence we decided to stick to a fixed set of questions based on our broad thematic concerns: livelihood enhancement, sustainability, equity and democratic decentralisation.

Third, our repeated attempts to get village information from government offices failed and to do a detailed one/two village study seemed out of the question.

The Programme Villages

The SLRC programme is being implemented in 18 villages—six villages in Merur Char Union, two villages in Sadhurpara Union, one village in Bagar Char Union (Bakshiganj Upazila of Jamalpur district) and nine villages in Erendabari Union (Fulchari Upazila of Gaibanda district). These villages are detached from their upazila headquarters by the rivers Jamuna and the Old Brahmaputra, respectively, and are more than 15 km away from the headquarters.

The landholding pattern in the project area is given in Table 7.1. About 35–40 per cent of the people inhabiting these chars are landless and were always so. They are the worst affected by the difficult environmental conditions. About 30 per cent of the people have homestead land where they also maintain a small vegetable garden. Approximately 20 per cent of the people have less than three acres of land and about 10 per cent have more than 10 acres of land. Though the landholding pattern is highly skewed in the region, there is not much difference in the level of vulnerability to river erosion (see Box 7.2). Once a char goes under water there is no difference between a landowner (especially the small and marginal farmers) and the landless.

Table 7.1 clearly shows that about half (45.4 per cent) the households surveyed are landless or have only homestead land (households with less than 0.05 acres of land have only homestead land). About two-thirds (66 per cent) are functionally landless as their landholding is less than 0.5 acres. Only about 10 per cent of the respondents are favourably placed in terms of land ownership.

Even landholders, as mentioned above, are vulnerable to the adverse environmental conditions in the chars (see Box 7.2). More than half of the total land owned by the surveyed households are under rivers and only about one-third is under cultivation. About 35 per cent of the total land is estimated to be cultivable land for agricultural and horticultural plants; about 15 per cent of the land is occupied by the dwellings and the remaining portion of the total area is either very recently

TABLE 7.1
Landholding Pattern in the SLRC Project Area

Land Owned by Households (in acres)	Erendabari Union		Merur Char Union		Total Project Area	
	No.	%	No.	%	No.	%
<0.05	693	46.4	639	44.4	1332	45.4
0.05–0.49	290	19.4	312	21.7	602	20.5
0.5–0.99	125	8.4	130	9	255	8.7
1–2.49	235	15.7	214	14.9	449	15.3
2.5–4.99	109	7.3	96	6.7	205	7
>5	43	2.9	48	3.3	91	3.1

Source: Anonymous, 2000.
Note: The baseline survey was carried out in 23 villages in the two unions of Merur
        Char and Erendabari. A total of 2,934 households (40 per cent of the total
        households in the two unions) were surveyed.

Box 7.2
Even the Landed are Vulnerable

In Chaumohan Char, one of the villages that we visited in Erendabari Union,
the people were ready to move to a new place as they feared that the entire char
might go completely under water within a month. There were about 1,000 households earlier in this char, but people started shifting gradually in small groups
over the last 10–12 years when the river started eroding the char. Now there are
only about 145 households left. Hussain, whose father owned about 500 bighas of
land (his share is 125 bighas), informed us that he also does not have any savings.
All his lands were gradually eroded by the river Jamuna and he kept shifting
his tin house to a safer place in the same char. Now he does not have any option
but to shift to a new char. But he has not thought about where to move. Similarly,
Alam, who once had 100 bighas of land, is landless today and has nowhere to go.
One of the GC staff who accompanied us to this village, and who belongs to the
same area, offered the people land in his village/char for setting up their home.
He also told them that they should inform GC in advance when they decided to
move so that GC could send the organisation's motorised boat for shifting the
people. The people were counting the days to when they had to move. The only
thing that was forcing them to delay the inevitable was the pain of leaving their
village.

---

formed charland (which may not yet be fit for cultivation) or the beds
of small rivers and drainage channels (Anonymous, 2003).

Table 7.2 depicts the high incidence of poverty in the project area.
About 65 per cent of the male household heads are engaged in day

TABLE 7.2

Occupational Structure in the SLRC Project Area

Occupation		Erendabari Union		Merur Char Union		Total Project Area	
		No.	%	No.	%	No.	%
Husband	Day labour	1,059	70.8	856	60	958	65.4
	Sharecropping	644	43.1	635	44.2	640	43.6
	Farmer	412	27.6	620	43.1	516	35.3
	Business	149	10	181	12.6	165	11.3
	Labour	14	0.9	30	2.1	44	1.5
	Traditional job	6	0.4	8	0.6	14	0.5
Wife	Day labour	25	1.7	82	5.7	54	3.7
	Business	49	3.3	64	4.4	56	3.8

Source: Anonymous, 2000.

labour, while about 47 per cent are engaged in sharecropping. Among women, only a small per cent (7.5) are engaged in activities other than household work. *This fact is significant from the point of view of the intervention, namely, that SLRC has focused on providing gainful income-generating activities for women in the project area.*

Other development indicators for this region are poor too. Boats are the only means of transport during the rainy season for the people living in these villages. Infrastructure services in these villages are very poor. Very few metalled roads and bridges exist and people have to either walk or use bicycles to travel during the dry season. Very few villages are electrified. The literacy rate is very poor (7–25 per cent for men and 1–15 per cent for women). Of the 12 villages we visited, only Merur Char has a high school and a health centre; it also happens to be the union headquarters. The rest of the villages only have a primary school (see Table 7.3).

Gono Chetona and NGO Interventions in the Chars[18]

Gono Chetona (GC) was formed in 1993. It grew out of Terre Des Hommes, France (TDH) which after working in the northern region of Bangladesh withdrew from the country in 1988.[19] Ira Rahman, the present executive director of GC like many other staff of the organisation, had worked with TDH. Ira Rahman herself worked as a social worker in the organisation and was responsible for running two

TABLE 7.3
Basic Information on Villages Visited in the SLRC Project Area

Name of the Village	Union	Total Population (1991)	Total Area of the Village (acres) in 1991	Educational and Health Care Facilities in 1991	Total Households (1991)	Total Beneficiary Households
Talpatti, Bhatiapara	Erendabari	1,019	593	P (3)	152	63
Algar Char	Erendabari	3,297	2,028	P (3)	507	48
Anandabari	Erendabari	744	675	P (1)	106	21
Char Chaumohan	Erendabari	2,246	1,354	P (1)	349	10
Char Harichandi (Fakir Para)	Erendabari	1,114	465	P (1)	155	45
Tinthopa	Erendabari	581	1,703	P (3)	86	20
Madarer Char	Merur Char	1,115	NA	P (1)	225	75
Airmary	Merur Char	1,357	NA	P (1)	261	64
Chiner Char	Merur Char	503	NA	P (1)	81	50
Kolkihara Bhati	Merur Char	2,352	NA	P (2)	418	85
Merur Char	Merur Char	2,226	NA	P (2), H (1), HC (1)	446	59
Baghaduba	Merur Char	429	NA	P (1)	77	32

Note: P = primary school, H = high school or madarsa school, HC = health centre.

orphanages; she also helped in the formation of women's groups. Prior to working with TDH, which she joined in 1978, Ira had also worked with BRAC for two years—1976–78. This was the period when a lot of educated people who wanted to bring about change in rural Bangladesh entered the voluntary sector. She and most other staff of GC have a long experience of working in the Brahmaputra–Jamuna charlands.

Prior to the SEMP, GC was engaged mainly with forming women's groups and working towards women's empowerment in the chars. Considering the very low literacy rate, especially among women, in the chars, the underprivileged status of women in households and the control over the free movement of women (because of the system of purdah), GC has sought to work with women. It has chosen to organise women by providing them with some livelihood support as it strongly believes that economic empowerment will lead to a general uplift of women. Gono Chetona has tried to involve the resource-poor households in starting new ventures like seed banks and in some cases has also helped them financially, for instance, in buying water pumps. The idea was to mobilise people, especially women, by giving them incentives. Some of the women's groups created by GC in the past have become strong enough to formulate their own plan of action, which GC passes on to donor agencies for support. From the nurseries that they run, some groups have provided GC with seeds to distribute among the poor free of cost. While working towards empowering women, GC takes care not to create a rift within the households and encourages a dialogue within the family.

Another area in which GC has been working (again through the women's groups) is disaster management and preparedness by creating awareness among the poor about how best to prepare oneself for emergencies like floods. It has also installed bore wells on raised platforms so that the people have some source of safe drinking water during the flood season. Gono Chetona has been very active in providing relief to victims of floods and erosion. It has in place rescue boats and flood protection centres and helps government agencies in distributing relief materials during disasters. In fact, most of its staff are still very 'relief minded' even though the focus of GC's work has now changed towards sustainable NRM.

The Intervention—Sustainable Livelihoods in Riverine Charlands

Considering the vulnerability of the chars and the past interventions that focused on disaster management and relief efforts, SLRC was a fresh programme where some effort was made to provide sustainable livelihoods to the people of the charlands. As has been mentioned earlier, the focus of the intervention is in alleviating poverty through initiating sustainable ecosystem management.

In the 18 programme villages there are 1,632 direct and 1,655 indirect programme beneficiary households.[20] The organisation has a policy that only the poor households can be beneficiaries and among them single women, widows, divorced women and handicapped women are given preference. Gono Chetona works mainly with women as it is the women who stay back in the villages while the men are away for most of the year (Anonymous, 2002).

The important components of the intervention may be identified as follows:

(a) Forming women's groups, initially for credit and savings and later involvement in the larger programme.

(b) Promoting sustainable livelihood options in the chars through a more productive use of homestead lands for vegetable gardens, or promoting non-land based livelihoods like animal husbandry and poultry farming, or introducing new crops and agricultural practices along with sanitation and health activities.

(c) Sustainable environment management such as planting a variety of trees for the greening of chars as well as economic benefits from increased vegetation (fruits, fodder and fuelwood) and prevention of erosion and alternative agricultural practices to reduce use of chemical fertilisers, etc.

(d) Flood management and disaster relief activities.

Each of these different components will be taken up for discussion in the following sections.

Formation of Women's Groups

As stated earlier, GC works through the women members of each beneficiary household, both for practical and ideological reasons. It has succeeded to a large extent in increasing the confidence levels of women and in improving the general health and well-being of women. A lot of women who we talked to categorically stated that the income that they earn through these new activities goes to them and they spend it for whatever they want. But more than the increased income, it is the feeling of empowerment that comes through very noticeably in these women. Earlier they had to scrounge for food or seek help from neighbours and relatives—something that increased their vulnerability to exploitation of various kinds. Now at least some of their basic subsistence needs are being met.

Some women are part of specific committees like the Environmental Protection Committee and the Disaster Preparedness Committee, but it is not clear what the duties and responsibilities of these committees are. Group meetings are used to inform members about the activities and to raise awareness about environmental issues. Some of the more active members of these groups are selected for training in running nurseries and for participation in integrated pest management (IPM) and poultry and cattle vaccination programmes. These women in turn train other members in the group meetings and also get some extra income from the new skills that they learn. While promoting the more active members in the villages to take up the leadership of these groups ostensibly seems to be a positive step, it might lead to unequal sharing of benefits in the absence of a close monitoring of the working of these groups. While it is easier for an NGO to promote leadership in the village from among the more active women, in this case the result has been that leaders or active members have cornered benefits for themselves and have later ignored the activities of the groups (see Box 7.4). Also, in terms of internal democracy within these groups, it appears that the decision-making is often restricted to the more prominent members and leaders and often the NGO staff sets the agenda for the group. It might be that in a time-bound project like SLRC, the implementing agency does not have enough time to identify and promote good leadership and monitor the working of groups and institutions to make them more democratic.

The women's groups formed in the villages also double up as savings groups. In each village there is more than one such group. They meet every fortnight, contribute a certain amount to the savings account and discuss other important matters. In our meetings with several such women's groups it was apparent that many of the members are not well informed about group action or managing accounts. One or a few members in these groups appeared to be the leaders—they were confident, forthcoming and aware of whatever was happening (see Box 7.3). In some groups, on the other hand, we had very lively discussions where almost all the women participated and put forward their opinions. So in terms of the performance of women's groups, the results have been mixed. Nonetheless, from many of the discussions we had the increase in confidence and self belief among the women was evident. This bears out the NGO's contention that their presence has helped these people find some hope and optimism. To a large extent, however, this also needs to be seen in the background of the complete absence of the state in rural Bangladesh, especially in the char villages.

Box 7.3
New Leadership Emerging

Mosammat Khairunisa is a young teenage girl in Baghaduba village. She is the leader of the group named Kariphool. She completed high school in a local madrasa. She told us that she was approached by GC and was convinced of the good work that GC proposed. But she found it very difficult to get a positive response from other women. She said that it took a lot of time and effort for her to convince the women to come together and form a group and take up the project activities. Mosammat is the only woman, not in her group alone but in the entire village, who appears to be aware and confident.

None of the groups have started giving loans to their members, though some of them have savings in excess of Taka 20,000. The logic is that the savings should first increase and then loans can be distributed. Gono Chetona, moreover, is not in favour of self-help groups giving out loans to members at this stage. This is partly because it feels that giving beneficiaries real benefits in terms of livelihood generation is a better way to help them and enlist their participation. This policy may also be linked to the fact that village (or group) membership is in a flux because of the unstable river regime. However, GC has been disbursing loans to some selected beneficiaries from the programme money and has 100 per cent success in loan recovery. These loans are given for very specific needs, like buying cattle, starting a new enterprise such as a nursery, etc. In some instances, we also found that prominent members of a group got more benefits from the programme than others allegedly because of their closeness to the NGO staff despite the fact that there were other more needy women (see Box 7.4).

While a number of women's groups have performed well, others have ceased to function (they no longer contribute money regularly or meet regularly) because they have not gained enough expertise or experience in handling transactions. They are dependent on GC staff to do such work for them. However, with GC scaling down its programme (it is now working with a skeletal staff) the future of these women's groups is in question. Moreover, though the activities introduced in the programme have helped the women in increasing their income in a small way, changing their socio-economic condition substantially will require structural changes in rural society. This can happen only through state support or conscienticising and mobilising these groups to fight for their rights. The latter does not seem to be the aim of either SLRC or the NGO programme.

Despite weaknesses in some of these groups, GC is going ahead and is in the process of forming a federation of these women's groups. Two central committees have already been formed in Bakshiganj and Fulchari sub-districts. The central committee has 21 members who

---

Box 7.4
Group Leader Corners the Benefit

Paroma, of village Merur Char, is the leader of a women's group named Protigya. She is very confident and knowledgeable and gave us all the information about the village. She did not appear to be very affected by the presence of men and spoke her mind. But all these qualities were not enough for her to keep her group going. Her group had stopped functioning for more than one year as the GC volunteer stopped coming to their village. However, she herself appears to have benefited from the programme. She got help from GC in setting up a nursery and tree plantation although she is from one of the better-off households in the village. Her husband works as a school teacher in a private school in the village.

are selected by leaders of 47 groups. These 21 members have selected nine members for the executive committee. The idea is to set up a common body of these women's groups which can work independently of any NGO and any other agency. The main role of this body will be advocacy. But this experiment has not shown very encouraging results as a lot of groundwork is still required to bring women under a single umbrella.[21]

Another factor which puts question marks on the sustainability of these new institutions is that they remain cut off from government officials and departments. Instead of the officers dealing directly with these groups, they prefer to work through the staff of GC. While GC maintains good working relations with government agencies, it is doubtful whether the women's groups have developed the confidence to deal directly with officials and also whether government officials have an open and healthy attitude towards the poor and inaccessible villagers. One also doubts the ability of these groups to engage in any way with the social forces at work in these villages, namely, the landed and political elites. On a more positive note, groupings like this, if nurtured and strengthened over time across the region can campaign for changes to be made at the policy level.

Promoting Sustainable Livelihoods in the Chars

The NGO has introduced a number of activities in the villages. The interventions have been aimed at introducing income-generating activities that landless women, who have no gainful employment in the villages, can take up. Moreover, the intention was to improve the living conditions of the poor who have very little control over the use of scarce natural resources in the chars.

Livelihood Activities

The most important and widely practised livelihood intervention has been the promotion of vegetable gardens in the homestead lands, animal husbandry and poultry farming. What is noticeable in these Bangladeshi villages is that almost all the households, including the landless, have some homestead land near their homes. There is, in fact, not much difference in the extent of homestead land possessed by small and big farmers. This land is used for many purposes, including as a vegetable garden. The NGO trained women selected from each group in horticultural and farming practices and also provided better and high-yielding varieties of vegetable seeds and tree saplings. Some poor women were also provided training in nursery management and given initial financial help in establishing nurseries. Tree saplings were distributed among beneficiaries and these have shown good survival rates. Most of the households have at least some fruit trees (jackfruits, banana and guava). This has given the women a source of income as they sell the vegetables, fruits, seeds, saplings and poultry and dairy products in the village and in the nearby marketplaces.

Women are now growing a variety of vegetables throughout the year which was not the case earlier.[22] With the increase in animal husbandry, the women have also been given training in making and using organic fertilisers. But the focus of all this is production for the market. All the women said that their income has increased many times from the sale of vegetables and fruits. The only limiting factor for them is scarcity of land. Most of these women have less than one bigha of homestead land where they grow vegetables and plant trees. But the other major problem, which should be resolved, is that of marketing. As it is a remote area, people find it difficult to access markets both because of poor communication links and low prices.

A number of beneficiary women were given loans to buy cows and goats. Rearing cattle for a year and selling them to butchers is a common practice in these parts. A few people also rear cows for milk, which they sell mostly in the village. Dairying activity was very rare in these parts and whoever reared cattle did so mainly to sell them to butchers. Otherwise, oxen are also kept by some farmers as draught animals in agricultural operations. The landless do not rear oxen as draught animals as they are not sure if they will get land for sharecropping every year. Rearing cows and buffaloes for milk is not a very profitable venture as marketing is a big problem. While milk is sold at Taka 10–12 in these villages, in the towns and cities it is sold at Taka 16–20. There are no milk cooperatives in the area. Rearing cattle for meat is a much more profitable venture. If a cow is bought for Taka 5,000, it can be fed and fattened in a year and sold at double the amount.[23] Since the dairy business is not that significant, there has not been much effort to introduce improved varieties of cows in this area. However, this is changing of late with GC giving loans for the dairy business in particular. There has been 100 per cent loan recovery by GC on the loans that were disbursed to the people. Goats and sheep are also reared in this region and fetch a good price in the market.

Cattle diseases in these parts are rampant and people lose their cattle from diseases and floods. Gono Chetona has given training to some women in the villages in veterinary services and its own staff provides veterinary services such as vaccinations and advice on the best types of cattle feed. In the absence of any extension service from government officials these services have been beneficial to the people. Rearing of cattle has increased in these villages and the people now take better care of their cattle.

Poultry farming is practised by almost all the households in the area. Local (*desi*) varieties of chicken, sparrow and duck are the most common bird species that are kept by people. Gono Chetona has trained some women from beneficiary households in each village to provide timely vaccination and poultry feed. These trained women charge a small amount for the service that they provide. Vaccines and medicines are bought from nearby towns.

Apiculture or bee-keeping has also been introduced as there is a lot of greenery and seasonal flowers in this area. Selected women have received training and a basic kit. An expert employed by GC visits regularly to oversee activities. This is a new activity with a lot of potential to generate income, but it will take some time to become popular among people. People have not yet taken up apiculture on their own although the initial cost of setting up a beehive is not much. This shows that either they do not have enough guidance/training or the profitability of this venture has not been demonstrated properly. Till now GC has not set up any marketing structure or producer's network that can give inputs on training, guidance and marketing of honey.

Tree plantations in some of the more stable char villages have also been promoted. Some villagers, whose lands are partly submerged under water during the flood season, are unable to grow crops because of sand deposition and have hence opted for tree plantations. Long duration trees are grown that also help in preventing river erosion. People believe that since their land was lying fallow, planting trees is a good investment as it does not require much initial investment and care (see Box 7.5). Even if the river eats up these lands they will not lose much. People also plant timber trees in their homestead gardens and on the edges of fields.

Box 7.5
Income Generation through Nurseries

Hunufa Begum and her husband Sona Mia are nursery owners in village Chiner Char. They have 22 decimels (one decimel is one-hundredth of an acre) of homestead land (that is raised and hence free from flood water) and have leased in another 8 decimals where they maintain a nursery. They sell saplings for Taka 3–8 and get a good income from this. Gono Chetona helped them initially with 1 kg seeds, fencing around the nursery and 1 pound of plastic sheet. Hunufa was also trained for seven days in the beginning and still goes for supplementary training. Earlier they used to grow *aus* and boro rice, wheat, jute and sugarcane on their land, but Sona Mia had to still migrate to earn some money. Five years back, they planted about 200 trees of mahogany, acacia and *sessum* in about 20 decimals of land that goes under water during the monsoon. This additional income has meant that Sona no longer has to work on others' land.

Similarly, Musiful Begum of the same village has planted trees on 8 decimals of land with the help of GC. Both see this as an investment which they will use after 20 years for their children's marriages. Dowry demands are quite high in these parts and the poor suffer a lot because of this practice.

Gono Chetona sees the activities in the programme as forming a package for strengthening the coping mechanisms of the people with respect to erosion and flooding. It has planned its interventions according to the three categories of land in the chars. Land that gets flooded often and is sandy (generally the two go together) is seen as suitable for planting trees that can tolerate seasonal inundation. Land that does not get flooded during normal floods or that gets flooded for a very short time is seen as suitable for improvement through improved methods of cultivation. And finally there are homestead land and kitchen gardens that normally do not get flooded or are raised so that they do not get flooded at all. These measures sit alongside other measures de-linked from vulnerable agriculture in the flood-affected zone: for example, dairy farming, cow fattening and poultry, apiculture, vegetable farming, etc. While elements of this coping mechanism strategy have been evolving through the years, their combination into a package is recent and has yet to be put into practice systematically.

As mentioned earlier, women in these villages spend most of their time alone as their spouses are away for work. These activities provide some income to women. However, it has in no way decreased migration as the income accrued from these new activities is not much and can only supplement existing incomes. Migration continues to remain the main livelihood option for a large majority of the people, but the women now have some income which they can spend in whatever way they feel necessary. Most of the women we talked with said that they now spend money in buying small things for the household, sending their children to school and have money in case of emergencies. But the most positive impact of increased agricultural production (vegetable and fruits) is that the poor have improved their nutritional intake. Many women informed us that neither they nor their children suffer from hunger now. The positives of these activities are that they are easily replicable even by non-beneficiaries. With expertise and extension services available in the village itself, not much investment is needed for others to take up these activities.

The chars also suffer during the dry months and in years of low rainfall. During the period from October–May, the rivers and drainage channels dry up and there is no source of water for irrigation. Dry spells during October–November adversely affect the aman paddy and during March–April the seed beds of sugarcane and jute. The problem now is of irrigation sources. The larger farmers dig their own bore wells and use diesel generator sets to pump water. But the small and marginal farmers do not have the money to invest in these. Moreover, most of the villages in the char area are not electrified and irrigation from diesel generator sets is very costly. For the poor and landless these dry periods are also difficult as there is no work for them.

Sanitation and Health

Due to the low literacy rate, high levels of poverty and lack of medical infrastructure, disease and poor health is common in these areas. Some of the commonly occurring diseases in these parts are malaria, cholera, typhoid, diarrhoea, asthma and skin diseases. Raising awareness about the need for clean and sanitary habits is also a very important activity of the programme. Sanitary latrines have been provided for all the beneficiary families. It was a 100 per cent subsidy programme; though the beneficiaries had to pay about 25 per cent of the cost, they got some other benefits to compensate them for the initial expenditure. This activity was aimed at improving the health and sanitation of the villages where disease can take epidemic proportions (especially during the monsoon and summer months). The privacy element was also an important consideration in promoting this activity. During the floods when the entire area goes under water for 10–30 days, access to a latrine is critical. Latrines have been built on the raised homestead area. The entire cost of buying the latrine and setting up a small tin room comes to about Taka 2,200. This cost is too high for the poor to bear and that is why the subsidy was provided. But the moot question is how this can be sustained. Who will provide sanitary latrines to all the residents of these rural areas? The NGO informed us that there is a cheaper version of the latrines (Taka 500) that they sell on a no loss no profit basis. The Environmental Protection Team set up by GC teaches people about cleanliness and hygiene. The villages and houses we visited appeared clean.

This programme was meant to be a pilot that could be replicable. But the non-beneficiaries, even the better-off ones, are not installing latrines in their homes. With the constant fear of erosion and having to shift to new homes, people do not find it very practical to invest so much money on latrines. In our visit to the various project villages, the areas where the programme was implemented were cleaner than the rest of the village. But in all the villages we visited, the first demand people had was that some medical facility should be set up close to their village. Gono Chetona has employed two paramedics who go visit villages and provide basic medical help and advice to people. But the job of these paramedics is to basically raise the awareness of the people about health and sanitation rather than to provide any medical services. It is the state that has to provide free and easily accessible medical services to the people.

Gono Chetona has also encouraged some people to grow locally available medicinal herbs and plants. These plants were earlier used by local traditional *vaidyas* to treat particular diseases. Gono Chetona has tried to teach/train the local people about how these plants are to be used and some people told us that their use is effective. Growing these plants for commercial purposes could also be started if proper markets are created.

The government bore wells provide drinking water to these villages. For its part, GC has installed several bore wells on raised platforms so that people can access safe drinking water even during periods of floods. Arsenic levels in groundwater are high in this area. The government of Bangladesh with the help of the World Health Organisation (WHO) has marked and isolated the bore wells that have arsenic-contaminated water.

Raising the homestead land was also a highly-subsidised package that cannot be replicated at least by the non-beneficiary poor. Each year floods reach the homestead land for at least a few days. During this period, people have to live on tree branches or rafts made of bamboo or banana stems. Not only can they not earn their daily wage, but there is no help on hand except for some NGOs who distribute relief. Raising homestead lands thus gives them respite from floods. It is only if the height of the flood is very high that vegetable gardens are inundated.

Though GC's efforts at promoting clean sanitary practices and alternative (indigenous based) medicine have shown some positive results in the villages where the programme is being implemented, what is actually needed is a more sustained effort by the state to provide primary health services in this area. The health and sanitary practices promoted through the programme can only be treated as part of a more comprehensive health package. Moreover, replicating this experiment over a much larger area can only happen with the involvement of state agencies.

Sustainable Environmental Management

The greening of the chars was one of the most prominent components of the programme. Most of the char villages have scarce vegetation.

Considering the temporariness of residence and impermanence of even land assets, investing in agricultural improvement is not the best option as the benefits accruing from it are uncertain. Also, with migration being the primary livelihood option for more than 50 per cent of the people, many of whom stay on leased land, improving one's own living condition by planting trees was not an option they seriously thought about. The SLRC programme sought to demonstrate that small investments on private lands can bring in both short-term as well as long-term benefits. The SLRC programme has demonstrated to the poor the advantages of maintaining a few trees and homestead garden and poultry farming and animal husbandry.

According to Ira Rahman, 'the programme encouraged the people to undertake activities which they should be doing in the first place in the charlands'. The SLRC programme did not introduce many innovative activities, but only helped improve and finance activities that boosted the morale of the people who have witnessed their homes and lands regularly going under water. For instance, the idea of raising homestead land to save oneself from floods was known to people, but they did not have the financial wherewithal to take it up.

Apart from new crops (especially marketable ones like vegetables), a number of trees have also been planted in the chars. Fruit trees like jackfruit, guava, papaya, *jamun*, lemon, banana and mango dot the landscape. Other species like sessum, eucalyptus, *bakain* (like neem), mahogany, bamboo and olive are also a very common sight. All these crops and trees are well suited to the agro-ecology of charlands. Apart from giving an extra income to the residents, these crops have also improved biomass availability.

People have started planting trees and are trying to adjust to their agro-ecological setting. With the setting up of nurseries and the presence of trained village women, it becomes less difficult for people to start all over again after shifting to a new place. The idea behind tree planting was to give people alternate livelihood options and also to 'green' the char villages which had a barren look (especially the newly-formed ones). This component of the programme has been very successful and has been replicated by non-beneficiaries too. But the confusion about the ownership of these lands might impede the greening of the chars in the long run.

Fodder/Animal Husbandry and Fuelwood

Though fodder and fuelwood should not be a problem in the chars, especially after the successful greening of the chars in the course of the project, people did complain about scarcity during floods. There are periods during the flood season when entire chars, including homestead lands, go under water. These are the times when collection of fuelwood and fodder become difficult. For the poor and landless, availability of fodder is a big problem even in normal circumstances as they do not have paddy stalks, the main source of fodder. There is very little common land in the chars and so cows and goats are mostly stall fed. There is a lot of private fallow land that is not fit for cultivation (in some of these sandy lands groundnut is grown during the winter season). In these lands a local succulent variety of grass grows in the wild. Some people also buy fodder from bigger farmers or from the market. Tree leaves, local varieties of grasses and stalks of crops like paddy are used for fodder. The stalks of jute, dried grasses and cow dung cakes are used as fuel by most people. But the absence of common grazing lands imposes limits on animal husbandry in these areas.

New improved chulhas that require little fuelwood have been introduced. They are energy efficient and even non-beneficiaries have adopted this on their own. It requires very little investment—only the wire mesh has to be bought from the market while the mud is locally available. The use of LPG is almost nil in these villages.

Use of Chemical Fertilisers/Danger to Biodiversity

With the introduction of HYV seeds and new varieties of vegetable crops, and the focus on increased yields and more intensive cropping, the use of chemical fertilisers has also seen an increase. The introduction of improved HYV paddy (developed by the Bangladesh Rice Research Institute) known locally as BRRI (pronounced biri) rice has brought in its own problems associated with use of chemical fertilisers. A scientific study of the impact of excessive use of chemical fertiliser on the biodiversity has not been done, but local people and NGO staff talk about how people use fertilisers excessively because they have not been given proper training on the exact quantities to use. They also believe that some of the local fauna is now disappearing. The programme has promoted the use of organic manure. The beneficiaries have been taught to produce organic manure with cow dung and household waste. However, eco-friendly farming practices promoted through the experiment focus only on the homestead gardens; similar experiments are not encouraged in the agricultural fields. Further, the ecological context places limits on animal husbandry in the region which in turn places limits on the quantity of organic manure produced in the chars.

There are some other practices promoted in the project that are difficult to replicate. Raising the homestead lands is only a protection against floods not against river erosion (it might at best delay the inevitable). It is surprising though that the project did not try to introduce the concept of building houses on bamboo stilts (which is a common practice in flood-affected areas of Assam in India). There are many project villages (namely, Anadabari, Char Choumohan, Madaner Char) that are under threat of erosion. Even char villages that are not under the threat of erosion (those which are connected with the main land) are constantly changing shape and size—some parts might go under water while a new area comes up. In this extremely vulnerable area, it is the poor who suffer the most. For them to find a new home is a big and difficult task.

Floods, Flood Control and Sustainable Interventions in the Charlands

Considering the bio-physical context and the dangers of structural measures (such as putting up embankments), the physical vulnerability of char dwellers can be addressed only in non-structural ways through flood warning and other mitigation measures, such as provision of adequate (temporary and permanent) relocation sites, accompanied by appropriate support for health and education and alternative sources of income. It should be recognised that the set of measures evolved by GC provide important protection to its beneficiaries against flooding. Its rescue boats, which also act as communication and transport hubs to some extent at other times, as well as its flood relief shelters are important in providing baseline protection against seasonal floods.[24] Similarly, elevating the homesteads as well as raising the kitchen garden lands are measures that extend the period that people can stay on inside their homes during the flood season. However, though they are closely related, it is important to separate the seasonal floods from the erosion process. Most of the measures are effectively planned for coping with the former. There is not much that can be done for the latter. In fact, nothing can be done to prevent erosion and river flooding.

There is a danger that whatever is being done towards improving the conditions of the char dwellers through these experiments might all go waste because of the developments within the water resources sector. Bangladesh had planned a large-scale project to embank all the rivers, but this project was shelved because of nation-wide protests from civil society organisations (Huq, 1997). Even GC does not favour flood protection measures in this area. The local residents too never listed this as one of the things that they needed, except in one village, Madaner Char. This village had one embankment built by Save the Children in 1989 that has now disappeared in the river. On the other hand, these char villages have to face increased problems of erosion and floods because of the presence of some protective embankments in the area. For instance, people of the Kolkihara Bhati complained that their land was facing excessive erosion and their village was cut in to two because of an embankment that was built about 25 years back to protect the nearby town of Dewanganj. The Old Brahmaputra changed its course and one of its branches has divided the village in to two parts. Any further attempts to strengthen or extend the embankments will only increase the vulnerability of the chars and undo any progress made by development programmes. If the rivers are embanked on both the banks it will result in raising the flood level in the rivers, which consequently will increase the flooding and erosion in the chars (or mid-channel islands). Though embanking the rivers through a holistic policy has been shelved, there has been an incremental increase in embankments along the Jamuna. The embankment arising from the construction of the Jamuna Bridge, other embankments arising from compartmentalisation, including the Jamalpur Priority Project, and the prospect of further construction of link roads to the Jamuna Bridge have already covered a long stretch of the river. These embankments have increased the vulnerability of millions of char dwellers. Further, char protection embankments are unlikely to be stable and successful considering that the chars are surrounded by fast moving channels of the river (Wood, 1997).

During our visit to these two unions we saw the construction of roads and bridges. These roads act as embankments and the bridges restrict the flow of the rivers. The rivers are trained to follow a fixed course that goes against nature. Rivers and drainage channels in this area are always changing their course and the experience in other parts of the subcontinent· shows that flood control measures, roads and bridges result in disastrous consequences. A number of developmental agencies have also in the past constructed roads and embankments in some areas with the same disastrous consequences.[25] This is the irony of the development process—the infrastructure required to get help to these remote ecosystems itself induces unsustainable development in the long run and aggravates the vulnerability of the region.

Community Involvement in the Project

As has been discussed earlier, the element of temporariness militates against the emergence of durable village institutions. Coupled with it the absence of men/household heads for most part of the year creates difficulties in community action around NRM. To counteract these factors, GC works through women's groups that it helped form in the beginning of the programme. Though most of its interventions have been at the household level, it is important to see whether the larger community—at the village or *para* (hamlet) level—has been involved in identifying the beneficiaries (village and households) and deciding on the nature of interventions.

How Participatory is the Intervention?

Gono Chetona conducted a baseline survey at the beginning of the project and it believes that this helped it get inputs for the intervention.

The GC staff also repeatedly emphasised that their knowledge of the area and long experience helped identify beneficiaries.[26] Among all the villages in the two unions (13 in Erendabari and 19 in Merur Char), more villages were taken up in Erendabari than in Merur Char as the former is more vulnerable to erosion. The staff of GC informed us that they did not choose the villages that were most prone to erosion, and which could go under the river in 5–10 years for their interventions. This argument appears a bit misplaced considering the uncertainty regarding the behaviour of rivers in the region (the absence of reliable data and study on rivers puts limits to forecasts about the swings of the rivers). In fact, there were five more villages under SLRC, namely, Bulbulichar, Ghatua, Kisamat, Dhanu, Poschim Jigabari and Dholi Patadhoa, but they have all been destroyed by river erosion (Montu, 2003: 11).

The baseline survey also helped identify the neediest who could be selected as beneficiaries of the project. Though all the beneficiaries are poor (landless with some homestead land or small peasants with less than half an acre of land), even among them first priority was given to widows and the disabled. In most villages, GC finally selected people living in compact blocks within a village so that the programme (especially the component of raising of homestead land) could be implemented efficiently and the *benefits easily visible*. Interviews with the local staff of the NGO revealed that the motivation level of the villagers also played a crucial role in their selection.

In practice, participation (in terms of beneficiary involvement in particular activities) seems to have been high. There has been a good uptake of practices promoted through the project, for instance, encouraging a much more scientific and intensive cultivation of homestead gardens, better upkeep of cattle, planting trees, improved chulhas, etc. These are low investment practices where the returns are quite high. The demonstration of these practices by encouraging the beneficiaries helped in its popularisation among non-beneficiaries in the project villages too. A number of women were trained to provide extension services to people at small charges.

But while GC itself appears to have acted in a fair and impartial manner in the selection of beneficiaries, it does not seem to have taken the entire village or hamlet community into confidence. Also, it was not the village community which decided who the beneficiaries should be. In programmes like SLRC, where most of the intervention is don through individuals, the role of the community is critical in preventin outsider discrimination and ill-feeling among the non-beneficiarie although community involvement has its own possible biases. More over, SLRC was also meant to be a pilot programme and the experi ments promoted through it were supposed to be disseminated t non-beneficiaries. The non-involvement of the larger community (a well as the fact that uptake requires some investment) at the village o char level has resulted in little uptake of the practices that were pro moted through the programme by the non-beneficiaries. Keeping i mind the goal of the project (helping the poor attain sustainable liveli hoods), the efforts that have been made to popularise the experiment among the non-beneficiaries even in project villages, as well as in othe villages do not seem commensurate to the level required.

Recently, the organisation has proposed the idea of developing som villages as 'eco-villages'. The organisation adopts a compact area c a village and tries to create an ideal village settlement. While the resi dents of these 'eco-villages' have clearly benefited from the attentio received, it is not clear as to whether the purpose of this experimen namely, showcasing the experiments for other villagers to follow sui has been fulfilled. There is a danger that these villages may becom favourite sites for outsiders to visit and may lead to a 'museumisatior of success, hence making participation a secondary concern.[27]

Having said that, the NGO has been flexible at times in listening t and responding to local people. Take, for example, the raising of home stead land. This was not a fully subsidised component; the house holds had to contribute 30 per cent of the cost, mostly in the form c labour input. Potentially this could be exclusionary. However, som of the poor beneficiaries entered into an agreement with neighbourin landowners (on whose land they were squatting) whereby they helpe the landowner raise his land on condition that he allowed them (th beneficiaries) to stay for 10 (or more) years. Some kind of a forma agreement was signed between them (though these are not legal docu ments). In some cases, an agreement was reached between some land less, small farmers and bigger farmers whereby the landless (who wer the beneficiaries of the programme) requested GC to raise a particula piece of land (owned by some big farmer), which everybody then used as grazing land.[28] Hence, the project offered the poor some powers to negotiate with the landowners. The landowners too agreed to this arrangement because they themselves did not have the means to invest. Moreover, their social standing improved by helping their needy neighbours. These win–win arrangements were not part of the original design, but were proposed by the beneficiaries. Gono Chetona was flexible and receptive enough to put these arrangements into practice. It also raised land on some common lands so that the non-beneficiaries too had some flood shelter. A number of other people also take shelter in the private (raised) homestead lands of the beneficiaries. Gono Chetona categorises them as indirect beneficiaries. It maintains that about a thousand non-beneficiaries took shelter in these raised platforms in the 2004 floods.

In other cases, however, participation is hampered by the cost question and the thinking or ideology of GC. The sanitary latrine case discussed earlier is a case in point. Costs have proved exclusionary. Perhaps it would have helped if GC had tried some other method of providing soft loans or partial subsidy instead of a 100 per cent subsidy. Otherwise, the non-beneficiaries will never have the wherewithal to replicate the experiments, especially the ones requiring investments like installing sanitary latrines. Some degree of innovation and flexibility would possibly have widened the beneficiary net. However, GC felt that it should be the responsibility of the state or donors to provide practically the full cost of these investments.

The programme, moreover, seems to have been designed by policy-makers and experts at the drawing board. Though SEMP is a continuation of the NEMAP process, which held several consultation workshops all over the country, the beneficiaries of the project area were never involved in the design process and phase of SEMP (or SLRC) itself. What is clearly visible is the very high involvement of the villagers in the implementation phase of the project. The villagers participated in the programme actively to the extent that they brought in their own innovations to get the maximum benefit. To the credit of GC, it has shown some flexibility in implementing the project though it was not built in the project design itself. But it is also clear that in the absence of political devolution and democratisation in these villages and lack of support, these experiments will only create isolated 'oases in deserts'.

Conclusion

There is no doubt that GC's interventions have made some impact in improving the coping mechanisms of the poorest sections of society in the chars and introducing methods of more intensive use of land and other resources. However, the moot question is whether this experiment can be replicated elsewhere over a larger area without as much financial or technical help. Gono Chetona, UNDP and other government agencies have to now test the lessons learnt through this pilot programme over a larger area, maybe with much less financial backing. Can an NGO sustain these activities with the limited resources at hand? Also, should an NGO do this—should it not lead to some initiative by the state? Gono Chetona has already started withdrawing from a number of activities. It does not seem to have even the resources to help the various women's groups and provide the necessary extension services that people might need. Very few women's groups have become strong enough to carry forward their own institutional responsibilities forget about liaison with government officers. Replicating the same experiment in a new village will at least require forming women's groups and training selected women in setting up nurseries and seed banks, animal husbandry and poultry farming and other such things. Does the NGO have the financial stability to continue taking up such activities? Is it not the responsibility of the state to take up and support similar initiatives on a larger scale?

Will GC itself replicate the same experiments and lessons in other places given its long-term commitment and presence in the area? Gono Chetona is also implementing other programmes (for instance Oxfam GB's River Basin Programme) and might also be part of DFID's Sustainable Livelihoods in Charlands Programme (which has already started). There does not seem to be much similarity in the two programmes and the approach of the two agencies—UNDP and DFID.

Moreover, they also do not seem to coordinate amongst themselves. While it is the local NGO that has a long experience of working in the area, the donor agencies are the ones trying to force the agenda. Talking with GC staff clearly brings out the fact that they want a project which has a strong livelihood component in it, which the DFID programme has only in name. Moreover, they are also not in favour of time-bound projects where the focus is more on completing individual components and less on bringing in long-term changes. They believe that for any major change to happen, work has to be done in a sustained manner over a period of time, but many donors are not interested in supporting long-term, process-based approaches and rather prefer implementation of a programme on a project mode. Despite these reservations, GC has gone ahead and taken up the implementation of the DFID programme. Obviously, its own institutional survival gets precedence over its long-term vision. Further, though GC has its own take on the various developmental issues and state policies it has shied away from lobbying with the government on changing the policy context.

During its work in the charlands, GC has had to encounter some opposition from the dominant groups and individuals in the area. Though GC has not compromised on its basic principles, it has not engaged directly with the elites and the dominant groups. It has chosen to quietly implement the programmes, probably because of the prevailing social and political situation in the area where the matbars and their political bosses hold complete sway over the region. It believes that change can be brought about by providing livelihood support to the poor, especially the women. Probably it is too much to expect an NGO to take on the all-powerful political and economic elites, who are also well connected with high-level officials and politicians. But, on the other hand, there have been some NGOs who have focused on mobilising the poor and challenged the dominant groups and state policies (for instance Swanirwar Bangladesh, Gono Shahajjo Shangstha, Nijera Kori). Gono Chetona, on the other hand, appears to believe that the solution to the problem of poverty is in providing livelihood support to the poor. It appears to be doing what the agencies of the state are supposed to do, but have failed to do. In other words, it remains a service delivery agency that seems to be doing its job well; it has tried to reach out to the remotest corner of the 'forgotten land'.

Moreover, the question of landlessness and land-grabbing in the chars are issues that will need to be resolved. The concentration of land resources in the hands of a few people does not bode well. These are the people who also grab any new charland that accrues to the government. To bring in significant change in the living conditions of more than half the population living in this region, the landholding pattern and land politics will need to be addressed. For this to happen, the state machinery will have to become more receptive to the needs of the poor. Or else, civil society organisations will need to mobilise resource-poor people to demand some basic rights. Till then any such intervention can only manage to marginally improve the living condition of a small section of the poor.

Notes

1. Chars or charlands are mid-channel land formations that are built up by accretion of sediment load carried by rivers.
2. One bigha is approximately one-fourth of a hectare.
3. Literally, society or community, it consists of a number of spatially contiguous households belonging to one or more patrilineage.

> Each shamaj formed a distinct and exclusive group the members of which had obligations of reciprocity to each other.... A critical function of the shamaj was to resolve conflicts among its mem-ber households.... These indigenous courts with their quasi-formal procedures were locally known as *shalish* or *bichar*.... The shamaj did not operate on the basis of any written rules or law. Instead, there were unwritten conventions. Refusal to accept the verdict of the shalish was met with imposing a social boycott, involving sanctions of exclusion and non-co-operation.... Such sanctions were usually more effective against the poor and weak compared to the rich and powerful.... Shamaj did not function as a political entity, but it did provide an arena where overt or covert political manoeuvres could be attempted. Typically, it was the richer peasantry who aspired to lead and expand their respective shamaj groups because of the legitimating attributes of the leader of the shamaj (Adnan, 1997).

4. Though Baqee's study is based on the chars of the river Padma, it is to a large extent representative of the settlement pattern in chars all over Bangladesh.
5. In anthropology, a patrilineage is a consanguineal male and female kin group each of whom is related to the common ancestor through male forebears.
6. Though the law of inheritance in Bangladesh gives the woman right over her father's/husband's property, under the living law (or Bengali rule of inheritance) a women's claim is to maintenance, not to a particular piece of land and its

products. Although women formally inherit land according to Islamic precepts, they do not take possession of it in any real sense. In return, the father and brothers have certain duties towards their daughters and sisters respectively (Rahman and Schendel, 1997).

7. In charlands, the environmentally-forced migrants look for help from relatives in establishing homesteads. The first ones they approach, and get help from, are relatives of the husband. In most instances, they get some land from the relatives or members of their own gusthis. But because of the continuous displacement forced by river erosion, a man might have already taken help from almost all his relatives, including distant agnates. It is then that the wife's entitlement basket is used to find feet in some place (Indra and Buchignani, 1997).

8. The discussion in this section is about the charlands of Jamuna and not specifically on the villages we visited. We did not spend too much time in the field studying this issue. This discussion is based on secondary literature. Competition and the resultant violence is higher in the settlement process in the chars of the Padma than in the chars in the upper reaches of the Jamuna.

9. We were lucky to meet most of the male members of the village because they come back to their villages during the flood season.

10. Zamindar: Legally, a proprietor of land, acting as an intermediary between the state and the tenants in rent collection. Taluqdar: The zamindars farmed out portions of their estates to independent or dependent taluqdars; they paid fixed amounts to the zamindars and pocketed the difference between their rent collection and fixed dues. Jotedars were independent ryots who acquired rights to hold land for the purpose of collecting rent or bringing land under cultivation by establishing tenants on it (Zaman, 1991: 676).

11. A union comprises 5–10 villages and the political institution representing these villages is the union parishad. There are 12 members in a union parishad, nine elected directly by the people and three selected, the latter reserved for women. The members select a chairman. Several unions comprise a sub-district (*upazila*) and there is no representative body at this level. A bureaucrat, *upazila nirbahi* officer (UNO), heads the bureaucratic structure and all the power is concentrated in his hands.

12. One such programme is UNDP's Sirajganj Local Governance Development Project started in 2000. One of the objectives of the programme was to 'secure the support of union parishads for local development initiatives in Sirajganj District and increase their ability to do so in an effective, sustainable and participatory manner'.

13. Thana literally means police station that in most South Asian countries is also an administrative unit.

14. Non-governmental organisations now distribute about 10 per cent of foreign aid in Bangladesh (Feldman, 1997: 57).

15. 'The reorganisation of social collectivity to ensure loan repayment serves as a mechanism of social control rather than an arena for building social solidarity and creating relations of social obligation and reciprocal exchange, since repayment by all group members is required before new loans are disbursed' (Feldman, 1997: 60).

16. This section is based largely on the 'Biodiversity Report' (Anonymous, 2003). The account is unique to the two unions and all the villages we visited have characteristics as described here.
17. We realised later that the NGO had a standard set of interventions in all the villages.
18. This section is largely based on conversations with Ira Rahman and other senior staff of GC.
19. Terre Des Hommes had a very narrow focus, namely, of rehabilitating infants and the unwanted children of the war in Kurigram.
20. Direct beneficiaries are those households that were selected by the NGO to receive financial and other benefits of the programme. Indirect beneficiary households, as defined by GC, are those that have benefited indirectly from the programme in terms of learning from the experiments promoted in the project, and they also find shelter during floods in the raised homestead lands of the beneficiaries.
21. We could not talk with any member of the executive committee as we came to know about this umbrella organisation when we finished our fieldwork. None of the women's groups ever mentioned the existence of such a network. This shows that the members of the respective groups are not aware of these committees, let alone its roles and functions.
22. Earlier they used to grow vegetables such as radish, brinjal, chilli, bitter gourd, *data* and bottle gourd. Now they are growing new vegetables such as lady finger, cabbage, cauliflower, tomato, carrots, long beans, *palak* (spinach) and *poi* (a leafy vegetable).
23. The poor of this region often opt for the cow-fattening business. They get a cow from a rich man for free and rear it for a year at the end of which they sell the cow for Taka 10,000 or more. The owner gets the original cost of the cow (Taka 5,000) and half of the profit (Taka 2,500) while the other half goes to the person who tended the cow.
24. It should be noted here that GC does not differentiate between beneficiaries and non-beneficiaries during the times of floods. Rescue boats and flood protection shelters are meant for anyone in need.
25. Personal communication with Ira Rahman (27 June 2005).
26. This is in contrast to the participatory approach in livelihood analysis that is ostensibly encouraged by UNDP (Carney et al., 1999: 16).
27. Others (Baviskar, 2002; Li, 1999) have also showed that development sites are chosen based on accessibility and willingness of the people to cooperate.
28. Though there is enough fodder for the cattle during most part of the year, the people face severe scarcity during the flood season when most parts of the char are under water.

8

Conclusion

Our interest in CBNRM in general and NGO-driven CBNRM in particular was sparked by the potential for 'alternative' development that both the idea of CBNRM and an earlier set of such experiments seemed to offer. We were, however, also cognisant of the fact that critiques existed that questioned the discourse and practice of CBNRM and the increasing 'mainstreaming' of NGO-driven CBNRM. While these critiques served as the background of our study, we felt the need for a more detailed enquiry (given the relative scarcity of such studies) that engaged more rigorously with both the making and working of CBNRM and captured more of the diversity of NGO/civil society initiatives.[1] In this enquiry into CBNRM in South Asia, therefore, we chose to focus on initiatives that have been implemented by civil society groups with or without support from government agencies and large donors. We chose to study six such initiatives in detail. The choice of case studies, although constrained by various factors, still gives us a reasonable spread in terms of socio-ecological conditions, policy contexts and approaches to CBNRM of the intervening agencies. Our chapter-length narratives of each case sought to both describe and analyse these initiatives within the broad framework outlined in the introductory chapter.

In this chapter, we shall attempt to generate some broader insights from these case studies through a comparative analysis of outcomes and processes that seem to have been at work in various locations. We end with a discussion of the broader implications of our analysis for NGO-driven CBNRM and CBNRM in general.

A Brief Profile of the Six Cases

Before we examine and understand the outcomes across the six cases,[2] it would be useful to summarise some of the key features of these cases. All six cases are located in ecologically vulnerable regions: two in the Himalayas, three in semi-arid parts of western India and one in the flood plains of the Brahmaputra–Jamuna. In all locations, agriculture is the primary occupation for most communities, although secondary occupations and the role of non-land activities vary, and in the case of the flood plains, remittances from migrant labour may be more important that returns from agriculture. Thus, there is generally a high level of dependence on natural resources. But the extent and nature of the commons differs: in the mountainous locations, forests lands are significant, whereas in the semi-arid locations the extent of common lands varies signficantly. Water is a commons everywhere, although its treatment is not uniform across the locations. In the floodplains of Bangladesh, the 'commons' are perhaps the common threat of flooding.

Concomitant with these variations in agro-climatic conditions, there are significant variations in the socio-economic context. In many cases, the village community is fragmented on the basis of class, caste and other primordial loyalties. Gender relations in general and in particular with regard to NRM remain tilted in favour of the men, more so in the villages in the plains than in the Himalayan communities. Class distinctions, in terms of access to land, are very sharp in four of the six cases, whereas perhaps somewhat less sharp in the Himalayan case studies.[3] Traditional village institutions remain dominated by village elites, although newer elites might be emerging in some cases. What this entails is that any development initiative for the village would disproportionately benefit these elites unless proper safeguards are put in place. The sense of a collective 'village' identity—and the extent to which such identity helped overcome internal differentiation—varies: only Hivre Bazar and the villages in the Doodha Toli region seem to have this sense of identity to some extent, and that too needs to be qualified. In the case of Hivre Bazar, the collective identity was actively fostered as part of the initiative and to a large extent deliberately underplayed the inequalities of access to land. Similarly, the 'village identity' in Doodha Toli did not prevent caste-based exclusions, for example, in the case of access to temples. On the whole then, while the large extent of and high dependence on common pool resources provided a material basis for CBNRM in all cases, the socio-economic context was generally much less conducive to community-based efforts and posed substantial challenges to the intervening agencies.

To these differing socio-ecological contexts came six organisations with varying backgrounds, styles and approaches. Three were external NGOs (Utthan, GC and TBS) that are undergoing a process of 'professionalisation', one a community-led effort (Hivre Bazar), one a grass-roots organisation (DTLVS) and one effort implemented by a state-supported scientific organisation (RNRRC). Of the six cases, three (TBS, GC, RNRRC) are aided by large donors, two (Hivre Bazar, Utthan) have drawn support from state programmes for their work in the study villages (though Utthan does receive funds from international donors as well) and one (DTLVS) has carried out its programme with small financial support coming from various channels. It should be noted that the durations of the interventions and their scope is also varied: DTLVS, TBS and Hivre Bazar have been functioning for one-and-a-half to two decades in the particular villages, whereas Utthan, GC and RNRRC have been working in the case study villages for only a few years (though both Utthan and GC have a longer presence in many of the other villages in the region). Similarly, the work of Utthan, GC, TBS and DTLVS spans several tens or hundreds of villages, whereas that of the Hivre Bazar community is naturally concentrated in its own village and that of RNRRC is focused on the particular set of villages that it has chosen for the CBNRM effort.

These organisations approached and implemented CBNRM in distinct ways.

(*a*) The group in Hivre Bazar adopted an integrated watershed development approach to the hilt that formed part of a wider effort at community development as a whole. They drew upon a watershed development funding programme, but complemented that with support from other state-funded development programmes, especially during drought years. The group worked through the formal gram panchayat as much as it worked through the informal gram sabha and informal village-level interactions.

(*b*) Utthan entered Nathugadh largely as a service delivery organisation, helping first implement a state (and international donor supported) drinking water and sanitation scheme and then a watershed programme. The watershed programme focused mainly on water harvesting. Most decisions were taken in committees specifically set up for these programmes, with limited linkage with the formal political structures. At times the general village body was also used to sort out issues.

(*c*) Tarun Bharat Sangh focused almost exclusively on setting up water-harvesting structures, wherever individuals or groups were willing to come forward and contribute to the activity. They drew support from various donors, although they also cooperated with a government programme to some extent. It used the so-called traditional gram sabha, which ended up being in practice, but not design, a gathering of male elders, to obtain initial sanction and support for their initiatives. Some forest conservation efforts had been initiated by the villagers themselves, which have continued as well.

(*d*) The RNRRC focused heavily on the process of involving the community and especially individual households in various technical efforts to boost on-farm and livestock productivity and generate some alternative livelihoods. They were supported by an international donor and also the state, which saw this as a pilot effort that would guide its larger CBNRM programme.

(*e*) The DTLVS focused on regeneration of forests, grasslands and drinking water sources as well as spreading renewable energy technologies; it also attempted some horticultural experiments. Its approach was based on spreading the message of forest-water-land conservation, making small interventions based on their understanding of local priorities and working with women's groups in some cases and informal consultative mechanisms at the village-level in others.

(f) Gono Chetona focused first on the question of providing some minimal protection to households against floods by raising homestead land and then on expanding the livelihood options of the poorer households, especially women, through various skill-building activities in horticulture, poultry farming and animal husbandry.

We shall now examine the outcomes, processes and possible influencing factors in more detail across the cases in a comparative framework. Unlike the individual chapters, where our attempt was to convey a connected 'story' of the interplay between interventions, context and outcomes, our discussion here is perforce sequential—considering one aspect at a time, while keeping in mind their interrelated nature. 'Bigger' questions related to the practice of CBNRM, in the context of existing critiques, are addressed after that.

Outcomes of the Interventions

We used a four-dimensional framework of livelihood enhancement, sustainability, equity and democratisation to characterise the outcomes in each intervention. We begin, therefore, with a summary of the achievements of all the interventions along each of these dimensions. This is not an attempt to characterise the interventions in terms of 'successes' and 'failures', or even to 'rank' them in some simplistic manner, but rather to indicate the extent to which they were able to make headway in certain commonly defined directions. As we explained in the introductory chapter, our framework simply makes explicit the concerns that underpin our notion of CBNRM. We recognise that this framework is to some extent 'external' and may not match entirely with the concerns of the intervening agency. We discuss this possibility in the next section, when we try to understand what factors shape the variations in the outcomes.

Livelihoods

The main focus of all the initiatives to various degrees, as we have seen in the individual chapters, is on livelihood enhancement. Most, if not all, of the initiatives seem to have resulted in significant livelihood benefits. These have been mainly of three types: (*i*) improvements in availability of livelihood support resources (fuelwood, fodder, drinking water), (*ii*) increased productivity (including diversification of cropping patterns) in agriculture and allied activities, and (*iii*) new sources of livelihood.

Gains in the form of enhanced livelihood support resources have been prominent in the DTLVS case and Hivre Bazar cases. The DTLVS' efforts at improving fuelwood and fodder availability have been very successful, as also their rejuvenation/protection of traditional drinking water sources—all from the commons. Fuelwood and fodder availability also got augmented in Hivre Bazar and RNRRC although in the former case it was mostly from increased agricultural residues from private lands. Drinking water availability and access also improved in Hivre Bazar, and to some extent in Utthan and TBS, where the semi-arid conditions made this a priority. The DTLVS also facilitated access to solar PVs for lighting—a crucial need in that remote region. The TBS case did not make any real dent in the availability of fuelwood resources to the women who are responsible for gathering them.

Gains in terms of increased agricultural production are visible in almost all cases, certainly wherever they were attempted. In Hivre Bazar, not only does part of the kharif crop get irrigated, but so does the rabi jowar. And there is also a summer crop of vegetables. Plot-level productivity has also increased. Similarly in the TBS case study, the construction of the johads has led to significant expansion in area irrigated and there has been an increase in wheat productivity. This is also the case in Nathugadh. In the RNRRC project, certain crop productivities have increased, though differentially across the watershed, although they are the consequence of the introduction of new varieties rather than increased water availability. The increases in incomes in the GC programme also came through agricultural diversification and promotion of home gardens. Although small in absolute terms, these increases were probably crucial to the beneficiaries. Additional gains from improved animal husbandry seem to have occurred primarily in Hivre Bazar, but all these sites also saw spill-over benefits of agricultural production increases in the form of increased wage employment for landless households. The one case where agricultural production gains were limited is DTLVS.

Finally, the CBNRM interventions in most cases have also led to new avenues of employment and income generation. For example, in Hivre Bazar, many households have benefited from the EGS because of the NGO's intervention. In the cases of Lingmuteychhu, solar driers made by SHGs will most likely earn them income in the future. In Hivre Bazar, some members of the BPL SHG have benefited from the loans provided to them for starting a new income-generating enterprise, mostly cattle rearing. In the case of the char villages too, GC successfully extended loans to some women to start new enterprises, such as plant nurseries, poultry farming and animal husbandry. On the other hand, the SHG formed by Utthan in Nathugadh failed and also did not benefit the Koli Patel women who were never given loans from the credit. Similarly, the SHGs started by DTLVS have stagnated somewhat.

Although livelihood gains are significant in every case, we also see the possible limits to how much of a contribution these gains can make under conditions of expanding populations, declining resources and terms of trade. In Gopalpura, the villagers believe that emigration has increased in recent years because of the reasons cited earlier. In the char villages of northern Bangladesh, though the intervention has provided important supplementary support, it has had little impact on the migratory pattern of landless families. Migration for work continues to remain the primary livelihood option for more than 50 per cent of the households in the region. Similarly, substantial migration to the plains continues in DTLVS' area, and the slow down, if any, has more to do with declining opportunities in the plains and some employment guarantee schemes in the hills. Even in Hivre Bazar, emigration has stopped or reversed not just because of watershed development and its impact on agricultural activity, but also because the entrepreneurial leadership has taken advantage of various state programmes to provide employment during drought years and improvement in the social atmosphere in the village.

Sustainability

The gains on the ecological sustainability front have been much more mixed. Whereas one-time resource regeneration has taken place in most cases, longer-term sustainability of the regenerated resources has been harder to address. The Hivre Bazar and DTLVS initiatives have probably been most successful on this front. In Hivre Bazar, not only has there been significant regeneration of the water table within the village, but mechanisms are in place so as to ensure that groundwater is not over-exploited. There is also an understanding that drinking water needs will get priority over irrigation during years of drought. In the Doodha Toli region, communities are collectively regulating regenerated forests and fodder sources.

In the case of Nathugadh, there is no sign of the community moving towards regulation of groundwater use. Despite the summer drinking water stress, there is no community understanding regarding which uses to prioritise at least in years of drought. In the case of Gopalpura also, the sustainability issue has not been prominently addressed: some johads have begun to silt up due to the lack of catchment area treatment and some are damaged but not repaired. There is a claim that water use is being regulated by banning water-intensive crops, but our investigations suggested that such crops cannot be grown in this particular site anyway. In Lingmuteychhu and in the charlands, the ecological sustainability component has been much less explicit, though in the latter there have been efforts to promote organic farming. In the case of Lingmuteychhu, the problem of soil erosion still seems to be largely unaddressed and the promotion of certain new varieties of crops might in fact increase fertiliser-intensive forms of agriculture. The question of forest use sustainability has also not been foregrounded. Some catchment area treatment issues are being dealt with gradually.

Not many interventions have focused on and hence made any contribution towards 'off-site' environmental concerns. The DTLVS initiative is the only exception. Here conscious efforts at generic recharge of water have been carried out, although on a limited scale and with limited local support. In Hivre Bazar, watershed development in the village has inadvertently benefited the village downstream (as acknowledged by people in that village). In the only case where upstream and downstream communities were both part of the intervention, that is, Lingmuteychhu, there has as yet been no progress on resolving upstream–downstream water sharing issues. In Gopalpura, forest cover within the village boundary has improved to some extent due to the successful community regulation, but at the expense of forest elsewhere and more arduous work for women. In the char lands, GC has consciously promoted organic methods in home gardens, but these methods have not spread to regular agriculture.

Are the interventions robust enough to sustain in the face of changes at a larger scale? Hivre Bazar appears to be paying some attention to these possibilities by trying to intervene in other villages before their sinking of bore wells begins to affect groundwater in Hivre Bazar itself. The DTLVS has also struggled successfully in the past against the forest department efforts to log forests being used by villagers. In the case of GC, the positive outcomes of the experiment might get watered down by developments at the larger level, namely, construction of roads and embankments in the region which might increase erosion and floods in the char villages.

Equity

When thinking about how equitable the impact of these interventions has been, one has to keep in mind the multifaceted nature of the concept and also the various inbuilt inequities in social relations that the interventions have to contend with. At one level, most interventions (where they have a choice) have tried to some extent to target relatively poorer or ecologically vulnerable areas or villages. One could even argue that by virtue of working in ecologically-vulnerable regions, they are targeting those communities that have been left out of mainstream development programmes. On the other hand, intra-village equity—either in economic terms or caste terms—has not been a strong dimension of most of these interventions, with the exception perhaps of GC's work. The benefits have often accrued in proportion to landholding. In some cases, subgroups have been left out. Some impact is visible on the gender front in a few of the cases, although much more needs to be done.

The equity of outcomes in economic terms is strongly related to the context in which they are implemented. Communities are strongly differentiated to begin with. Watershed interventions naturally tend to favour the landed and the richer amongst them—those who own more land and those who own land in the lower reaches. This was the case in Hivre Bazar, Gopalpura and Nathugadh. Interestingly, even domestic benefits such as drinking water can get inequitably distributed when user costs are sought to be recovered, such as in the case of the landless Koli Patels in Nathugadh. Traditional resource rights might be originally inequitably distributed—as was the case with water rights in Lingmuteychhu. Caste relations can tend to exclude some from the decision-making mechanism. Equally important, lack of regulation of resource use can have cross-sectoral implications (irrigation overuse means drinking water decline)—something that the leadership in Hivre Bazar has anticipated and addressed, but has not been addressed in Nathugadh. Even when the context is relatively less differentiated, the outcomes can strengthen or maintain inequities—as happened in the Doodha Toli region—unless strategies are carefully and consciously chosen and implemented. Gono Chetona very consciously targeted poorer households and women, and the results are there to see. In Hivre Bazar, there was a conscious attempt to redress the imbalance to some extent through special activities such as loans for the landless. Attempts were also made to make water accessible to the poorer farmers through water sharing arrangements. Others only chose to address equity issues at the inter-village level, if at all, by targeting poorer or less-endowed villages, as for instance the focus on Nabchey village in Lingmuteychhu indicates.

It is not easy to separate out caste inequities (in the Indian cases), because in many cases economic inequities are strongly correlated with caste. However, one sees that the interventions do not address questions of caste head-on and in the process caste-based discrimination may get reproduced. For instance, in DTLVS' programme, traditional drinking water sources were rejuvenated using the labour of all households, but access to these sources continued to be on caste lines.

In Gopalpura, caste issues were never brought to the table; caste-based discrimination continues and developmental programmes (implemented both by TBS and the state) continue to favour the Meenas. Caste-based discrimination also got reproduced in the women's SHG formed by Utthan in Nathugadh. On the other hand, in Hivre Bazar, the leadership has made concerted efforts at reducing the social discriminations in the village to bring about unity in the village though the land question has not been addressed.

In the case of gender equity, again GC's intervention addressed the issue most explicitly by involving the women of the households in income-generating activities so as to improve their economic status. The DTLVS also tried to address some issues of importance to women, such as improving availability of fuelwood and fodder. In other cases, gender concerns have been addressed largely by trying to provide some alternative forms of employment or sources of credit to women. For example, in the Lingmuteychhu Watershed, the women's SHGs in Limbukha has started marketing solar driers meant to dry meat, a possible source of revenue if the idea is taken to its logical end. In Hivre Bazar, the women members of the BPL SHG have availed of loans to buy goats. It is unclear, however, to what extent these SHGs are sustainable. Many women in Hivre Bazar complained that they were not able to make their contributions over the last few years due to drought. Moreover, the long-term vision of SHGs acting as a platform for employment diversification has not really occurred because women are mostly busy with agricultural work and hence do not have spare time. What SHGs have achieved, at least nominally, is provide women with a platform from which they *might be* increasingly able to play a more important role in the matters of the village. In practice, however, this does not seem to have happened. For example, in the case of DTLVS' initiative, after a promising start for the MMDs, which not only functioned as chit funds but also took on some NRM responsibilities, they have lost steam. Today, much of the decision-making remains in the hands of the male villagers. So too is the case in Lingmuteychhu, despite the fact that women appear to play a much more active role in the public domain.

In some cases, it appears as if women's work has, in fact, increased due to ecological rejuvenation. In most watershed experiments (TBS,

Utthan and Hivre Bazar), the work burden has increased with the intensification of agricultural activity—with no consequent improvement in their social status. In the case of the TBS, the ban on tree felling has imposed a further burden on women who have to travel longer distances to collect fuelwood.

Democratic Decentralisation

We have examined the question of democratic decentralisation at three levels: have people participated, have the communities as a whole gained greater control over the management of the natural resources they use (or increased their capacity to hold the state/NGOs accountable) and has there been an internal democratisation of the process of community decision-making about these resources.

The concept of participation itself has several layers or dimensions (see, for example, Cohen and Uphoff, 1980). Perhaps most important is the distinction between participation in implementation and participation in design. Equally important is the question of 'who participates'. In all the six cases, mobilising the villagers has been a major component of the activities and the project implementers have generated very substantial levels of participation in implementation. In most cases, villagers or certain groups have also had a significant say in the design of at least some of the interventions. For instance, the villagers in Hivre Bazar collectively decided that private grasslands would be treated under the AGY but then would be available as a commons.[4] In the area where DTLVS works, we see different activities (afforestation, wall building, etc.) being taken up in different villages depending upon priorities expressed by the villagers. In Nathugadh, not only did the selection of sites for check dams have the full involvement of the farmers, but the idea of providing farm ponds to farmers who did not benefit from the check dams emerged through these discussions as well. The RNRRC conducted PRAs to elicit villagers' priorities and designed its programme on the basis of this information.

It must be noted, however, that the 'breadth' of the participation has been uneven. While the intervening agencies always talk about 'community mobilisation' and 'villager participation', at least in two cases—Nathugadh and Gopalpura—there have been clear exclusions of certain groups. This has been sometimes in spite of NGO efforts and sometimes due to inadequate attention to this issue, as we shall discuss in the next section.

In terms of decentralisation, the gains seem surprisingly limited. Although Hivre Bazar, largely due to the visionary leadership of its sarpanch, managed to extract maximum benefits from the state AGY programme and also attract funds from programmes, there is no sign that this has gotten institutionalised; in fact, the AGY programme has collapsed and the state is back to implementing more routine (that is, bureaucratically-controlled) watershed development programmes. The DTLVS' efforts made little dent in the level of community control vis-à-vis the state; in fact, existing community institutions such as VPs have continued to be eroded by increased state interference and that too in spite of the state of Uttarakhand being created in response to grass-roots agitations for autonomy for the hills. In Rajasthan, the TBS, in spite of its strident anti-state rhetoric, has not managed to increase villager's control over forest or water resources significantly, while Utthan does not seem to have gone into this aspect at all in Nathugadh (although the NGO has played an important policy advocacy role in other programmes). Similarly, decentralisation was simply not on the agenda of the sustainable livelihood programme in the charlands (GC).

With regard to internal democratisation, the achievements are even more limited. Again, Hivre Bazar is a bit of an exception, because they have a functioning gram sabha and the leadership seems to encourage debate and discussion. The DTLVS had the opportunity of such democratisation, but failed to link its MMDs with the actual resource use and developmental decision-making or to get involved with existing institutions such as VPs or gram panchayats. Elsewhere, internal democratisation has not been on the agenda, even though the need for it is fairly obvious, as in the case of Gopalpura (TBS), where participation in the gram sabha is marred by the observance of ritual purity between the Meenas and the Balais and also by the imposition of cultural norms that prohibit women's participation in the public space. In Nathugadh, as mentioned earlier, while all the men of the

Kanbi Patel caste were co-opted in the decision-making process, the women and the landless Koli Patels remained excluded, despite some efforts by Utthan.

Summary of the Outcomes

If one tries to summarise the gains or headway made along the four dimensions of concern that we have identified, one may say that the gains along the livelihood enhancement dimension are prominent, whereas those along the sustainability and equity dimension mixed or limited. Participation levels in implementation are uniformly high—much higher probably than many state-implemented programmes—but transferring real control over resources to communities and internal democratisation of community-level decision-making processes has not happened to a significant extent. Why variations occur, generally in spite of efforts at times by the intervening agency, and what role is played in each location by the local and larger context and NGO understandings of CBNRM is something that we now try to explore.

Understanding Outcomes: Strategies, Visions and Micro–Macro Factors

How does one understand these varied outcomes? Each chapter, whether explicitly or implicitly, highlights the complex interplay of intervention strategies and the micro-context in which such interventions are made. Most analyses of CBNRM-type interventions tend to stop at this level— they ask whether 'improved crop varieties make a significant contribution to livelihood enhancement', or 'SHGs work better than PRIs' or 'success is more likely in homogeneous communities than in heterogeneous communities', or 'sustainable management is more likely with indigenous technologies as against modern technologies' and so on. We believe, however, that the choices of strategies and the decisions to intervene through a particular programme or approach are themselves deeply embedded in the way CBNRM itself may be envisaged, that is, in perceptions of what is to be achieved and what constraints the organisation is operating under (and the constraints themselves may be the outcome of strategic decisions taken earlier). While the factors clearly interact,[5] for the sake of exposition we shall first discuss what appear to be the 'proximate' factors: the micro-context, and the technologies, institutions and processes adopted by the intervening agency in this context. We shall then go into the deeper question of what governed the choices of strategies.

Constraints Imposed by the Micro-context

At the outset, it is important to recognise the significant and often constraining role played by the micro-context. To begin with, these initiatives are all located in ecologically-vulnerable and marginal parts of South Asia. This was particularly true of the GC initiative, where the threat of floods probably looms large over everything else. Furthermore, economic, caste and gender divides were deeply embedded in the social fabric in most sites, although they may have been slightly less of a barrier to community mobilisation in the case of the Doodha Toli region of the Himalayas. Certain differences were particular to a site—such as the peculiar situation of the 'bhagiyas' in Nathugadh. But many differences—for example, major variations in landholding and even access to water—occurred across most sites. In Hivre Bazar, Nathugadh, Gopalpura, Lingmuteychhu and the charlands, for example, landholding and/or access to water inequalities tended to create barriers to the intervention in the first place and to aggravate the differential impacts of community-level interventions. Moreover, the notions of equity and sustainability are not particularly strong in many communities. For instance, there was strong resistance to the notion of community regulation of individual groundwater use in Nathugadh. In most sites, the concept of equity that people subscribed to was the limited one of 'something for everyone'. Caste-based discrimination in access to drinking water continues in an otherwise more 'communitarian' Uttarakhand. To an extent, as we shall see next, the intervening agencies tried to tune their intervention to the context, although not always successfully.

Strategies of Intervention and Community Building

We described the outcomes along four distinct dimensions. But the processes that lead to those outcomes are not separable, that is, one intervention typically has impacts along all dimensions. Broadly speaking, it appears that 'how' the intervention takes place, both in the technical and the social sense—is the key to the kinds of outcomes one observes in a particular context.

In the semi-arid context, it was natural that the emphasis would be more on water, on drought proofing or increasing irrigation or drinking water. Thus, one sees all three groups that work in this region (Hivre Bazar, TBS and Utthan) giving priority to this aspect. But the manner in which they did it differs significantly. Tarun Bharat Sangh focused only on building water-harvesting structures. It did not carry out comprehensive watershed development, particularly catchment area treatments that would have reduced siltation in the johads and hence enhanced their life. And this 'technical' choice also had implications for the 'social' process—building johads required the cooperation of the (male) farmers on whose land the johad was being built or at most those who were going to benefit from the johad, whereas treating the entire catchment would have required involving the entire village community and thereby opened up possibilities for negotiations between different groups. Utthan went a step further by adopting a watershed approach. But in practice, the focus was only on putting up water-harvesting structures—check dams, farm ponds and well recharge. The earlier work on drinking water and the later work on watershed development were not well integrated and the increase in water availability could not be harnessed to relieve summer drinking water stress because the additional water was simply used for agriculture. It may be noted here that drinking water stress is an impact that is felt more by the women than by male landholders alone. Hivre Bazar implemented a comprehensive watershed treatment plan that included not just constructing water-harvesting structures and treating private and public lands, but also regulating water use. In other words, water-harvesting structures provide immediate livelihood enhancement, but inherently also tend to undermine the sustainability and equity dimensions. Even watershed development, if defined solely as treating the entire catchment to ensure a combination of recharge and water harvesting, has inbuilt inequity and unsustainability dimensions that need specific attention (Joy et al., 2004). This aspect seems to have been adequately recognised only in Hivre Bazar, where an attempt was made to make the recharged water more widely available, although even here the effects of closure of the common lands were felt most sharply by the women in the landless households.

Similar differences exist in the mountain regions. Whereas DTLVS was able to mobilise women from all households to participate in the regeneration of oak forests (that are crucial to the fuelwood and fodder collection that is done by the women), RNRRC's focus on agricultural technology innovations meant that the 'community mobilisation' actually took place much more at the individual level. Community mobilisation of the village in the latter case was limited to a few villages where community forestry was practiced.

One can look at the problem from the social process side and see similar links between intervention and outcome. As suggested earlier, the efforts invested in building institutions of collective action varied significantly across the interventions. In interventions where households are targeted, the role of collective action or decision-making is inherently limited. In GC's work, beneficiaries were chosen from a compact block in each village and in group meetings, but the focus of the initiative was on individual households and the choice of households was largely based upon a baseline survey conducted earlier. Even if women's groups were formed, they were really not 'self-help' groups, but simply focal points for easy dissemination and training. In Lingmuteychhu too, women's groups were mostly for rotating capital and starting new income-generating programmes and less for initiatives that required collective action.

Some of the groups, such as RNRRC, have argued that it chose to focus on individually-beneficial activities so as to earn the trust and good will of the community before it moved to more complicated community-level issues such as water sharing. However, experience from other areas suggests that even confidence-building measures can and perhaps should be done at a collective level to the extent possible— such as the idea of setting up village development funds (*gram kosh*) or common infrastructure that is used by Seva Mandir in Rajasthan or Aga Khan Rural Support Programme (AKRSP) in Pakistan respectively (Ballabh, 2004; Wood and Shakil, 2003).[6] The experience of Hivre Bazar points in the same direction: strategic interventions that require collective action, like improving the village school, building a temple and *akhara*, are useful and perhaps necessary before getting into the more complicated question of watershed development—though undertaking watershed development is no guarantee at all that distributive issues will be tackled.

There is, however, a need to distinguish between community-building/collective action and redressing inequalities. While it is true that giving non-land based benefits to the landless reinforces at one level the divide between the landed and the landless, watershed development initiatives (even if they included the landless) also do not necessarily tackle the inequality question head-on as the Hivre Bazar example illustrates. Interestingly, none of the cases studied sought to directly replicate the more innovative approach of giving equal rights in resources regenerated, such as the equal water rights concept in Sukhomajri and Pani Panchayat, and thereby mobilise the landless in managing the common resource.

Other interventions such as SHGs were also set up to tackle win–win situations in terms of rotating some capital or getting access to a market—for homogeneous sub-groups within the larger village community. And they were generally not directly linked with NRM, whether it was the women's groups in Hivre Bazar or those in Nathugadh. It has been argued in the literature, and could be argued by the implementers in these cases, that SHG formation and management provides an arena where historically backward groups, such as women, can experience a sense of solidarity and develop the skills necessary to engage in higher level decision-making (such as in the gram sabha). But we did not see such linkages developing in these cases. Even in the DTLVS' work, where the women's groups were involved in reforestation, they did not step into domains where resource management structures already existed, namely, the male-dominated VPs.

In none of the cases have new broad-based institutions been consciously set up and nurtured, and only in a few cases was the involvement of existing democratic institutions such as panchayats sought. In Nathugadh, the watershed guidelines require the setting up of a broad-based watershed committee, but this requirement could not be properly operationalised. The RNRRC has also made some efforts to involve the local bodies, namely, the geogs, and ultimately aims to have these bodies become responsible for the interventions In Hivre Bazar's case, the panchayat was fully involved. But in most other cases, the NGOs have stayed away from the existing local bodies. There may be good reason for this difference—the conditions in Hivre Bazar favoured gram panchayat involvement more than anywhere else, not least because the NGO leader was also the panchayat head, and also because the boundary of the gram panchayat coincides with that of the revenue village and forms a semi-autonomous upstream area un-affected by other micro-watersheds. In most other cases, the local bodies cover too large an area and were seen as more concerned with implementing developmental programmes of the government in a haphazard manner rather than providing a platform for CBNRM.

Finally, none of the formal institutions—whether SHGs or water-shed committees or gram panchayats—would function or last without being part of a wider cultural process of building (or rebuilding) the 'community' in a form that is on the one hand conducive to collective action, but on the other hand also infused with some environmental and social concerns. This process was most consciously attempted in Hivre Bazar. According to Popatrao Pawar, the sarpanch, the main problem confronting the village of Hivre Bazar prior to his arrival was factionalism and in-fighting and other social problems such as alcoholism. From the outset, he was very conscious of the need to over-come factionalism and he took some very strategic steps to reduce inter-group tensions. This has also presumably limited the extent to which he can push the equity dimension. At the same time, he always kept as a larger goal the building of a community that was infused with certain concerns, whether explicit or implicit, that led to addressing the problem of alcoholism, limiting groundwater exploitation and ameliorating the inequitable impacts of the watershed development approach. But such a process of community-building has taken them more than a decade and is in some sense continuously going on, not stopped. The DTLVS attempted some amount of awareness building on environmental issues, but with a small organisation attempting to work across several villages, its work did not or could not go deeper into issues in any one village. In contrast, although TBS has invoked the notion of village community repeatedly, it seems to have taken this as a given, as pre-existing, and hence has made little contribution to the rebuilding of the contours and content of the community. In RNRRC's work, although there was a significant emphasis on 'process', it has seemed limited thus far to ensuring greater 'participation' and not on building of community; whereas in the other two cases (Utthan and GC), the attention given to this process was also limited, at least partly because of their project mode of operation, as we shall discuss next.

'Capacity' of the Intervening Agency

To what extent are these processes and outcomes influenced by the capacity of the intervening agency? Capacity can be defined in various ways and a full discussion of this aspect is perhaps beyond the scope of this study, since we did not gather detailed information on the organisation's capacity in each case. But some exploration of this issue is possible.

There are certainly differences in the capacities of the different intervening agencies. In terms of skills of the staff, clearly RNRRC staff were strong in their 'technical' knowledge, but had limited training in or exposure to social issues and processes of mobilisation. On the other hand, some of the staff of Utthan and GC had some training in social work, whereas in the case of Hivre Bazar and DTLVS, the 'staff' were simply volunteers from the local village/villages with almost no exposure to the kinds of debates and experiences that their leaders (Bharati of DTLVS or Popatrao Pawar in Hivre Bazar) had and continued to get. In the case of TBS, the staff were drawn from the local communities, but were full-time employees of the organisation. But it did not appear that they had received much training or exposure to issues concerning rural development or social work. Similarly, there are differences in the access to financial resources that the different organisations have had. After factoring out the scale of operations, one still sees that the support for DTLVS has been very modest, whereas the support available to the Hivre Bazar group, and to Utthan, TBS and GC, probably fall in the mid-range. The RNRRC has perhaps been most heavily supported of all the implementers in our cases.

To a large extent, however, these capacities are not simply 'givens', but rather the outcome of larger choices made by the organisation in terms of at what scale and over what time frame it wants to intervene, using what mixture of local and external personnel, taking support from (and hence accepting conditionalities from) which donors or government programmes and so on. For instance, working intensively in only one village (Hivre Bazar) over a long time period requires limited staff as compared to any attempt to work across many villages. And working across many villages and implementing government pro- grammes (Utthan, GC) necessarily means recruiting a large number of staff and working with project deadlines and accepting the con- sequent 'routinisation' of rural development work that is almost bound to follow. And the fact that smaller organisations working within a relatively limited region (but longer time frame) seem to have made a somewhat deeper or more sustained impact on certain dimen- sions (and Utthan had a similar impact when working intensively in the Bhal region prior to its entry into watershed development pro- grammes) seems to suggest that formal skills and resources may have relatively little to do with the outcomes and that, in fact, the capacity to engage in community-building and work on a need-based approach may probably decline with the expansion and formalisation of the intervening agency, as we believe has happened in the case of Utthan's work in Nathugadh.

State Policy and Donor Influences

In the introductory chapter, we had highlighted the ubiquitous role that NGOs garnered for themselves or were given as CBNRM was mainstreamed in different ways by governments across South Asia. It seemed, therefore, that the state was providing enough space and indeed positive encouragement for a community-based approach to NRM to be implemented widely. Nevertheless, an examination of the specifics of different decentralisation programmes suggested that there is substantial variation across sectors and across countries in the actual quality and extent of policy support for this approach. We, therefore, kept open in our analysis the possibility that NGO-led interventions might, on the ground, face difficulties in bringing about a genuine transition to community-based resource management. It would be important to examine whether this factor might explain some of the variations in outcomes that one observed across the six cases.

The case studies suggest that the relationship between the state, CBNRM programmes and individual NGOs and villages is complicated. On the one hand, the state has provided some active support (to a certain kind of CBNRM) to NGOs in several cases, ranging from state funding through the AGY and EGS (in Hivre Bazar) or the watershed development and drinking water programmes (in Nathugadh) to actually treating the initiative as a pilot scheme for a widespread shift to CBNRM (in Lingmuteychhu). Gono Chetona's work is also supported entirely through a state-approved and UNDP-funded programme. The DTLVS is not directly involved in state programmes, but it drew upon the state's renewable energy initiatives during the early work on solar PV lighting and has received small amounts of support indirectly from the central government. Tarun Bharat Sangh is the only case where the organisation draws almost all funds from non-governmental donors, but even here it is important to note that it did figure as an NGO partner in the state-implemented PAWDI programme.

On the ground, however, there has been a mixture of mostly collaboration or non-interference and some tension with state agencies. The Hivre Bazar group managed to get the cooperation of state agencies—with each 'successful' implementation gaining them the confidence of the bureaucrats. Utthan faced some delays in release of state funds that led to misunderstanding with the villagers, although this was eventually rectified. In GC's work, the state does not figure very much—GC's work appears to proceed largely independently of state structures at the regional or local level, but is funded. The DTLVS did initially confront the forest department over certain policies, but later the department became its major client for sale of seedlings from its village nurseries and DTLVS' efforts at reforestation and water recharge has faced no problems from state agencies. Perhaps the only case where friction with the state has been persistent is the case of TBS—mainly over the question of whether state permission is required to build johads. But even here, on the ground in Gopalpura, there seem to have been significant phases or points of collaboration—as witnessed by the involvement of the TBS in the state's PAWDI programme.

From a larger policy perspective, the role of the state in facilitating CBNRM seems to have been more mixed. In the case of India, while the state has tried to restructure its watershed development programmes to make them more 'participatory', it has provided little by way of concrete policy support for (say) enabling regulation of groundwater use (even where state funds have been invested for recharge) or promoting sustainable agriculture. Even in the Bhutanese case, proactive adoption of CBNRM as a state policy has not yet translated into any revisions of (say) water rights or forest rights that might be essential for more equitable and sustainable resource management. At the same time, the state has tried to implement its own sectoral decentralisation programmes that often make only cosmetic changes but co-opt some of the villagers in the process. For instance, in the forestry context, Uttarakhand is witnessing a systematic policy of reducing the autonomy of the historical VPs and the typically top–down implementation of JFM or so-called 'participatory watershed programmes' through the gram panchayats as simply another opportunity for partaking in state/donor largesse. And sometimes, state programmes can work at cross-purposes to on-the-ground efforts, as in Bangladesh, where the flood control programmes seem oblivious to the efforts put in by GC in micro-level rehabilitation and flood-proofing.

Conversely, in the case of Hivre Bazar, the design of a state programme not only created the space for NGO participation, but actively structured their efforts in the direction of sustainable resource use (through its emphasis on suspension of open grazing and firewood hacking), social goals (by asking villagers to control alcoholism and implement birth control), and decentralisation (by giving a major role to the village general body). One could also argue that the structural conditions favourable for a more democratic implementation of watershed development are in place in Maharashtra due to its more decentralised panchayati raj structure. It is, therefore, ironical that AGY was subsequently stifled for funds and watershed development programmes were brought back into the normal bureaucratic channels.

Linked to the issue of state policy are the policies of donor agencies. In fact, in most cases, one cannot really separate the two, because the donor funding is routed through the state or at least has state approval. But funding, in general, and donor funding, in particular, seems to push the CBNRM concept much more into 'project' mode, thereby imposing constraints in various subtle ways. Generally speaking, projects have short time frames during which community-building and post-implementation follow-up get short shrift. Gono Chetona, for instance, strongly believes that it is constrained by the limited time it has within the project period and limited financial backing during the post-project period to carry out its plan to continue working with the existing women's groups and form new ones (and also to build a network of existing women's groups). Utthan could not systematically take up, because of bureaucratic arbitrariness, watershed development in villages neighbouring Nathugadh (which it believed was necessary to maximise the benefits of the soil and water conservation efforts). The haphazard and 'spending' oriented approach in various donor-supported programmes in Uttarakhand has already been noted—DTLVS' decision of not participating in such programmes still did not help its cause as it tended to get sidelined. The TBS' efforts to garner financial support from other donors also appears to have pushed it towards building more physical structures across a large landscape rather than focusing on engaging in more intensive and sustained work in fewer sites (Kumar and Kandpal, 2003). Even AGY failed to get integrated into state policy and remained a programme that was slowly strangled by the bureaucracy, although the innovative leadership in Hivre Bazar ensured that it derived maximal gains from the scheme. Perhaps the only case where projectisation may not have been significant is that of RNRRC, where the donor has provided ample time and scope for innovation and bringing about community involvement.

One need not take the implementing agencies as simply 'recipients' of higher-level policy. One could also ask whether, knowing that state and donor policies directly and indirectly influences the possibilities of success at the micro-level, these agencies actively sought to modify these constraints in the projects and programmes they were involved in or whether they have engaged with programmes or policies that hold some positive potential. On the whole, there appears to have been

limited engagement. Popatrao Pawar from Hivre Bazar has been actively involved in trying to keep the AGY programme alive, but with little
success. Utthan has been active in state-level policy issues, but it appears to have focused more on the drinking water programme than
on watershed development or groundwater regulation, and the engagement seems to be at the level of ensuring better programmes rather
than major changes in resource governance. The TBS has agitated against
the state government (and mobilised support from several other civil
society groups) in the name of 'autonomy of local communities' to
build johads, but has not actively engaged with panchayati raj institutions.[7] The DTLVS has stayed away from the ongoing battle between
village VPs and the state of Uttarakhand, possibly because of its very
limited resources. It does not appear that RNRRC or GC have lobbied
their respective governments for any major policy changes either.

In short, the state has overtly supported CBNRM, but fallen short
of providing the substantive and multidimensional policy changes that
would be required to internalise and support the concept. Donor support in the form of financial resources comes at the cost of increased
project expectations and implementation scale in a short time frame.
Most of the implementing organisations seem to have gone along or
tolerated these constraints rather than actively and consistently loos-
ened them up. For at least some of them, these are not seen as major
constraints because their own perceptions of what they want to achieve
is perhaps more in line with state and donor perceptions.

Visions—Plurality or Limitations?

The differences we observed in the strategies pursued across the six
cases are stark enough to call for a deeper analysis. It seems that these
differences cannot simply be explained in terms of variations in the
local context or even variations in the capacity of the intervening
agency alone. The fact that, for instance, in a semi-arid context some
organisation chooses to focus on only constructing water-harvesting
structures on common or private land whereas another focuses on
comprehensive watershed development (with government funds
invested only in common land treatment) and also imposes some regulation of water use suggests that significantly different ideas of CBNRM

are at work. The fact that, in a Himalayan context, one organisation chooses to focus substantially on mobilising all the women in a village for reforestation work whereas another (working in a matrilineal society) focuses more on technical interventions on individual farm plots (even if as a means to gaining local acceptance) suggests significantly different ideas of how communities might be mobilised and to what end. And the apparent 'shortfall' in achievements on certain dimensions such as equity or democratisation across most interventions naturally also raises the question as to whether we have imposed a totally 'external' notion of what CBNRM is supposed to be for. We shall try to outline next what we believe are the visions underpinning the interventions and highlight the key differences and similarities across cases and how they might relate to the choice of strategies. We are aware of the difficulties of this task. First, some organisations such as Hivre Bazar and DTLVS were much more forthcoming (and maybe more articulate than others) about their ideologies, whereas others were less so. Second, our own efforts at eliciting an understanding of the organisation's visions varied somewhat from case to case, although we did try to fill gaps during our feedback workshop and through further correspondence. Third, it is always hard to separate the organisation's actual vision from the way it got (or did not get) operationalised in a particular context or case. We have tried to deduce their vision from not just their achievements in these locations, but also the activities undertaken (regardless of whether they succeeded in them), including activities in locations other than just those studied by us.

Differences in vision could be at two levels: the normative concerns that drive the agency's work and their conceptualisation of what CBNRM is and how it might enable them to address these concerns. Our understanding of the normative concerns underpinning the organisations involved in the six cases is summarised in Table 8.1.

It is apparent that there is a fairly uniform concern for enhancing livelihoods, but much more variation along other dimensions. Concern for sustainability are most sharply articulated in the case of DTLVS, with a broad focus on '*jal, jangal, jameen*' (water, forests, land conservation) and in Hivre Bazar, even though limited to local resource use. Interestingly, TBS seems to have a concern for sustainability, including downstream concerns, if one goes by its efforts in setting up

TABLE 8.1
Differences in Normative Concerns Underpinning the CBNRM Initiatives

Case	Livelihood Enhancement	Ecological Sustainability	Equity	Democratic Decentralisation
Hivre Bazar	Meeting both domestic and production needs, beyond subsistence	Ensuring water availability for the future	Mitigating inequitable impacts to the extent possible	Community's implementation or control over implementation of development/NRM programmes; building a strong and self-reliant 'community'
Utthan	Meeting first domestic and then production needs, beyond subsistence	Visualised in terms of self-reliance, mainly local resources	Reducing gender and caste inequities; mainstreaming women's concerns	People's participation in programme implementation; especially that of marginal groups
TBS	Meeting the water requirements for domestic and production needs	Ensuring forest conservation and water availability locally and downstream water availability	Silent	Community's implementation or control of development/NRM programmes
RNRRC	Meeting production needs, beyond subsistence	Not clear/silent	Tackling inter-village inequities, no direct targeting of poorer households	People's participation in programme implementation
DTLVS	First meeting domestic needs and then production needs, beyond subsistence if necessary	Ensuring conservation of forest and water availability for local and off-site stake-holders, including wildlife	Reducing gender inequities/empowering women	Community's empowerment to control development/NRM programmes; building a self-reliant community
GC	Meeting the need for supplementary income, after securing lives against floods	Reducing vulnerability to floods, ensuring sustainability of the agricultural practices introduced	Reducing absolute poverty and gender-based inequity	People's participation in programme implementation

the Aravari River Parliament and its publicity materials. On the other hand, Utthan and the RNRRC do not seem to have a well thought-out position in this regard, although Utthan has tried to focus on the concept of self-reliance in its wider drinking water programmes. And GC's idea of sustainability is a very specific one—that of ensuring sustainability of livelihoods in the face of recurring natural disasters (floods).

On the dimension of democratic decentralisation, there seem to be clear differences between some (GC and RNRRC) who are mainly focused on increasing people's say in governmental programmes, and others (Hivre Bazar, DTLVS, TBS and Utthan) who feel the need for increasing the 'sovereignty' of the community vis-à-vis the state. There are cross-overs, of course. The RNRRC, for example, while not lobbying for more control to communities are interested in increasing the role of local bodies. Utthan sees the process of making marginal voices heard a critical part of increasing people's participation as a whole. On the other hand, the Hivre Bazar group does not really see itself in opposition to the state, but sees the need for having a community that is strong enough to make the state implement its programmes 'properly'. Nevertheless, this broad distinction between those focusing on 'better, more participatory' programme implementation and those who are seeking to go beyond participation to more devolution of resource control to local communities is clear. And in this notion of devolution, the local community is often viewed as homogeneous, apolitical and somehow distinct from the highly politicised bodies of local self-governance. In other words, while decentralisation is a goal for some of the intervening agencies, internal democratisation of the decentralised resource management institution is not generally on the agenda—certainly not for TBS, marginally so for DTLVS.

The differences in vision get more sharply etched and linked to the choices of strategies and thence the outcomes when one examines how the idea of CBNRM is articulated in relation to these concerns. If the role of the local community and the role of natural resources is central to the broad idea of CBNRM, then how this role is visualised varies significantly across initiatives. Further, if one says that the *process* of a transition in resource management towards more community-based NRM cannot take place without some intervening agency (individual or group, local or external) and without reference to what the state is expected to do or not do (because the state has traditionally implemented top–down NRM, but also promoted rural development), then one sees additional differences in the visions of CBNRM. All these aspects in some sense also relate to the 'problem perception', what the intervening agency sees as the core issue that calls for community-based NRM actions. We have tried to sketch out in Table 8.2 the basic features of the thinking of each intervening agency across these dimensions.

There are clearly major differences in the way the problem is perceived and what role different actors and components of CBNRM are supposed to play in solving this problem. At one extreme, GC is really not focusing on a problem of resource degradation, but that of poverty of certain households under conditions of extreme vulnerability to natural hazards. In that sense, it might be inappropriate to impose on GC's work criteria that are related to CBNRM rather than simply to poverty alleviation and flood management. The RNRRC, though recognising watershed level problems, has in the short-run focused mostly on individual poverty that may be the result of various NRM and non-NRM factors, such as the productivity of individual lands or livestock or lack of access to credit. Collective action thus far, if at all required, is mainly for learning or sharing of monetary capital. While the link between access to water and production is understood, the question of water sharing has not been tackled head-on as yet though the formation of a watershed committee is a start. For Utthan, the main purpose of collective action is to make the community's (and particularly women's) voice heard in the process of implementation of state programmes—whether NRM-related or otherwise. But one is not very sure whether Utthan thought through the significance of the need to encourage collective decision-making not only when selecting sites or beneficiaries, but also in maintaining the physical structures, of coordinating catchment area treatment along with check dam construction and of eventually regulating resource use.

Amongst those who recognise that collective action for resource regeneration and regulation is at the core of CBNRM, there seems to be a belief that there is something called a 'local community' that is homogeneous in its interests and distribution of power and can

TABLE 8.2
Differences in Problem Perception and Notion of CBNRM

Case	Perception of the 'Problem'	Role and Scope of NRM	Role and Form of Community	Role of Intervening Agency	Role of the State/Its Form at the Local Level
Hivre Bazar	Lack of collective action, regulation and community identity	Integrated soil and water conservation is central to drought-proofing and production increase, but needs to be supplemented with credit and market interventions	Central role for entire community in the form of collective action for resource regeneration and regulation; smaller groups for secondary activities such as SHGs; recognises that the community may be fragmented	To catalyse community building, provide technical support, develop long-term leadership	State should provide enabling environment; local-level bodies should be the nodal point, but accountable to the general body
Utthan	Lack of voice for rural communities versus the state and especially the marginal groups within that; lack of room for indigenous innovation in a bureaucratised development framework	Water harvesting will provide self-reliant improvement in livelihoods	As key participants/ partners in the state-led development process; recognises that community is fragmented	To help marginal groups (especially women) to voice their concerns in government programmes; to implement CBNRM itself; to build local capacity by involving them as NGO staff	State should provide enabling environment, reduce the bureaucratisation of its programmes

TBS	Lack of voice for rural communities versus the state, lack of room for indigenous innovation in a bureaucratised development framework	Water-harvesting structures are central; one-time resource augmentation will be automatically sustained, although regulation was also attempted	Central, as key decision-makers; assumes traditional community exists and is sufficient	To implement CBNRM as widely as possible on its own; to build local capacity by involving them as NGO staff	State should stay out of village-level resource management completely
RNRRC	Lack of people's input into technical solutions, lack of sensitivity in line departments to local needs, lack of a watershed approach	Augmenting resource productivity is central, but may have to be supplemented by credit and market linkage	Making choices about interventions, more active participation in local bodies, collective management of watershed in the form of small groups that provide mutual support and learning; assumes traditional community exists and is sufficient	To implement CBNRM on a pilot-scale; to improve capacity of local bodies and line departments to address NRM	State should support the transition to CBNRM, but line departments will remain in the picture to support local bodies

(Table 8.2 continued)

(*Table 8.2 continued*)

Case	Perception of the 'Problem'	Role and Scope of NRM	Role and Form of Community	Role of Intervening Agency	Role of the State/Its Form at the Local Level
DTLVS	Lack of bridge between local communities and the state, lack of voice for women, lack of the spirit of environmental conservation	Integrated NRM is central and should be the basis for future development	Central, as the key decision-makers and regulators; assumes traditional community exists by and large	To spread the message of environmental conservation, to facilitate CBNRM through strategic interventions or demonstrations	State policy should be sensitive to local needs; local bodies should be 'autonomous' of state politics
GC	Lack of access to financial and technical resources for coping with floods; lack of voice for women; lack of skills for income generation amongst the landless	Natural variability/ hazards needs to be adapted to; some focus on productivity improvements, but substantial dependence on remittance income will continue	Limited role in the form of small groups that provide mutual support and learning; no assumptions about traditional community	To channel resources towards marginal groups and build their capacity	State should stay in the background as a facilitator of NGO-led efforts; although it would like to see improvements in the functioning of current institutions of local governance

(therefore or otherwise) come together quite easily for sustainable resource management. That there are often deep divisions and inequities within the so-called local community that can either hinder collective action or, by creating communities through certain exclusions, generate outcomes that are lopsided in many ways is not recognised—certainly not in TBS. Nor is it recognised that certain interventions lend themselves easily to being 'privatised', such as johads on private lands, and that even a supposedly tightly-knit community cannot ensure that upstream johads get repaired to help downstream farmers. Equally unrecognised in this assumption of a homogeneous and pre-existing community is the fact that the existence of a sense of community does not by itself ensure sustainable use of resources. For instance, the Kanbi Patels of Nathugadh are a tightly-knit group, but they do not see the need for regulation of groundwater use.

Several interveners also seem ambivalent as to how central a role NRM can play in meeting the livelihood needs of the poor. The perception seems to be that the existing pattern of land ownership cannot be changed, that access to or benefits from the commons is also constrained by such patterns, and so poverty alleviation is best pursued outside the natural resource context.

Common to most initiatives is the perception that the state should stay hands-off, even though it is also expected to fund such initiatives. This perception is probably rooted in the history of top–down state-led development of the past. But it often ignores the presence of the state in the midst of the village community itself, in the form of local self-governance institutions as well as the frontline staff of various line departments and the programmes they try to implement. Only in the Bhutan case, because of state backing, does the mandate include reorientation of the line department staff and involvement of local self-governance institutions.

It is in these inter-related combinations of underlying concerns and understandings of community-based resource management that one finds much of the explanation for the strategies chosen and the resultant outcomes. For instance, absence of an explicit commitment to sustainability led to the absence of efforts towards a priori negotiations on water use regulation[8] in Utthan's work in Nathugadh and a rather limited effort in Gopalpura. And not thinking through the ecological implications led to the non-treatment of catchment areas in both places. Interventions focusing on agricultural plots necessarily leave out the landless or the marginal as well to a large extent, but that is seen as an acceptable approach by the RNRRC, given its focus on farmer participation in agricultural innovation. Individual households are targeted by GC because it is primarily concerned with reducing individual poverty and does not see a major link between collective management of resources and poverty alleviation. The ubiquitous presence, but confused role (credit, training and exposure, or resource management), and indifferent functioning of SHGs is a result of the easy adoption of an institutional quick-fix without interrogating its role and relevance. Tarun Bharat Sangh works with the traditional gram sabha, although this more often than not excludes both lower castes and women, because its concern for local-level democratisation is limited and it does not perceive the link between inclusive decision-making and the sustainability and equity of outcomes.

Linking Visions to Strategies and Outcomes

Admittedly, there is no linear relationship between visions, strategies and outcomes. Nevertheless, what emerges on the whole from this analysis is not really a plurality of concerns and innovative visualisations of CBNRM strategies, but a relatively limited range. The primary interest seems to be in livelihood enhancement, with some priority for the poor, through some form of NRM. Community-based management often gets translated as participation of individual villagers or groups in externally implemented programmes or, alternatively, reification of exclusive notions of community. Natural resource management often gets translated as one-time improvement in the productivity of some disconnected set of resources. The need for integrated resource management strategies, and the concomitant need to carefully build a collective decision-making culture and an ethos of sustainable use while negotiating and loosening up traditional hierarchies, exclusions and inequities in resource control and access do not seem to have been recognised in most cases. Agitating for changes in macro structures so as to facilitate such a transition is, therefore, simply not in the picture by and large. Implementing programmes within limited time frames is seen as constricting but acceptable to most implementers.

In making this critical statement, we do not believe that we are holding up the cases to some external and unrealistic standard. First, as our review showed, the idea of CBNRM emerged most strongly from concerns of decentralisation and sustainable resource management. Second, seeing CBNRM as an opportunity for bringing about such changes was the hallmark of the earlier pioneering experiments such as Sukhomajri (Chopra et al., 1988), Pani Panchayat (Salunkhe et al., 2000), Ralegaon Siddhi (Antia and Kadekodi, 2002) or even the failed effort at equitable water sharing in the Upper Andhi Khola Irrigation Scheme (van Etten et al., 2002). Third, these and other experiments also showed that it is possible to go beyond the limits imposed by the existing socio-economic context by using some innovative strategies. For instance, Sukhomajri was able to put in place a system of water rights that was independent of land rights. Examples of community regulation of groundwater use also exist. Fourth, it seems possible to combine aspects of our case study initiatives into a more broad-based but realistic notion of CBNRM, although local dynamics will determine its feasibility to some extent.

The case of Hivre Bazar highlights the core tensions or tradeoffs involved in implementing this broader, multidimensional notion in a context characterised by economic, caste and gender inequities and changing perceptions about the role of natural resources. On the one hand, material conditions were favourable. The village had sub-watersheds within the village boundary, upstream of and not influenced by activities in other villages. The boundaries of the village coincided with the boundaries of the official gram panchayat—something that is not the case in many other regions. Within this favourable context, an enterprising and far-sighted leadership made use of a supportive government programme to push a relatively progressive agenda. But to do so, its *mantra* for bringing disparate factions together was the 'ideal village'. The discourse of an ideal village both created but also limited the space for bringing difficult issues to the table, for example, the inequality in landholdings. Potentially disruptive issues were postponed at times and the efforts taken to redress imbalances were also limited, although significant. Notions of sustainability or environmental conservation also had to be defined in relatively simple terms.

Modes of intervention are also potentially important. The group in Hivre Bazar is from the village itself and the founder of DTLVS is also from that locality. Both groups have worked for more than a decade and have functioned in a non-professional manner (both NGOs are tiny in terms of budgets and staff).[9] The other interventions are not entirely 'external'—at least Utthan and TBS have drawn staff from the villages in which they are working. Indeed, Utthan began as a movement for empowering women and Dalits. But the approach and discourse in Nathugadh—in spite of Utthan's best intentions—had more the character of professional social work than of a move-ment. As organisations move towards expanding their programmes, it seems that their ability to devote the time and energy required for the slow and difficult process of shaping local processes gets inevitably reduced.

Summary and the Need for Rethinking the Way Forward

We started our enquiry into NGO-driven CBNRM in South Asia keeping in mind the substantive critiques of both the concept and practice of CBNRM and NGO-driven development more generally. The currently in favour idea of CBNRM emerged as a response to critiques of centralised resource management. Following an initial set of experiments that highlighted the benefits and feasibility of a community-based approach, the concept of CBNRM got mainstreamed to various degrees in governmental and donor policy in South Asian countries, leading to the coexistence of multiple approaches. These ranged from a replication of the individual experiments by civil society groups to state-led sectoral decentralisation programmes to attempts at political devolution of resource control to lower levels of government.

Non-governmental organisations have come to play a prominent if not ubiquitous, role in not only the civil society replications, but also the decentralisation programmes. The evolution of CBNRM in South Asia is, therefore, intertwined with the changes in the NGO sector. In South Asia and other parts of the world, NGOs had emerged as a growing force in the 1970s and were seen by many as a possible force advocating an alternative vision of development that would also pressurise the state to democratise.

After the initial hype about both concepts (CBNRM and NGO-based development), several critiques have emerged. The simplified notion of community underpinning the concept of CBNRM has been a main point of criticism. Several other critiques, such as the need to get away from a simplistic opposition between state and community, have also been tabled. Similarly, it has been pointed out that NGOs may easily fall prey to hegemonic discourses of development, the micro-politics of development practice and the shortcomings of projectisation.

Despite these critiques of NGO-led CBNRM, we felt a significant lacuna existed in terms of our understanding of this phenomenon. First, a number of 'innovative' experiments dotted the landscape, experiments that did not appear to blindly accept the terms of reference of 'mainstream' development discourse of the 1970s and 1980s or its more recent neo-liberal manifestations. Second, we felt that there was a scarcity of studies that analysed the diverse forms of NGO-led CBNRM in a comparative framework. Third, most studies tended to take a somewhat unidimensional perspective on what is expected from CBNRM.

Our study was aimed at filling this gap through a critical enquiry into mostly NGO-led CBNRM initiatives. Our substantive concerns of livelihood enhancement, sustainability, equity and democratic decentralisation provided a multidimensional lens through which to examine and compare different cases. The main question that we wanted to answer is to what extent does CBNRM continue to offer a vision for the future and what role (if any) could NGO-led CBNRM play. We sought to answer this question by examining the outcomes of the interventions studied and the strategies used to achieve them, and trace the role played by the micro and macro context on the one hand, and the visions and understanding of CBNRM that the intervening agencies brought with them on the other hand.

Our study clearly reveals that these CBNRM efforts have generally made significant contributions to livelihood enhancement, but, contrary to the aim of CBNRM, have made only limited gains in terms of collective action for sustainable and equitable access to benefits and continuing resource use, and in terms of democratic decentralisation. The explanations for limited gains are multiple and interlinked. First, regardless of what they want to do, the implementers have to confront the socio-ecological reality of the region, which almost always is that of fragmented communities (or communities in flux) with unequal dependence on and access to land and other natural resources and with great gender imbalances. This rural reality also includes a local-level discourse that is generally bereft of environmental and social issues. Second, the macro-level policies, although superficially supportive of CBNRM, are not conducive to seriously addressing these concerns. Finally, however, the differences in outcomes is largely due to the limited visions of many of the implementers, in terms of both what they are concerned about and how they see CBNRM addressing these concerns.

Our historical review of the concept of CBNRM presented in the introductory chapter had suggested that this concept represented a powerful convergence of different concerns and motivations: simplistically put, whether one was concerned about the environment or about rural poverty or about democratic decentralisation or about preserving indigenous knowledge or tribal cultures, one saw CBNRM as the way forward to address any or all of these concerns. Although we then identified different streams within the CBNRM umbrella, by choosing to focus on the NGO-led stream, we believed that we would be studying initiatives with a broad-based set of concerns and understanding of issues. Yet, our case studies show that many groups do not share such broad-based concerns or at least have not actively explored practical ways of translating these concerns for sustainability, equity and democratisation systematically and simultaneously into practice. What is further disturbing is that such explorations had actually begun with the first generation of CBNRM experiments. What one is seeing now seems to be a narrowing of the practice of CBNRM. This narrow practice, when coupled with a similar absence of broader concerns and facilitating policies on the part of the state, represents a narrowing of the whole discourse of CBNRM. The co-option of the rhetoric of CBNRM by the state and donor agencies and its promotion through large-scale heavily-funded but badly designed and short-term projects rather than through long-term changes in governance structures reduces the space available to independent initiatives that may be trying to promote more democratic and responsible resource management. Many NGOs have observed this co-option, but rather than resist have chosen to become part of official (state-sanctioned and donor-supported) programmes. Perhaps they believed that they would have the freedom to innovate within these programmes. Perhaps they were tempted by the resources that these programmes made available. Perhaps some of them were never exposed to these wider issues. Whatever the reasons, the outcomes are not entirely encouraging.

There are, of course, some limitations to our study. First, even for the cases that we studied, the particular village(s) chosen may not be representative of the work of an agency that may be working across several tens or even hundreds of villages. In particular, Utthan feels that Nathugadh was its first attempt at watershed development and that it has internalised the lessons from Nathugadh in its subsequent work. Second, the initiatives being compared are at different stages in their lives—some are as old as 25 years, others just at the end of the first project cycle. The younger ones may legitimately say that they are still refining their strategies and options. Third, the chosen cases are not necessarily typical of NGO-led CBNRM in the entire South Asian region, especially since we have not covered Nepal, Pakistan and Sri Lanka. The indication from the literature from these countries is, however, that the approach to CBNRM is fairly similar in these countries as well.[10] Despite these limitations, we feel comfortable in making some broad comments.

First, the role of NGOs in CBNRM needs to be re-examined. Our study suggests that bringing about successful CBNRM requires enormous attention to and involvement in processes at the micro level. This militates against the idea that NGOs can be service providers, replacing line departments as the agency through which widespread impacts can be achieved through some kind of 'standard' CBNRM model. They can at best play a complementary role, carrying out experiments that can then be promoted though policy-level changes as well as grass-roots-level learning. The challenge for NGOs is not the replication of a simple model across hundreds of sites, but rather the changing of the discourse of development in a few of them without investing huge financial resources.

Second, and a corollary to the first, is that the idea of promoting CBNRM through 'projects' or 'programmes'—whether NGO-implemented or department-implemented—needs to be questioned. This relates to the question of what is visualised under CBNRM. If one is not just talking of some kind of participatory, natural resource-based livelihood enhancement, but a concept that embodies issues of sustainability, equity and democratic decentralisation, it requires serious changes in the structure of governance in the region. By all counts, most current efforts have fallen seriously short of any meaningful changes. If some governments, such as Bhutan, have already committed themselves to such changes in principle (no doubt in a context of a centralised state), NGOs with experience in such efforts might have a significant role to play in identifying the kinds of policy changes that are necessary. But in cases where the state has gone ahead with a rhetorically sound, but practically highly-flawed (and heavily-funded) programmes (as in Uttarakhand), to even get the government to re-think will require a long engagement with the discourse at both the state and local level—as communities will have to be weaned off state largesse. Getting the state to refocus on what it should do best, such as reallocating access to resources—land, water, forests—will be the challenge. This will require cooperation amongst NGOs to resist rather than compete for the carrot of donor-funded and state-approved programmes that notionally involve NGOs.

Third, there is the question of the role of local state institutions in CBNRM. The view, often held by many NGOs, is that of a corrupt, overly-politicised state apparatus. That might, in fact, be the case in many South Asian countries. Yet, there is a difference between wanting the state to reform and wishing it away. Moreover, in some South Asian countries, there are institutions of democratic decentralisation already in place and relatively significant packages of devolution. One of the problems confronting these institutions, of course, is the lack of wherewithal to seriously address NRM concerns. Can the NGO-led 'innovations' include ways of closer linkage with existing institutions of democratic decentralisation? Or, alternatively, can the debate be shifted to restructuring state institutions in a way that will eventually make them more appropriate vehicles for decentralised natural resource governance?

These questions are partly rhetorical. But even if the direction suggested by them is deemed desirable, moving in that direction is filled with complexities. Our study reinforces many of the critiques of NGO initiatives laid out at the outset: hegemonic discourses of development, micro-politics of community-based development and increasing project mode of functioning, along no doubt with a number of situational obstacles. Moreover, in a context where the political economy of development is seeing NGO-driven CBNRM increasingly as a means of sanitised 'good governance' or a way to reform a poorly functioning state, is it short-sighted to expect such changes to emerge from within this 'community of actors'? At the least, what it suggests is the need to expand the terms of reference when examining CBNRM into the realm of social and political movements that may be embracing CBNRM with different visions in mind, visions that perhaps are more cognisant of alternative imaginations that address concerns of democratisation and sustainable and equitable development.

Notes

1. Though our initial intent was to focus on NGO-driven CBNRM, as stated in the 'Introduction', there is a large diversity of organisations that can be called NGOs. In the case of Hivre Bazar, though an NGO was formed, it is actually the village panchayat that is the NGO. In the case of the RNRRC in Bhutan, it is a research organisation funded by the state and hence falls more broadly into the category of a civil society organisation.
2. Throughout this chapter, when referring to the 'cases', we have tried to use the village or site name when discussing characteristics of the site or the community, and the name or acronym of the intervening agency when we are referring to the interventions or the logic behind them. However, in the case of Hivre Bazar, where the NGO (Yashwant Krishi Gram and Panlot Vikas Sanstha) consists entirely of people from that village, we have tended to use the village name to refer to the intervening agency.
3. In the case of Bhutan, class differences become more ostensible if rights to water are taken into account.

4. As mentioned earlier, given that the 'NGO' in Hivre Bazar is a subset of the village community, it is almost impossible to really distinguish the role played by villagers at large vis-à-vis the role played by the 'intervening agency' in a particular decision.
5. For example, choices of strategies may to a varying extent be tuned to local conditions. Conversely, regardless of the strategy adopted, local conditions might overwhelmingly determine a particular outcome.
6. Indeed, Seva Mandir has argued that even starting with a sub-group (such as farmers who benefit from a water-harvesting structure) is perhaps not appropriate when the goal is to democratise local governance of the commons.
7. Something that another major NGO working in south-east Rajasthan (Seva Mandir) has done much more consciously, although it was fully aware of the pitfalls of these institutions, because it believes in democratic decentralisation of resource governance (Ballabh, 2004).
8. As we have been at pains to point out in the chapter on Utthan, this is not at all to suggest that regulation of groundwater use would have been automatically achieved, but that there were no attempts to even raise this issue.
9. Indeed, the DTLVS consciously refused funds from various sources because they did not want to get 'NGO-ised'.
10. The exception might be the community forestry effort in Nepal where the state created a relatively radical legislation and created a process of transferring significant control over forests to local forest user groups.

Bibliography

Adnan, Shapan, 1997, 'Class, Caste and *Shamaj* Relations among the Peasantry in Bangladesh: Mechanisms of Stability and Change in the Daripalla Villages, 1975–86', in Jan Breman, Peter Kloos and Ashwani Saith (eds), *The Village in Asia Revisited*, Oxford University Press, New Delhi, pp. 277–310.

Agarwal, Anil and Sunita Narain, 1989, *Towards Green Villages: A Strategy for Environmentally Sound and Participatory Rural Development*, Centre for Science and Environment, New Delhi.

———, 1997, *Dying Wisdom: Rise, Fall and Potential of India's Traditional Water Harvesting Systems*, Centre for Science and Environment, New Delhi.

Agrawal, Arun and C.C. Gibson, 1999, 'Enchantment and Disenchantment: The Role of Community in Natural Resource Conservation', *World Development*, 27(4): 629–49.

Agrawal, Arun, 2001a, 'Common Property Institutions and Sustainable Governance of Resources', *World Development*, 29(10): 1649–72.

———, 2001b, 'The Regulatory Community: Decentralization and the Environment in the Van Panchayats (Forest Councils) of Kumaon, India', *Mountain Research and Development*, 21(3): 208–11.

Agrawal, Arun and Elinor Ostrom, 2001, 'Collective Action, Property Rights, and Decentralization in Resource Use in India and Nepal', *Politics & Society*, 29(4): 485–514.

Agrawal, Nand Kishore, 1999a, 'Equity Issues in Natural Resource Management: A Case Study of Bharuch District in Gujarat', Aga Khan Rural Support Programme (India), Ahmedabad.

———, 1999b, 'Institutional and Vegetational Issues in Joint Forest Management: A Case Study of Khaidipada Village in Bharuch, Gujarat', Aga Khan Rural Support Programme (India), Ahmedabad.

Agrawal, Rakesh, 1999, 'Van Panchayats in Uttarakhand', *Economic and Political Weekly*, 34(39): 2779–81.

Ahmed, Qazi Faruque, 1997, 'People's Participation: Democratisation of the Development Process', *Discourse*, 1(1: Summer): 55–61.

Alvares, Claude A., 1991, *Decolonising History: Technology and Culture in India, China and the West from 1492 to the Present Day*, The Other India Press, Goa.

———, 1979, *Homo Faber: Technology and Culture in India, China and the West, 1500–1972*, Allied Publishers, Bombay.

Anderson, Paul Nicholas, 2000, 'Commodification, Conservation and Community: An Analysis and a Case Study of South India', Edinburgh Papers in South Asian Studies No. 14, Centre for South Asian Studies, Department of History, University of Edinburgh, Edinburgh.

Anonymous, 1995a, *National Environment Management Action Plan: Summary*, Ministry of Environment and Forests, Bangladesh, Dhaka.

———, 1995b, 'A Study on Status of Natural Resources in Coastal Areas of Gujarat', Utthan, Ahmedabad.

———, 1998a, *Ripples of the Society: People's Movements in Watershed Development in India*, Gandhi Peace Foundation and PWMTA Programme, FAO, New Delhi.

———, 1998b, *Participatory Rural Appraisal in Nathugadh Village*, Utthan, Ahmedabad.

———, 1999, *Bhutan 2020: A Vision for Peace, Prosperity and Happiness*, Planning Commission, Royal Government of Bhutan, Thimphu.

———, 2000, 'Baseline Survey: Sustainable Livelihood in Riverine Charlands', Gono Chetona, Dhaka.

———, 2002, 'Sustainable.Livelihood in Riverine Charlands: Consolidated Two Year Report (2000–2001)', Project Report, Gono Chetona, Dhaka.

———, 2003, 'Charland Biodiversity Assessment—Final Report', Gono Chetona, Dhaka.

———, n.d.a, 'Adgaon at a Glance', A Report on Impact of Adgaon Work, Marathwada Sheti Sahayak Mandal, Aurangabad.

———, n.d.b, 'Participatory Forest Management for Local Use', mimeo.

———, n.d.c, 'Ralegaon Siddhi: An Experience in Watershed Development', mimeo.

Antia, N.H. and G.K. Kadekodi, 2002, 'Dynamics of Rural Development: Lessons from Ralegaon Siddhi', Project Report submitted to the Ministry of Rural Development, Government of India, Foundation for Research in Community Health (FRCH), Pune.

Apte, Tejaswini, 2001, 'Water Rights, Land Reform and Community Participation: The Pani Panchayat Model for Sustainable Water Management', mimeo, Kalpavriksh, Pune.

Baker, J.M., 1997, 'Common Property Resource Theory and the Kuhl Irrigation Systems of Himachal Pradesh', *Human Organization*, 56(2): 199–208.

Baland, Jean-Marie and Jean-Philippe Platteau, 1996, *Halting Degradation of Natural Resources: Is There a Role for Rural Communities?*, Oxford University Press, Oxford.

———, 1999, 'The Ambiguous Impact of Inequality on Local Resource Management', *World Development*, 27(5): 773–88.

Ballabh, Pankaj (ed.), 2004, *Land, Community and Governance*, National Foundation for India, New Delhi and Seva Mandir, Udaipur.

Ballabh, V. and K. Singh, 1988, 'Van (Forest) Panchayats in Uttar Pradesh Hills: A Critical Analysis', CPRM Discussion Paper Series, Institute of Rural Management, Anand.

Baqee, Abdul, 1998, *Peopling in the Land of Allah Jaane: Power, Peopling and Environment: The Case of Char-lands of Bangladesh*, The University Press Limited, Dhaka.

Barot, Nafisa and Ashok Chatterjee, 2004, 'Vision, Plan and Reality: Challenges and Experiences from Gujarat, India', Paper presented in the Dakar Forum (November 2004), Dakar.

Baviskar, Amita, 2002, 'Between Micro-politics and Administrative Imperatives: Decentralization and the Watershed Mission in Madhya Pradesh, India', Workshop on Decentralization and the Environment, 18–22 February, World Resource Institute, Bellagio, Italy.

Bawa, Kamal S., Sharachchandra Lélé, K.S. Murali and Balachander Ganesan, 1999, 'Extraction of Non-Timber Forest Products in Biligiri Rangan Hills, India: Monitoring a Community-Based Project,' in Anonymous (ed.), *Measuring Conservation Impact: An Interdisciplinary Approach to Project Monitoring and Evaluation*, Biodiversity Support Program, World Wildlife Fund, Washington, DC, pp. 89–102.

Bebbington, Anthony J., 1997, 'Reinventing NGOs and Rethinking Alternatives in the Andes', *Annals of the American Academy of Political and Social Science*, 554 (November): 117–35.

Belbase, Narayan and Dhrubesh Chandra Regmi, 2002, *Potential for Conflict: Community Forestry and Decentralisation Legislation in Nepal*, ICIMOD, Kathmandu.

Berkes, Fikret (ed.), 1989, *Common Property Resources: Ecology and Community-Based Sustainable Development*, Belhaven Press, London.

Berreman, Gerald D., 1963, *Hindus of the Himalayas*, University of California Press, Berkeley.

———, 1977, 'Demography, Domestic Economy and Change in the Western Himalayas', *Eastern Anthropologist*, 30(2): 157–92.

———, 1978, 'Ecology, Demography and Domestic Strategies in the Western Himalayas', *Journal of Anthropological Research*, 34(3): 326–68.

Bhatt, Yogesh Kumar and Shibha G. Iyer, 1994, 'A Study of Community Participation in Rural Water Supply Project: Utthan–Mahiti Experiment in the Village of Bhal at Dhanduka', Project Report, Indian Institute of Management, Ahmedabad.

Bhatta, Binod, 2002a, 'Access and Equity Issues in High Mountain Regions: Implications of a Community Forestry Programme', in *Policy Analysis of Nepal's Community Forestry Programme: A Compendium of Research Papers*, Winrock International, Nepal, Kathmandu, pp. 1–32.

———, 2002b, 'Access and Equity Issues in Terai Community Forestry Programme', in *Policy Analysis of Nepal's Community Forestry Programme: A Compendium of Research Papers*, Winrock International, Nepal, Kathmandu, pp. 93–132.

Bhujel, Aita Kumar, n.d., 'Farm Household Categorization of Dompola and Matalungchu Villages in Lingmuteychhu Watershed', mimeo.

Bookbinder, Marnie P., Eric Dinerstein, Arun Rijal, Hank Cauley and Arup Rajouria, 1998, 'Ecotourism's Support of Biodiversity Conservation', *Conservation Biology*, 12(6: December): 1399.

Brahme, S., 1983, *Drought in Maharashtra 1972*, Gokhale Institute of Politics and Economics, Pune.

Bromley, Daniel W. (ed.), 1992, *Making the Commons Work: Theory, Practice and Policy*, Institute for Contemporary Studies, San Francisco.

Carney, Diana, Michael Drinkwater, Tamara Rusinow, Koof Neefjes, Samir Wanmali and Naresh Singh, 1999, *Livelihoods Approaches Compared: A Brief Comparison of the Livelihoods Approaches of the UK Department for International Development (DFID), CARE, Oxfam and United Nations Development Program (UNDP)*, Department for International Development (DFID), London.

Census of India, 1981, *District Census Handbook, Paudi Garhwal District, Uttar Pradesh*, Office of the Registrar General and Census Commissioner of India, New Delhi.

——, 1991, *District Census Handbook, Bhavnagar District, Gujarat*, Office of the Registrar General and Census Commissioner of India, New Delhi.

——, 2001a, *District Census Handbook, Ahmednagar District, Maharashtra*, Office of the Registrar General and Census Commissioner of India, New Delhi.

——, 2001b, *District Census Handbook, Alwar District, Rajasthan*, Office of the Registrar General and Census Commissioner of India, New Delhi.

——, 2001c, *District Census Handbook, Paudi Garhwal District, Uttaranchal*, Office of the Registrar General and Census Commissioner of India, New Delhi. [Uttaranchal was renamed Uttarakhand in January 2007.]

Chakraborty, Rabindra Nath, 1998, 'Problems of Intra and Inter Group Equity in Community Forestry: Evidence from the Terai Region and Nepal', Workshop on Participatory Natural Resource Management, Mansfield College, University of Oxford, Oxford.

Chambers, Robert, 1983, *Rural Development: Putting the Last First*, Longman, London.

Chambers, Robert, Arnold Pacey and Lori Ann Thrupp (eds), 1989, *Farmer First: Farmer Innovation and Agricultural Research*, Intermediate Technology Publications, London.

Chatterjee, Partha, 1998, 'Community in the East', *Economic and Political Weekly*, 33(6): 277–82.

Chatterji, Angana P., 1996, *Community Forest Management in Arabari: Understanding Sociocultural and Subsistence Issues*, Society for Promotion of Wastelands Development, New Delhi.

Chopra, Kanchan, G.K. Kadekodi and M.N. Murty, 1988, *Sukhomajri and Dhamala Watersheds in Haryana: A Participatory Approach to Management*, Institute of Economic Growth, New Delhi.

Cohen, J. and N. Uphoff, 1980, 'Participation's Place in Rural Development: Seeking to Clarify through Specificity', *World Development*, 8: 213–35.

Conroy, Czech, 2001, 'Factors Influencing the Initiation and Effectiveness of Community Forest Management: Hypotheses and Experiences in Orissa', Project Report No. 5, Society for Promotion of Wasteland Development, New Delhi.

Conroy, Czech, Ajai Rai, Neera M. Singh and Man-Kwan Chan, 1998, 'Conflicts Affecting Participatory Forest Management: Experiences from Orissa, India', Workshop on Participatory Natural Resource Management, Mansfield College, University of Oxford, Oxford.

Corbridge, S. and S. Jewitt, 1997, 'From Forest Struggles to Forest Citizens? Joint Forest Management in the Unquiet Woods of India's Jharkhand', *Environment and Planning*, 29(12): 2145–64.

Correa, M., 1996, 'No Role for Women—Karnataka's Joint Forest Management Programmes', *Economic and Political Weekly*, 31(23): 1382–83.

Coward, E.W. (ed.), 1980, *Irrigation and Agricultural Development in Asia: Perspectives from Social Sciences*, Cornell University Press, Ithaca.

CWR (Centre for Water Resources), 2000, *Final Report of Monitoring and Evaluation of the Tank Modernisation Project (with EEC Assistance), Phase II and Phase II—Extension, Volume I & II*, Centre for Water Resources, Anna University, Chennai.

de Graverol, Gail, 2003, 'The Relationship between Caste and Tribes in a Former Kingdom of Rajasthan: Kingship and Tribal Sovereignty: The Case Study of Minas in the ancient Princely State of Amber', Working Paper No. 129, Institute of Development Studies, Jaipur.

Deb, D. and K.C. Malhotra, 1993, 'People's Participation: The Evolution of Joint Forest Management in South West Bengal', in S.B. Roy and A.K. Ghosh (eds), *People of India: Bio-cultural Dimensions*, Inter-India Publications, New Delhi, pp. 329–42.

DHAN Foundation, 2004, 'Study on Customary Rights and Their Relation to Modern Tank Management in Tamil Nadu, India', DHAN Foundation (Development in Humane Action), Madurai.

Dixit, A., 2000, 'Water as an Agent of Social and Economic Change in Nepal', in Peter Mollinga (ed.), *Water for Food and Rural Development: Approaches and Initiatives in South Asia*, Sage Publications, New Delhi.

Dougill, A.J., J.G. Soussan, E. Kiff, O. Springate-Baginski, N.P. Yadav, O.P. Dev and A.P. Hurford, 2001, 'Impacts of Community Forestry on Farming System Sustainability in the Middle Hills of Nepal', *Land Degradation & Development*, 12(3): 261–76.

DRDS (Department of Research and Development Services), 2002, 'Community Based Natural Resource Management in Bhutan: A Framework', Department of Research and Development Services, Ministry of Agriculture, Royal Government of Bhutan, Thimpu.

Edmunds, David, Eva Wollenberg, Antonio P. Contreras, Liu Dachang, Govind Kelkar, Dev Nathan, Madhu Sarin and Neera M. Singh, 2003, 'Introduction', in David Edmunds and Eva Wollenberg (eds), *Local Forest Management: The Impact of Devolution Policies*, Earthscan, London, pp. 1–19.

Farrington, J., C. Turton and A.J. James (eds), 1999, *Participatory Watershed Development: Challenges for the Twenty-First Century*, Oxford University Press, New Delhi.

Feldman, Shelley, 1997, 'NGOs and Civil Society: (Un)stated Contradictions', *Annals of the American Academy of Political and Social Science*, 554(November): 46–65.

Ferguson, James, 1990, *The Anti-Politics Machine: 'Development', Depoliticization, and Bureaucratic Power in Lesotho*, University of Minnesota Press, Minneapolis.

Fisher, William F., 1997, 'Doing Good? The Politics and Antipolitics of NGO Practices', *Annual Review of Anthropology*, 26: 439–64.

Gadgil, Madhav and Ramachandra Guha, 1992, *This Fissured Land: An Ecological History of India*, Oxford University Press, New Delhi.

Gain, Philip, 2002, *The Last Forests of Bangldesh*, second edn, Society for Environment and Human Development, Dhaka.

GEC (Gujarat Ecology Commission), 2006a, 'Agroclimatic Regions of Gujarat', Gujarat Ecology Commission, http://www.gec.gov.in/envis/SoER_Table_htm/AgrCliReg.htm, accessed on 18 June 2006.

———, 2006b, 'District Wise Number and Area of Operational Holders', Gujarat Ecology Commision, http://www.gec.gov.in/envis/SoER_Table_htm/DisWisOpeHolOth.htm, accessed on 18 June 2006.

———, 2006c, 'Source Wise Area Irrigated in Gujarat State—1999–2000', Gujarat Ecology Commission, http://www.gec.gov.in/envis/SoER_Table_htm/SouWisAreIrr.htm, accessed on 18 June 2006.

Gidwani, Vinay, 2001, 'Labored Landscapes: Agro-ecological Change in Central Gujarat, India', in Arun Agrawal and K. Sivaramakrishnan (eds), *Social Nature: Resources, Representations, and Rule in India*, Oxford University Press, New Delhi, pp. 216–47.

Gilmour, Don, 2003, 'Retrospective and Prospective View of Community Forestry in Nepal,' *Journal of Forest and Livelihood*, 2(2): 5–7.

Gloekler, A. and J. Seeley, 2003, 'Gender and AKRSP—Mainstreamed or Sidelined?', Paper presented at the conferences on 'Lessons in Development—The AKRSP Experiences', organised by Aga Khan Rural Support Programme at Islamabad, 15–16 December, also available at http://www.livelihoods.org/lessons/project_summaries/comdev1_projsum.html.

GOI, 2001, *Guidelines for Watershed Development (Revised-2001)*, Department of Land Resources, Ministry of Rural Development, Government of India, New Delhi.

Guha, Ramachandra, 1985, 'Scientific Forestry and Social Change in Uttarakhand', *Economic and Political Weekly*, 20(45–47: Special Number, November): 1939–51.

———, 1989, *The Unquiet Woods: Ecological Change and Peasant Resistance in the Himalayas*, Oxford University Press, New Delhi.

Gunjal, R.S. and V.H. Deshmukh, 1998, 'Watershed Management and Conservation: A Case Study of Ralegaon Siddhi', in *International Conference—Watershed Management & Conservation*, Central Board of Irrigation and Power, New Delhi.

Gupta, Tilak D., 1999, *From Subsistence Agriculture to Irrigated Farming: Experiences of Community Managed Lift Irrigation in Bihar and Orissa Plateau*, PRADAN, New Delhi.

Harriss, John and Hamza Alavi (eds), 1989, *Sociology of Developing Societies, South Asia*, Macmillan, London.

Hazare, Anna, Ganesh Pangare and Vasudha Lokur, 1996, *Adarsh Gaon Yojana: Government Participation in a People's Program*, Hind Swaraj Trust, Pune.

Hirway, Indira, 1998, 'Paradigms of Development: Issues in Industrial Policy in India', QEH Working Paper Series No. 22, University of Oxford, Oxford, also available at http://www.qeh.ox.ac.uk/pdf/qehwp/qehwps22.pdf (18 June 2006).

Hooja, Rakesh, Ganesh Pangare and K.V. Raju (eds), 2002, *Users in Water Management: The Andhra Model and Its Replicability in India*, Rawat Publications, New Delhi.

Huq, Saleemul, 1997, 'NGOs' Advocacy against Structural Approach to Flood Control: The Case of the Flood Action Plan (FAP) in Bangladesh', *Discourse*, 1(1: Summer): 75–87.

Husain, Tariq, 1992, 'The Aga Khan Rural Support Programme: An Approach to Village Management Systems in Northern Pakistan', in Narpat S. Jodha, Mahesh Banskota and Tej Pratap (eds), *Sustainable Mountain Agriculture: Farmer Strategies and Innovative Approaches*, Oxford & IBH Publishing, New Delhi, pp. 671–709.

Hussein, M.H. and S. Plateau, 2003, 'Micro-Finance', Paper presented in 'Lessons in Development—The AKRSP Experience', organised by the Aga Khan Rural Support Programme at Islamabad, 15–16 December, also available at http://www.livelihoods.org/lessons/project_summaries/comdev1_projsum.html.

IFAD (International Fund for Agricultural Development), 2003, 'HLFFDP Interim Evaluation Report', International Fund for Agricultural Development, Kathmandu.

Indra, Doreen Marie and Norman Buchignani, 1997, 'Rural Landlessness, Extended Entitlements and Inter-household Relations in South Asia: A Bangladesh Case', *The Journal of Peasant Studies*, 24(3 April): 25–64.

Jairath, Jasveen, 1999, 'Participatory Irrigation Management: Experiments in Andhra Pradesh', *Economic and Political Weekly*, 34(40): 2834–37.

Jamal, T., R. Hooja and G. Pangare, 2002, 'Community Water Harvesting: Process Documentation of the Activities and Approach of Tarun Bharat Sangha', IndiaPIM, New Delhi.

Jeffery, Roger and Nandini Sundar (eds), 1999, *New Moral Economy of India's Forests*, Sage Publications, New Delhi.

Joshi, L.K. and Rakesh Hooja (eds), 2000, *Participatory Irrigation Management: Paradigm for the 21st Century*, Rawat Publiations, Jaipur.

Joshi, Niraj, Ashok Pingle and Biswaranjan Pattnaik, 2004, 'Impact of AKRSP (I)'s Intervention in Drought Coping by People: The Case of Surendranagar Programme Area', Aga Khan Rural Support Programme, Ahmedabad.

Joshi, Satish, 1988, 'Integrated Community and Area Development Project—Depur, Dholpur & Shyampur: A Case Study', Seva Mandir, Udaipur.

Joy, K.J., Suhas Paranjape, A.K. Kiran Kumar, Rohini Lele and Raju Adagale, 2004, 'Watershed Development Review: Issues and Prospects', CISED Technical Report, Centre for Interdisciplinary Studies in Environment and Development, Bangalore.

Kadekodi, G.K., Kanchan Chopra and P.R. Mishra, 1991, *Chakriya Vikas Pranali: A New Way Forward for India's Rural People*, Centre for Science and Environment, New Delhi.

Kamat, Sangeeta, 2002, *Development Hegemony: NGOs and the State in India*, Oxford University Press, New Delhi.

Karim, Mahbubul, 1999, 'Breaking Boundaries to Participation in Local Governance: Empowering the People's Organizations (POs) in Bangladesh', *Discourse*, 3(1: Summer): 71–88.

Kerr, John, Ganesh Pangare and Vasudha L. Pangare, 2002, 'Watershed Development Projects in India: An Evaluation', Research Report No. 127, International Food Policy Research Institute, Washington, DC.

Khan, Niaz Ahmed, Junaid Kabir Choudhury and Khawja Shamsul Huda, 2004, *An Overview of Social Forestry in Bangladesh*, Bangladesh Forest Department, Dhaka.

Khan, Waheeduddin and R.N. Tripathy, 1976, *Plan for Integrated Rural Development in Paudi Garhwal*, National Institute of Community Development, Hyderabad.

Kohlin, Gunnar, 1998, *The Value of Social Forestry in Orissa, India*, Economiska Studier, Utgivna Av Nationalekonomiska Institutionen, Handelshogskolan Vid Goteborgs Universitet, Goteborg.

Kolavalli, S., 1995, 'Joint Forest Management: Superior Property Rights', *Economic and Political Weekly*, 30(30): 1933–38.

Korten, David C., 1986, 'Introduction: Community-Based Resource Management', in David C. Korten (ed.), *Community Management: Asia Experience and Perspectives*, Kumarian Press, West Hartford, CT, pp. 1–15.

Kothari, Rajni, 1988, *State against Democracy*, Ajanta Publications, New Delhi.

———, 1989, *Politics and the People: In Search of a Humane India*, Ajanta Publications, New Delhi.

Krishna, Sumi, 1996, *Environmental Politics: People's Lives and Development Choices*, Sage Publications, New Delhi.

Kumar, Nalini, N.C. Saxena, Yoginder K. Alagh and Kinsuk Mitra, 1999, 'Alleviating Poverty through Participatory Forestry Development: An Evaluation of India's Forest Development and World Bank Assistance', Operations Evaluation Department, World Bank, Washington, DC.

Kumar, P. and B.M. Kandpal, 2003, 'Project on Reviving and Constructing Small Water Harvesting Structures in Rajasthan', SIDA Evaluation No. 03/40, Department of Asia, Swedish International Development Agency, Stockholm.

Kumar, Rajeev, 1993, *Integrated Land Management: Scarcity to Surplus*, Society for Promotion of Wastelands Development, New Delhi.

Lélé, Sharachchandra, 1999, 'Institutional Issues in (J)FM(&R)', in *Proceeding of the National Workshop on Joint Forest Management*, VIKSAT, Gujarat Forest Department and Aga Khan Foundation, Ahmedabad.

———, 2005, 'Decentralising Governance of Natural Resources in India: A Review', Country review report, UNDP Dryland Development Centre, Nairobi.

Lélé, Sharachchandra, A.K. Kiran Kumar and Pravin Shivashankar, 2005, 'Joint Forest Planning and Management in the Eastern Plains Region of Karnataka: A Rapid Assessment', CISED Technical Report, Centre for Interdisciplinary Studies in Environment and Development, Bangalore.

Lélé, Sharachchandra, Kamal S. Bawa and C. Made Gowda, 2004, 'Ex-Post Evaluation of the Impact of an Enterprise-Based Conservation Project in BRT Wildlife Sanctuary, India', Paper presented in 'The Commons in an Age of Global Transition: Challenges, Risks and Opportunities, 8th Biennial Conference of the International Association for the Study of Common Property', at Oaxaca, Mexico, 9–13 August 2004.

Lélé, Sharachchandra, K.S. Murali and Kamaljit S. Bawa, 1998, 'Community Enterprise for Conservation in India: Biligiri Rangaswamy Temple Sanctuary', in Ashish Kothari, Neema Pathak, R.V. Anuradha and Bansuri Taneja (eds), *Communities and Conservation: Natural Resource Management in South and Central Asia*, Sage Publications, New Delhi, pp. 449–66.

Li, Tania Murray, 1996, 'Images of Community: Discourse and Strategy in Property Relations', *Development and Change*, 27: 501–27.

———, 1999, 'Compromising Power: Development, Culture and Rule in Indonesia', *Cultural Anthropology*, 14(3): 1–28.

Manor, James, 1999, *The Political Economy of Democratic Decentralization*, The World Bank, Washington, DC.

———, 2004, 'User Committees: A Potentially Damaging Second Wave of Decentralisation?', *European Journal of Development Research*, Spring, 16(1): 192–213.

Mawdsley, Emma, 1998, 'After Chipko: From Environment to Region in Uttaranchal', *Journal of Peasant Studies*, 25(4): 36–54.

McCully, Patrick, 2002, 'Harvesting Rain, Transforming Lives: India's Stellar Water-Harvesting Movement Inspires Hope', *World Rivers Review*, December: 8–9.

McKay, Bonnie J. and James M. Acheson, 1987, *The Question of the Commons: The Culture and Ecology of Communal Resources*, University of Arizona Press, Tucson.

Mehta, Ajay S., 1996, 'The Micro-Politics of Development: An Anatomy of Change in Two Villages', Working Paper Series, Seva Mandir, Udaipur.

Mehta, Manjari, 1996, '"Our Lives are No Different from that of Our Buffaloes": Agricultural Change and Gendered Spaces in a Central Himalayan Valley', in Dianee Rocheleau, Barbara Thomas-Slayter and Esther Wangari (eds), *Feminist Political Ecology: Global Issues and Local Experience*, Routledge, London, pp. 180–210.

Meinzen-Dick, Ruth and K. Palanisami, 2001, 'Tank Performance and Multiple Users in Tamil Nadu, South India', *Irrigation and Drainage Systems*, 15(2): 173–95.

Mencher, Joan, 1999, 'NGOs: Are They a Force for Change?', *Economic and Political Weekly*, 34(30): 2081–86.

Mishra, Anupam, 1993, *Aaj Bhi Khare hain Taalab* (in Hindi), Gandhi Peace Foundation, New Delhi.

———, 1995, *Rajasthan ki Rajat Bundein* (in Hindi), Gandhi Peace Foundation, New Delhi.

Mollinga, Peter, 2002, 'Power in Motion: A Critical Assessment of Canal Irrigation Reforms in India', in Rakesh Hooja, Ganesh Pangare and K.V. Raju (eds), *Users in Water Management: The Andhra Model and Its Replicability in India*, Rawat Publications, New Delhi, pp. 265–81.

Montu, Rafiqul Islam, 2003, 'The Life in Char Lands: A Factual Narrative of the People and Habitat of Dewanganj, Bakshiganj and Phulchari Upazilas', *Ecofile*, October–December, 2003, pp. 7–17.

Mosse, David, 1997, 'The Symbolic Making of a Common Property Resource: History, Ecology and Locality in a Tank-irrigated Landscape in South India', *Development and Change*, 28: 467–504.

———, 1999, 'Colonial and Contemporary Ideologies of "Community Management": The Case of Tank Irrigation Development in South India', *Modern Asian Studies*, 33(2): 303–38.

———, 2003a, 'Good Policy is Unimplementable? Reflections on the Ethnography of Aid Policy and Practice', EIDOS Workshop on Order and Disjuncture: The Organisation of Aid and Development, School of Oriental and African Studies, London.

———, 2003b, *The Rule of Water: Statecraft, Ecology, and Collective Action in South India*, Oxford University Press, New Delhi.

———, 2005, *Cultivating Development: An Ethnography of Aid Policy and Practice*, Vistaar, New Delhi.

Mukundan, T., 1988, 'The Ery System of South India', *PPST Bulletin*, 16 (September): 1–37.

Nadkarni, M.V., Syed Ajmal Pasha and L.S. Prabhakar, 1989, *Political Economy of Forest Use and Management*, Sage Publications, New Delhi.

Najam, Adil, 2003, 'Working with Government: Partnerships of Necessity', Department for International Development (DFID), http://www.akrsplessons.org/downloads/thematic_papers/Working_with_Government-LessonsLearntExcercise.pdf, accessed on 14 June 2005.

Narain, Iqbal and P.C. Mathur, 1990, 'The Thousand Year Raj: Regional Isolation and Rajput Hinduism in Rajasthan Before and After 1947', in M.S.A. Rao (ed.), *Dominance and State Power in Modern India: Decline of a Social Order*, Oxford University Press, Bombay, pp. 1–58.

Negi, G.C.S. and Varun Joshi, 2004, 'Rainfall and Spring Discharge in Two Drainage Catchments in the Western Himalaya', *The Environmentalist*, 24: 19–28.

Ostrom, Elinor, 1990, *Governing the Commons: The Evolution of Institutions for Collective Actions*, Cambridge University Press, Cambridge.

———, 1992, 'The Rudiments of a Theory of the Origins, Survival and Performance of Common-Property Institutions', in Daniel Bromley (ed.), *Making the Commons Work: Theory, Practice, and Policy*, Institute for Contemporary Studies Press, San Francisco, pp. 293–318.

Pahadi, Ramesh, 2004, '*Jal, Jangal our Jameen ke Saath Jan to Jodaa Gayaa Hai: Bharati* (People have been Linked with Water, Forests and Land, says Bharati)', *Jal Sanskriti* (*Water Culture*, in Hindi): 9, 16.

Pangare, Ganesh and Vasudha Lokur, 1996, *The Good Society: The Pani Panchayat Model of Sustainable Water Management*, INTACH, New Delhi.

Paranjape, Suhas, K.J. Joy and Seema Kulkarni, 1997, 'Report of the Evaluation Team for CARITAS supported UVM Project "Community based Eco-regeneration in Dhar and Bagdunda Areas of South Rajasthan"', SOPPECOM, Pune.

Parthasarathy, R., Vinayak Dave and Mahesh Chhatrola, 1994, 'Lift Irrigation Scheme in Zadka Village: Documentation of the Process of Implementation by AKRSP and Some Issues', Report of Process Documentation Research, Gujarat Institute of Development Research, Ahmedabad.

Patel, G.D., 1957, *The Land Problem of Reorganised Bombay State*, N.M. Tripathi, Bombay.

Pathak, Akhileshwar, 1994, *Contested Domains: The State, Peasants and Forests in Contemporary India*, Sage Publications, New Delhi.

Pathak, Neema and Vivek Gour-Broome, 2000, *Tribal Self-Rule and Natural Resource Management: Community-based Conservation at Mendha-Lekha, Maharashtra, India*, Kalpavriksh and International Institute for Environment and Development, New Delhi.

Pattanaik, M., 2002, 'Community Forest Management in Orissa', *Community Forestry*, 1(1&2): 4–8.

Pradhan, Prachanda, 2003, 'Farmer Managed Irrigation Systems: A Mode of Water Governance', *Water Nepal*, 9/10(1/2).

Pradhan, Prachanda and Upendra Gautam, 2002, 'Water Users Association: Towards Diversified Activities', in Prachanda Pradhan and Upendra Gautam (eds), *Proceedings of the Second International Seminar*, Farmer Managed Irrigation Systems Promotion Trust, Kathmandu.

Rahman, A. Atiq, Saleemul Haque, Amala Reddy, Mozharul Alam and Sughra Arasta Kabir (eds), 2001, *State of the Environment—Bangladesh, 2001*, United Nations Environment Programme, Dhaka.

Rahman, Md. Mahmubar and Willem van Schendel, 1997, 'Gender and the Inheritance of Land: Living Law in Bangladesh', in Jan Breman, Peter Kloos and Ashwani Saith (eds), *The Village in Asia Revisited*, Oxford University Press, New Delhi, pp. 237–76.

Rahman, Md. Makhluqur, 2000, 'Growth and Development of Thana/Upazila Administration in Bangladesh', *The Journal of Local Government*, 29(2): 58–76.

Rahul, 1997, 'Masquerading as the Masses', *Economic and Political Weekly*, 32(7): 341–42.

Ramakrishnan, Rajesh, Manish Dubey, Rajiv K. Raman, Pari Baumann and John Farrington, 2002, 'Panchayati Raj and Natural Resources Management: How to Decentralise Management over Natural Resources', National Synthesis Report, Overseas Development Institute, London (along with Taru Leading Edge [New Delhi and Hyderabad], Centre for Budget and Policy Studies [Bangalore], Centre for World Solidarity [Hyderabad] and Sanket [Bhopal]).

Ravindranath, N.H., K.S. Murali and K.C. Malhotra (eds), 2000, *Joint Forest Management and Community Forestry in India: An Ecological and Institutional Assessment*, Oxford & IBH, New Delhi.

Reddy, Amulya K.N., 1999, 'Rural Energy: Goals, Strategies and Politics', *Economic and Political Weekly*, 34(49: 4 December): 3435–45.

———, 2004, 'Science and Technology for Rural India', *Current Science*, 87(10: 10 October).

Reddy, Jagannadha, Jayappa, Padma and Baliah, 2006, 'Sanghams: Experiences, Expectations', Deccan Development Society, http://www.ddsindia.com/www/default.asp, accessed on 26 July 2006.

Reddy, Somashekara, 1991, *Forfeited Treasure: A Study on the Status of Irrigation Tanks in Karnataka*, Prarambha, Bangalore.

Ribot, Jesse, 2002, *Democratic Decentralisation of Natural Resources: Institutionalizing Popular Participation*, World Resource Institute, Washington, DC.

RNRRC (Renewable Natural Resource Research Centre), 2000a, 'Community Based Natural Resource Management Research in Lingmuteychhu Watershed: Annexure 1—Technical Report', Renewable Natural Resource Research Centre, Bajothang, Bhutan.

———, 2000b, 'Community Based Natural Resource Management Research in the Lingmuteychhu Watershed: A Process Document', Renewable Natural Resource Research Centre, Bajothang, Bhutan.

———, 2000c, 'Farm Household Categorisation in the Lingmuteychhu Watershed, Wangdue Phodrang, Bhutan', Renewable Natural Resource Research Centre, Bajothang, Bhutan.

RNRRC and Sustainable Soil Fertility and Plant Nutrition Management (SSF & PNM) Project, 2001, 'Management and Use of Farm-Yard Manure in the Lingmuteychhu Watershed: Results from a Household Survey', Department of Research and Development Services, Royal Government of Bhutan, Thimpu.

Robbins, Paul, 2001, 'Pastoralism and Community in Rajasthan: Interrogating Categories of Arid Lands Development', in K. Sivaramakrishnan and Arun Agrawal (eds), *Social Nature: Resources, Representations, and Rule in India*, Oxford University Press, New Delhi, pp. 191–215.

Rossi, Benedetta, 2004, 'Revisiting Foucauldian Approaches: Power Dynamics in Development Projects', *Journal of Development Studies*, 40(6 August): 1–29.

Roy, Prodipto, Joya Roy and Rajive Jain, 2001, 'The Potential of Chakriya Vikas Pranali for the Greening of Jharkhand—A Participatory Appraisal Sponsored by the Ford Foundation', CENDIT.

Rudolph, Lloyd I. and Susanne Hoeber Rudolph, 1987, *In Pursuit of Lakshmi: The Political Economy of the Indian State*, University of Chicago Press, Chicago.

Runnalls, David, 1995, *The Story of Pakistan's NCS: An Analysis of its Evolution*, IUCN, Karachi.

Rutten, Mario and Pravin J. Patel, 2002, 'Twice Migrants and Linkages with Central Gujarat: Patidars in East Africa and Britain', in Ghanshyam Shah, Mario Rutten and Hein Streefkerk (eds), *Development and Deprivation in Gujarat: In Honour of Jan Breman*, Sage Publications, New Delhi, pp. 17–36.

Salunkhe, Vilasrao, Vasudha Lokur and Ganesh Pangare, 2000, '*Pani Panchayat* Model of Water Management', in S.N. Chary and V. Vyasulu (eds), *Environmental Management: An Indian Perspective*, Macmillan, New Delhi, pp. 187–205.

Saqui, Q. Md. Afsar Hossain, 2001, 'Village Government in Transition', *The Journal of Local Government*, 30(1: January–June): 13–53.

Saqui, Q. Md. Afsar Hossain and Jamshed Ahmed, 2000, 'Decentralization of Power and the Role of Local Government in Bangladesh', *The Journal of Local Government*, 29(2: July–December): 1–25.

Sarin, Madhu, 1995, 'Regenerating India's Forests—Reconciling Gender Equity with Joint Forest Management', *IDS Bulletin*, 26(1): 83–91.

———, 2001a, 'Empowerment and Disempowerment of Forest Women in Uttarakhand, India', *Gender, Technology and Development*, 5(3): 341–64.

———, 2001b, 'From Right Holders to "Beneficiaries"? Community Forest Management, *Van Panchayats* and Village Forest Joint Management in Uttarakhand', mimeo.

Sarin, Madhu, Neera M. Singh, Nandini Sundar and Ranu K. Bhogal, 2003a, 'Devolution as a Threat to Democratic Decision-making in Forestry? Findings from Three States in India', in David Edmunds and Eva Wollenberg (eds), *Local Forest Management: The Impact of Devolution Policies*, Earthscan, London, pp. 55–126.

———, 2003b, 'Devolution as a Threat to Democratic Decision-making in Forestry? Findings from Three States in India', ODI Working Paper No. 197, Overseas Development Institute, London.

Sarkar, Sarbani, Neena Singh, Saloni Suri and Ashish Kothari (eds), 1995, *Joint Management of Protected Areas in India*, Indian Institute of Public Administration, New Delhi.

Saxena, N.C., Madhu Sarin, R.V. Singh and Tushaar Shah, 1997, 'Independent Study of Implementation Experience in Kanara Circle', Review committee report, Karnataka Forest Department, Bangalore.

Schumacher, E.F., 1975, *Small is Beautiful: Economics as if People Mattered*, Harper & Row, New York.

Seckler, David and Deep Joshi, 1981, 'Sukhomajri: A Rural Development Program in India', mimeo.

Sengupta, Nirmal, 1991, *Managing Common Property: Irrigation in India and the Philippines*, Sage Publications, New Delhi.

Shah, Amita, 1996, 'Watershed Project in a Rainfed Region: A Case Study of AKRSP's Intervention in Jalandhar', Report submitted to AKRSP, Gujarat Institute of Development Research, Ahmedabad.

Shah, Amita, Vinayak Dave and Mahesh Chhatrola, 1994a, 'Initiating Social Forestry under Conflicting Environment: AKRSP's Experiences in a Rainfed Region', Report of the Process Documentation Research, submitted to AKRSP, Gujarat Institute of Development Research, Ahmedabad.

———, 1994b, 'Introducing Watershed Project in a Drought Prone Region: Issues for Developmental Intervention', Report of the Process Documentation Research, submitted to AKRSP, Gujarat Institute of Development Research, Ahmedabad.

Shah, Esha, 2004, 'Technological Vulnerability and Farmer's Suicide', Paper presented at the 'Annual Conference of European Association for Studies in Science and Technology', organised by EASST, 26–28 August, Paris.

———, 2005, 'Local and Global Elites Join Hands: Development and Diffusion of Genetically Modified Bt Cotton Technology in Gujarat, India', *Economic and Political Weekly*, 40(43): 4629–39.

Shah, Ghanshyam and Mario Rutten, 2002, 'Capitalist Development and Jan Breman's Study of the Labouring Class in Gujarat', in Ghanshyam Shah, Mario Rutten and Hein Streefkerk (eds), *Development and Deprivation in Gujarat: In Honour of Jan Breman*, Sage Publications, New Delhi, pp. 17–36.

Shah, Parmesh and Meera Kaul Shah, 1999, 'Institutional Strengthening for Watershed Development: The Case of Aga Khan Rural Support Programme (AKRSP) in India', in Fiona Hinchcliff, John Thompson, Jules Pretty, Irene Guijit and Parmesh Shah (eds), *Fertile Ground: The Impacts of Participatory Watershed Management*, IT Publications/IIED, London, pp. 309–23.

Shankari, Uma, 1991, 'Tanks: Major Problems in Minor Irrigation', *Economic and Political Weekly*, 26(39): A-115–A-125.

Sharma, K.L., 1998, *Caste, Feudalism and Peasantry: The Social Formation of Shekhawati*, Manohar, New Delhi.

Shiva, Vandana, 1991, *Ecology and the Politics of Survival: Conflicts over Natural Resources in India*, Sage Publications, New Delhi.

Singh, Rajendra, 1995, *Tarun Bharat Sangh: Ek Dashak, Ek Nazar me* (in Hindi), Tarun Bharat Sangh, Alwar, Rajasthan.

Sinha, Frances and Sanjay Sinha, 1996, *From Indifference to Active Participation: Six Case Studies of Natural Resource Development through Social Organisation*, EDA Rural Systems, India.

SPWD, 1984, 'Hill Resource Development and Community Management: Lessons Learnt from Micro-watershed Management from Cases of Sukhomajri and Dasholi Gram Swarajya Mandal', Society for Promotion of Wastelands Development, New Delhi.

Sundar, Nandini, Roger Jeffery and Neil Thin, 2001, *Branching Out: Joint Forest Management in India*, Oxford University Press, New Delhi.

Thompson, John and Ian Scoones, 1994, 'Challenging the Populist Perspective: Rural People's Knowledge, Agricultural Research, and Extension Practice', *Agricultural and Human Values*, 11(2 & 3): 58–76.

Tiwari, S., 2002, 'Access, Exclusion and Equity Issues in Community Management of Forests: An Analysis of Status of Community Forestry in the Mid-Hills of Nepal', in Anonymous (ed.), *Policy Analysis of Nepal's Community Forestry Programme: A Compendium of Research Papers*, Winrock International, Nepal, Kathmandu, pp. 33–92.

Underwood, Barry, 1997, 'Village Institutions and Federations: An Overview of AKRSP(I)'s Work in the Area of Human Resources Development', Aga Khan Rural Support Programme, Ahmedabad.

UNDP, 1997, 'Programme Support Document: Sustainable Environment Management Programme', United Nations Development Programme, Dhaka.

Upadhyaya, H.K., 2002, 'Strengthening and Sustaining Community Forest User Groups in Nepal: Required Support Services and Proposed Arrangements for Delivery', in Anonymous (ed.), *Policy Analysis of Nepal's Community Forestry Programme: A Compendium of Research Papers*, Winrock International, Nepal, Kathmandu, pp. 133–56.

Uphoff, Norman, 1998, 'Community-Based Natural Resource Management: Connecting Micro and Macro Processes and People with Their Environments', Paper presented at the 'International Workshop on Community-Based Natural Resource Management', organised by the Economic Development Institute, World Bank, 10–14 May, Washington, DC.

Van den Brand, L. and K. Jamtsho, 2002, 'Water Management in Small Farmer Managed Irrigation Schemes in the Lingmuteychhu Watershed in Bhutan', Liquid Gold Paper No. 7, Irrigation and Water Engineering Group, Department of Environmental Sciences, Wageningen University, Wageningen.

van Etten, Jacobijn, Barbara van Koppen and Shuku Pun, 2002, 'Do Equal Land and Water Rights Benefit the Poor? Targeted Irrigation Development: A Case of the Andhi Khola Irrigation Scheme in Nepal', IWMI Working Paper No. 38, International Water Management Institute, Colombo.

van Koppen, Barbara, Rashmi K. Nagar and Shilpa Vasavada, 2001, 'Gender and Irrigation in India: The Women's Irrigation Group of Jambar, South Gujarat', Working Paper No. 10, International Water Management Institute, Colombo.

Vohra, Gautam, 1990, *Altering Structures: Innovative Experiments at the Grassroots*, Tata Institute of Social Sciences (TISS), Bombay.

Wade, Robert, 1988, *Village Republics: Economic Conditions for Collective Actions in South India*, Cambridge University Press, Cambridge.

Warghade, Sureshchandra, 2003, *Hivare Bazarchi Yashogatha* (The Success Story of Hivre Bazar), second edn, Hivre Bazar Parivar Prakashan, Hivre Bazar (Ahmadnagar).

Weisgrau, Maxine K., 1997, *Interpreting Development: Local Histories, Local Strategies*, University Press of America, Lanham, Maryland.

Wiest, Raymond E., 1988, 'Domestic Group Dynamics in the Resettlement Process Related to Riverbank Erosion in Bangladesh', Paper presented at the 'International Symposium on the Impact of Riverbank Erosion, Flood Hazard and Problem of Population Displacement', at Dhaka.

Wilkinson, Susan, 1990, 'A Community Based Approach to Development', Utthan–Mahiti, Ahmedabad.

Wood, Geoff, 1997, 'Playing with Rivers: Democracy and the Water Sector in Bangladesh', *Discourse*, Summer, 1(1): 24–54.

Wood, Geoff and Abdul Malik, 2003, 'Poverty and Livelihoods', Paper presented at the conference on 'Lessons in Development—The AKRSP Experience', organised by the Aga Khan Rural Support Programme, 15–16 December 2003, Islamabad. Also available at http://www.akrsplessons.org/downloads/case_studies/povert_and_livelihoods.pdf.

Wood, Geoff and Sofia Shakil, 2003, 'Collective Action: From Outside to Inside', Paper presented at the conference on 'Lessons in Development—The AKRSP Experience', organised by the Aga Khan Rural Support Programme, 15–16 December 2003, Islamabad. Also available at http://www.akrsplessons.org/downloads/thematic_papers/CollectiveAction_Lesson LearntExcercise.pdf.

Yadav, Ram P. and Ambika Dhakal, 2000, 'Leasehold Forestry for Poor: An Innovative Pro-poor Programme in the Hills of Nepal—Policy Outlook', Winrock International Series No. 6, Winrock International.

Yoder, Robert, 1994, 'Organisation and Management by Farmers in the Chattis Mauja Irrigation System, Nepal', Research Report No. 11, International Irrigation Management Institute, Colombo.

Zaman, M.Q., 1991, 'Social Structure and Process in Char Land Settlement in the Brahmaputra–Jamuna Floodplain', *Man* (New Series), 26(4 December): 673–90.

Zohir, Sajjad, 2004, 'NGO Sector in Bangladesh: An Overview', *Economic and Political Weekly*, 39(36): 4109–13.

Zwarteween, Margareet and Nita Neupane, 1996, 'Free Riders or Victims: Women's Non-participation in Irrigation Management in Nepal's Chattis Mauja Irrigation Scheme', Research Report No. 7, International Irrigation Management Institute, Colombo.

Index

by, 298; in development, 8, 9, 327n, 328n; formation in Nathugadh, 96–101; homogenous, in Doodha Toli areas, 236–37; involvement in projects in Bangladesh, 278–82; mobilization, 206–7, 228, 298, 301; notion of, 112; participation, 9; in Gujarat, 82; sense of, in Paudi Garhwal, 200; space, and natural resource management in Gopalpura, 148, 153; traditional, in Gopalpura, 137–38; formation of, na-tural resource management 137–38; impact of state-initiated structural changes on, 137

community-based natural resource management (CBNRM), in Bangladesh, 242; in Bhutan, 159–63, 178, 189, 192–93; devolution and democratisation, 189–92; process and intervention in, 170–75; 'community' in, 18–20, 314, 319–20; concept and practice of, 1–2, 9, 322, 324; in Central Himalayas, 196–238; critique of, 2–3; definition of, 13–14; difference in problem perception and notion of, 316–18; discourse of, 325; emergence of, 4–9; in Hivre Bazar, 34; implementation of, 3; intervention types, 293, 300; initiatives, 313; in Nathugadh, 96, 110; mainstreaming, 9–13; state's support to, 311; supra-local institutions in, 60; under rural development policy, 1; vision of, 315

'Community-based Rural Drinking Water Project', Nepal, 10

community forest management (CFM), in Nepal, 9, 10; in Orissa, 17

community forestry management groups (CFMG), Bhutan, 174

Community Forestry Project (1981–88), Bangladesh, 10

Community Water Supply and Sanitation Programme (CWSS), Nepal, 10

contour-bunding, initiative in Nabhy Lingmuteychhu, 186

cotton, cultivation in Nathugadh, 87, 88, 103

cultural plurality, 7

Cypressus, 174

dadhiyas (daily labourers), in Nathugadh, Gujarat, 86, 97

dairy industry, in Nathugadh, 88; promotion of, in Hivre Bazar, 47

Dalits, in Hivre Bazar, 34, 55, 62, 64; movement for empowerment of, 322

dams, movements against big, 7

Darbar caste, in Nathugadh, 81, 84; tenancy under, 96–97

Dasholi Gram Swarajya Mandal, in Kumaon region, 205

Dasholi Gram Vikas Mandal, Kumaon region, generating extra income for villagers, 233

Dawakha Community Forest, Bhutan, 174

decentralisation, administrative, 14–15; democratic, 21, 232, 291, 298–300, 314, 323–24, 326; gains from, 299; policy, 9

deodar trees, 197

Department for International Development (DFID), Sustainable Livelihoods Programmes in Charlands, 282–83

development(al), activities, mobilisation for implementation of, 4; discourse and practices of, under DTLVS, 237; efficiency, 1; failure of government's approach to, 6; planning, 7; process, 7; projects, failure of, 6; 'rationalism', 7

devolution, notion of, 15, 314

Devooben, 79, 83, 108

DHAN Foundation, irrigation tank rejuvenation in Tamil Nadu, 13

Dhondiyalso patti, 209–10

diamond cutting and polishing industry, migrants from Nathugadh in, 88

Fruit, productivity in Hivre Bazar, 60;
tree plantation in Lingmuteychhu,
Bhutan 175, 178–79
Fuelwood, 174, 216–17; DTLVS' programme of, 292; and fodder collection in Bangladesh, 275; for
subsistence use, 203

Gandhi, Indira, 122
Gandhi, Mahatma, 232
Gandhians, on community devel-
opment, 8
Environmentalism, 232–33
Ganga–Brahmaputra–Meghna catch-
ment areas, runoff from, 245
Gangachukha Canal, Bhutan, 186
gaon talav, in Nathugadh, 89
Garhwal, Chipko movement in, 196;
see also Chipko movement
Gaselo chir pine plantation, Bhutan, 174
gauchar (grazing), 118; see also grazing
gender, equity and empowerment, 82,
297; imbalance, 324; issues, 232; in
Nathugadh, 107–8; relations, 288;
and forest conservation, 143–44; in
Gopalpura, 125; in Lingmuyeychhu,
169
Ghimeray, Mahesh, 181–82
Ghogha Regional Water Supply and
Sanitation Programme (GRWSSP),
89–90
Ghogha taluka, Gujarat, 84, 85; water
and sanitation programme in, 81
Girdhari ka nala, Gopalpura, 132
gliricidia trees, 44
Gohil, Bhavsingh, 76
Gohil dynasty, Gujarat, 76
Gohil Rajputs, 76
Gomukh Trust, in Pune District,
Maharashtra, 30
Gono Chetona (GC), Bangladesh, 242–
43, 289, 295, 315, 318; beneficiaries
of, 279, 281, 303; CBNRM initiatives
in, 313; DFID programme of, 283;
flood management and disaster re-
lief activities of, 262–63, 270, 276–78,

291, 301; formation of, 260; on
gender equity, 297; increase in income under programmes of, 267–68,
270–71, 292–93; interventions in
charlands, 260, 262–63, 278–82;
limited finances for, 310; loan provision to women, 266, 268; natural
resource management programme
of, 262; relationship with government, 267; survey by, 278–79; sustainable environment management
programme of, 24, 263, 273–76, 314;
sustainable livelihood programme of,
243–44, 263, 267–73; training to
women, 279; and women's groups,
262–67
Gono Shahajjo Shangstha (GSS),
Bangladesh, 253, 283
Gopalpura, Rajasthan, 118–20, 308;
abstaining from consumption of
alcohol and meat in 151–52; agriculture in 127, 135–36, 140; barley
cultivation in, 128; castes in, 121;
community-based natural resource
management in, 115–54; community,
state intervention and marginalisation of local, 115; contours of
community in, 144–52; credit relations in, 141; forest conservation in,
142–44, 151, 295; gender relations
in, 125; gram sabha in, 148–51, 153;
land, reforms in, 122; types in, 133,
134; use in, 133; landowners in,
121–24; livelihood strategies in,
134–37; location of, 119; migration
from, 136–37; population of, 119;
samaj seva in, 120–21; social ostracisation in, 143; streams in, 120; sustainability issue in, 294; tenancy
pattern in, 122; traditional community in, 137–38; vegetable cultivation in, 136; water-harvesting
structure in, 126, 128–32, 135, 141,
144, 152; wells in, 138–40; wheat crop
in, 127–28, 141
Goryala Bandh, Gopalpura, 132

160, 180; land, access to, 189; types of, 163; livelihood gains in, 175–78, 188; livestock economy in, 165–66, 173, 176; location of, 161–62; matrilineal norms in, 169; paddy cultivation in, 164, 172, 177–78; reforestation programme in, 174; self-help groups (SHGs) in, 175–77, 179; sharecropping in, 168; soil erosion problem in 185, 294; villages in, 163, 164; wage labour in, 169; water conservation, concern over, 187; women, groups in, 303; position in, 169–70

livelihood, activities in riverine charlands of Bangladesh, 267–71; animal husbandry and poultry farming, 268; dairy business, 269; tree plantation, 269–70; vegetable growing by women, 267–68; in Doodha Toli area, focus on improving, 237; enhancement, 21, 232–33, 291–94, 312, 323, 326; in Hivre Bazar, 49–55; in Lingmuteychhu, constraints/gains 175–78, 190; issues in Nathugadh, 94, 101–4; options, for local communities in Gopalpura, 116

livestock rearing, in Doodha Toli area, 212; in Dumlot Malla village, 212; in Hivre Bazar, 46–47, 53–54; in Jandriya Malla village, 212; in Jandriya Talla village, 212; in Lingmuteychhu, in Bhutan, 165–66, 173, 176; in Nabchey, Bhutan, 173; productivity programme, 290

Lobesa, Bhutan, 174

local self-governments, in India, 12

local state institutions, 314, 312

Maharashtra, 14, 29–30, 54; CBNRM in, 30; rains in, 29; see also Hivre Bazar

Mahendra Balai ki medh, Gopalpura, 130

Mahi water supply scheme, in Nathugadh, 90, 104, 109

Mahila mangal dals (MMDs), 218, 229, 297, 299, 240n, 241n; in Doodha Toli

area, and DTLVS, 207–8, 223–26; in Jandriya villages, 216

Mahila Samakhya, programme of Uttarakhand government, 224–25

Mahiti (NGO), in Bhal region, Saurashtra, 72, 77; formation of, 78

malgujari tanks, in Maharashtra, 14

Manavlok (NGO), in Beed district, Maharashtra, 30

Mandalwas, Tarun Bharat Sangh's work in, 117

Manga Baba, 121–22, 124–25, 127, 145

Maratha kingdom, in Maharashtra, 35

Maratha-Kunbi caste, in Hivre Bazar, 34, 42

marginalised groups, say in natural resource management (NRM), 5

markets, access to, 304; in charlands in Bangladesh, 268; network in Rajasthan, 135; people's relationship within Hivre Bazar, 59

matbar-dominated chars, 246–48

Matalungchu village, Bhutan, 163; community forestry in, 173, 185; benefits from forest trees in, 175; improved varieties of paddy seeds in, 172; soil erosion in, 185

Matulumchu village, Bhutan, community forestry in, 179

medhbandis, 116, 147, 154–55n

medicinal plants, in central Himalayas, 234–35; herbs and plants, programme by Gono Chetona, 272–73

Meena caste, in Gopalpura, 121, 148, 155–56nn; and Balais, 122–23; and Balai and Banjaras, 126; as farmers, 122–23, 148; landowning caste, 121; dominant position of, 153; programmes favouring, 297

Menon, Ajit, 28, 159

Merur Char Union, Bangladesh, land-holding pattern in, 258–59; landless population in, 258; literacy rate in, 260; occupational structure in, 260; paddy cultivation in. 257; poverty in, 259; socio-economic conditions in, 254–58; SLRC programme in, 258

About the Authors

Ajit Menon is Associate Professor at the Madras Institute of Development Studies. He was formerly a Fellow at the Centre for Interdisciplinary Studies in Environment and Development (CISED), Bangalore. His research interests include the political economy of natural resource management, environmental politics and decentralised governance in India.

Praveen Singh is Visiting Fellow at CISED. A historian, his area of interest is the history of flood control and agro-ecological change in the Indian floodplains during the colonial and post-independence period.

Esha Shah is Research Fellow with the Institute of Development Studies, University of Sussex. She was formerly a Fellow at CISED. She has authored a book *Social Designs: Tank Irrigation Technology and Agrarian Transformation in Karnataka*. Esha is an environmental engineer turned social scientist whose main research interest involves anthropology and history of science and technology. Her recent research activities include politics of knowledge generation, risk and uncertainty of new and emerging technologies in developing societies, and history of the green revolution.

Sharachchandra Lélé is Coordinator and Senior Fellow at CISED which he co-founded. He has held senior positions at the Pacific Institute of SIDES and Tata Energy Research Institute, and was a Bullard Fellow at Harvard University. He is also a founder-member of the Indian Society for Ecological Economics and has served on its Executive Committee. His work spans conceptual issues in sustainable development and sustainability, and institutional, economic, ecological and technological issues in forest, energy and water resource management.

Suhas Paranjape is Visiting Fellow at CISED and Senior Fellow with the Society for Promoting Participative Ecosystem Management (SOPPECOM), Pune. He is also founding member of the Lok Vidnyan Sanghatana. His interests cover the area of water, energy and biomass resources. He has co-authored a number of books in these areas: *Sustainable Technology: Making the Sardar Sarovar Project Viable*; *Banking on Biomass: A New Strategy for Sustainable Prosperity Based on Renewable Energy and Dispersed Industrialisation*; *Watershed Based Development: A Source Book* and the forthcoming *Water Conflicts in India: A Million Revolts in the Making*.

K.J. Joy is Senior Fellow with SOPPECOM. An activist-researcher, Joy has been a Fullbright Scholar at University of California, Berkeley, and a Visiting Fellow at CISED. His research interests centre around people's institutions for natural resource management. He has co-authored a number of books on water, watershed and energy issues, including *Sustainable Technology: Making the Sardar Sarovar Project Viable*; *Banking on Biomass: A New Strategy for Sustainable Prosperity Based on Renewable Energy and Dispersed Industrialisation*; *Watershed Based Development: A Source Book* and the forthcoming *Water Conflicts in India: A Million Revolts in the Making*.

Cover photographs (clockwise from top):

1. *Impounded water in upper watershed in Hivre Bazar.*
2. *A char village in Bangladesh.*
3. *Regenerating oak forest in Jandriya Talla, Uttarakhand.*